THE ROAR AND THE SILENCE

Wilbur S. Shepperson Series in History and Humanities

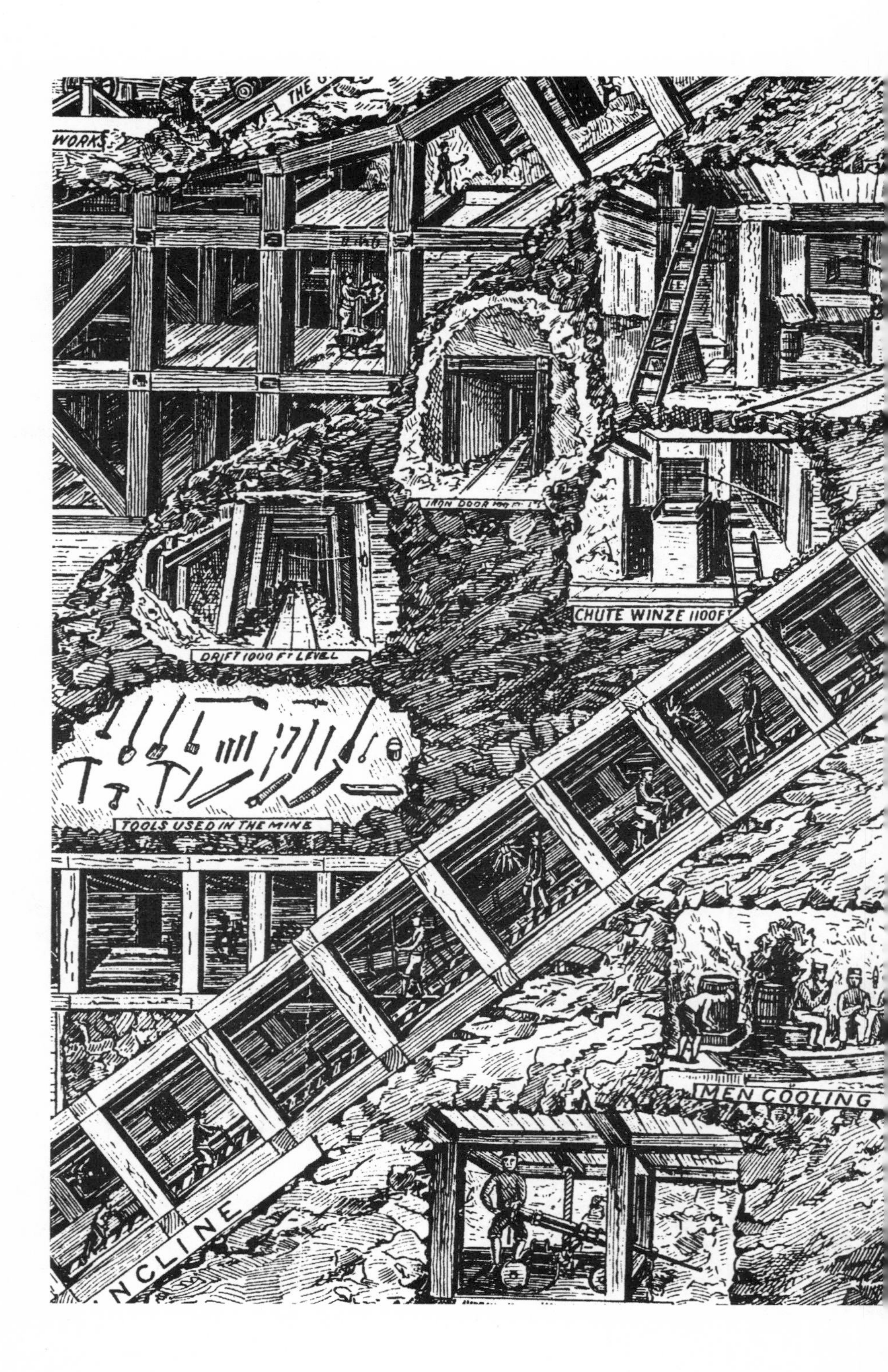
WORKS
IRON DOOR
CHUTE WINZE 1100FT
DRIFT 1000 FT LEVEL
TOOLS USED IN THE MINE
MEN COOLING

RONALD M. JAMES

THE ROAR AND THE SILENCE

A History of Virginia City and the Comstock Lode

University of Nevada Press
Reno & Las Vegas

CONTENTS

List of Illustrations xi

Acknowledgments xv

Introduction xix

1 A Glimmer of Opportunity: The Setting 1

2 The First Boom: Building the Community 21

3 The First Boom: Building the Mines 45

4 Grief, Depression, and Disasters: Successes in the Midst of Failures 70

5 A Time of Bonanza 91

6 The Workers: Labor in an Industrialized Community 119

7 The International Community: Ethnicity Celebrated 143

8 The Moral Options: Sinners 167

9 The Moral Options: Saints 191

10 Princes and Paupers: Contrasts in Class 215

11 Over Time: Bonanza and *Borrasca* (1877–1942) 235

12 The Sequel to the Big Bonanza: Tourism and Television 258

Afterword 275

Notes 277

Bibliography and Bibliographical Essay 323

Index 343

ILLUSTRATIONS

Figures

6.1 Ages of Men in Storey County, Nevada: 1860, 1870, and 1880 138
7.1 Ages of Men Born in the United States in Storey County, Nevada: 1860, 1870, and 1880 163
7.2 Ages of Irishmen in Storey County, Nevada: 1860, 1870, and 1880 164
10.1 Distribution of Wealth in 1870: Men and Women of Storey County, Nevada 224
11.1 Population Distribution for Storey County, Nevada, 1880 239
11.2 Population Distribution for Storey County, Nevada, 1900 246
11.3 Population Distribution for Storey County, Nevada, 1910 247

Illustrations

The Great Strike of June 1859 9
Henry "Pancake" Comstock 15
Teamsters on the Comstock, 1878 28
Camel Transportation on the Comstock, 1877 29
Virginia City, 1860 31
Gold Hill During Its Earliest Period of Development 38
Virginia City, 1860, by Sir Richard Burton 43
Two Approaches to Milling, 1875 47
The Washoe Pan Process 49
California Pan Mill, 1878 50
Interior of a Mill, 1878 52
Philipp Deidesheimer and His Wife 56
William Stewart 63

Jacob Van Bokkelin 73
William Sharon 78
Gold Hill's Crown Point Trestle 82
The Great Seal of Nevada, 1866 83
A Priest Comforting Families During the Yellow Jacket Disaster of 1869 87
Entrance to Sutro Tunnel 89
Virginia City, 1878 93
Street Scene, 1865 100
John Mackay 102
James Fair 104
Virginia City Mine Headframe, 1878 109
Miners Reporting for Duty, 1877 123
Newspaper Touring Party, 1878 124
Miners Laboring 3,000 Feet Below the Surface, 1878 127
Miners Recovering from Hot Work Deep Underground 128
Prospectors near Gold Hill 132
Visitors at a Mill, 1878 135
Gold Hill with Fort Homestead, 1878 148
Virginia City's Chinese, 1877 151
Clarence Sands, an African American, and Other School Graduates, 1883 155
Northern Paiutes Gathering for a Card Game, ca. 1880 158
Virginia City, ca. 1880 174
Virginia City, 1878, and the Nevada Brewery 183
The Fourth Ward School 192
The Miners' Dressing Room 194
Methodist Church 200
Presbyterian Church 202
Virginia City View, ca. 1880 210
Virginia City and the Combination Building, 1878 220
Organ Grinder on Virginia City's C Street 223
Virginia City, ca. 1880 233
Charles Butter's Cyanide Mill 243
"The Seniors Visit Virginia City" 249
Parade Float 250
St. Mary in the Mountains, ca. 1925 253
Hoisting a Bucket at Gold Hill, ca. 1925 255
Train Leaving Gold Hill, ca. 1925 256
Bonanza Map of Virginia City 262

Virginia City's Camel Races 264
Aerial View of Gold Hill, 1979 270

Maps

1 Virginia City, Nevada: African American Dwellings, 1863 98
2 Virginia City, Nevada: African American Dwellings, 1873–1874 99
3 Virginia City, Nevada: African American Dwellings, 1880 153
4 Virginia City, Nevada: Barbary Coast, 1890 179

Tables

1.1 Bullion Production from Gold Canyon, 1850–1857 5
2.1 Occupations for Comstock Men, 1860 26
2.2 Nativity of People Living on the Comstock, 1860 35
4.1 Production of Comstock Mines During the 1860s 74
5.1 Occupations for Comstock Men, 1860 and 1870 92
5.2 Occupations for Comstock Women, 1870 94
5.3 Ethnicities on the Comstock, 1860 and 1870 95
6.1 Industrial Workers of Storey County, Nevada, 1860 to 1880 139
8.1 Crime Statistics for Virginia City, June 1863 and June 1880 173
11.1 Population Size for Storey County, Males and Females 245

ACKNOWLEDGMENTS

Certain fundamental assumptions warrant underscoring, and many people deserve acknowledgment for this volume. The works of two humanists partly inspired this project. In his eloquent study of method, Christopher Lloyd points out that all the humanities examine the past, since none of them can study the future and the present is but the razor-thin line that separates what has been from what will be.[1] Lloyd proposes merging all the methods of the humanities to form one discipline. Clearly this is not likely to happen, but it is possible to employ diverse tools to understand the past. This book proposes the use of a methodology that crosses academic barriers. It asserts that these divisions are meaningless when the objective is to understand humanity. Historians study the past using the written record, archaeologists examine it with material culture, folklorists peer through the crystal of oral tradition, geographers consider human behavior in the context of space, and so on—but what we all really care about is understanding what happened and why. We may specialize in our methods, but the common goal we share is ultimately more important.

The work of Eric Hobsbawm has also been particularly useful here. The approach he employs in his impressive multiple-volume treatment of European history seems fitting for the Comstock. Hobsbawm demonstrates that almost any aspect of society has its antithesis.[2] In a place as full of contradictions as Virginia City, his study of opposites finds a natural home.

The many people who assisted me with information and ideas are acknowledged in the notes, at the points at which their specific help was particularly useful. A few of those people deserve additional mention here. Andria Daley-Taylor, chair of the Comstock Historic District Commission, encouraged me in 1991 to begin applying myself seriously to the writing of this work, and she has never stopped in her support of this project or me. Bernadette Smith Francke, former inspector/clerk of the commission, tirelessly provided information from the primary record as she conducted her own research on the Comstock. Her support of me personally has also been extremely important.

Bernadette's successor, Kelly J. Dixon, continues this tradition, and I appreciate her willingness to read the manuscript and offer suggestions. Other colleagues who have read part or all of the manuscript and have never faltered in their support are Jean and Robert E. Diamond, Nevada State Archives and Records administrator Guy Louis Rocha, Chris Platt, Terry Springer Farley, David and Anne Harvey, William Chrystal, William Kersten, and Sue Fawn Chung of the University of Nevada, Las Vegas. At the University of Nevada, Reno, former chair of the Department of History Jerome Edwards, current chair C. Elizabeth Raymond, and Donald L. Hardesty and Kenneth Fliess of the Department of Anthropology, have consistently been helpful. In addition, my coworkers at the Nevada Historic Preservation Office and at the Department of Museums, Library, and Arts have always encouraged me, and I appreciate their friendship. The same can be said for the staff of the Nevada Historical Society; the staff of Special Collections, University of Nevada, Reno; and the staff of the University of Nevada Press. It is appropriate to note here that chapter 1 appeared in a different form as an article in the *Nevada Historical Society Quarterly*.

Margaret Lowther and her staff in the Storey County Recorder's Office are a joy to work with, or simply to visit. They have always been extremely helpful. I would match their enthusiasm and knowledge of the past against those of any historian I know. This is true as well of hundreds of Comstock residents, many of whom offered suggestions and insights for this project and have never ceased in their encouragement of my work. My deepest thanks goes to my many friends who live in the mining district. I also wish to express my deep appreciation and respect for John and Kathy McCarthy of Virginia City, who have always been supportive of my work and who granted permission to use several of the images in this book.

My mother, my brother and his family, my wife, Susan, and my son, Reed, have continued their support for my projects, and Susan was kind enough to read this material and make numerous suggestions. She has always been there when I needed her. Of particular importance on the domestic front is Chloe, who joined our household in November 1994.

While I had previously worked on the project from 6:00 to 7:00 in the morning, Chloe recognized the importance of greater speed and proceeded to awaken me between 4:00 and 5:00 A.M. The additional time spent on writing sped this book to completion. By rights this publication should be dedicated to Chloe, but she is just a cat and doesn't know the difference.

INTRODUCTION

My heart gave a skip of exaltation as first I saw [Virginia City] lying sprawled there in its canyons and along the scarred mountainside—the greatest mining camp ever in America! . . . It was not long before I imbibed the [folklore] and history of the camp from hospitable old-timers.
—Wells Drury, upon arriving at the Comstock Lode in 1874

Virginia City clings to the steep side of Mount Davidson. It is an improbable town site. Before the 1859 strike that spawned the city, placer miners worked the sand and gravel of Gold Canyon far below, living in tents and shacks. They settled in enclaves where nature provided water, for drinking and washing sand away from gold, and cottonwoods, for shade and a break from the wind. Those early prospectors could not have envisioned the future Virginia City, looming far above. J. Ross Browne, one of the first authors to describe the community, observed that the climate was one of "hurricanes and snow; [its] water, a dilution of arsenic, plumbago, and copperas; [its] wood, none at all except sage-brush." He went on to point out that no one has "title to property . . . [but that there is] no property worth having."[1]

One hundred years later, American television exploited the history of Virginia City and the Comstock Lode almost as if that heritage itself were a body of ore. The *Bonanza* series conjured up an image of a mining boomtown situated on conveniently level ground. In the television Virginia City, unlike the real place, there were no buildings balanced on 40-percent grades, with extra stories built on the downslope side and foundations scraped into the mountain on the other side. Wagons rolled effortlessly into TV's Virginia City, once again denying the improbable location and nature of the West's premier mining town.

Coincidentally, tourists who come to visit the site of one of the world's richest ore strikes also struggle with the reality of Virginia City's peculiar disposition. Directed by signs along the main thoroughfare to parking downhill, flatlanders from throughout the world ascend the steep grades to view the silent remnants of the nineteenth-century mining boomtown. With chests heaving from exertion, they come to realize that if the silver deposit had not been discovered, no one would have planned a town on this mountain. The city stands nearly a mile and a half above sea level, on ground so steep that in the nineteenth century runaway wagons became a daily—and unremarkable—occurrence.

Still, as any miner knows, it is not possible to establish mines where people would most like to live. The discovery of ore dictates the location of the mine, and nature sometimes deposits that ore inconveniently. Such was the case with Virginia City. Historians should never lose sight of this fundamental fact concerning the town that was often called the Queen of the Comstock Mining District: its location, perched high on a steep, desolate mountain that was inaccessible to the rest of the world, shaped its development and its nature. First of all, then, Virginia City is a product of its place.

While the land furnished the bedrock upon which miners and entrepreneurs built the mining district, it also served as fertile soil for the growth of myth, a process that began long before television. The real Comstock may provide an irrefutable basis for good history, but since the earliest days legend has given resident, visitor, and those far away a prism that transmuted the appearance of the mining district, challenging and continually altering perceptions. While one might prefer to disregard the Comstock myth as an annoying distraction, it has become part of the place's reality, warranting its own study and appreciation during the course of any effort to come to terms with the district. The myth of the Comstock adds a second element essential to an understanding of the place.

A third critical aspect revolves around Virginia City's international context. Separated from the rest of the world by mountains and desert, hundreds of miles from major metropolitan centers, the town was nevertheless part of a global community. Its citizens arrived from everywhere. As one of its alumni, Samuel Clemens, pointed out, "all the peoples of the earth had representative adventurers in the Silverland."[2] And when the bonanza days were over, the Comstock had given back such notables to the global community as Clemens in the guise of Mark Twain, as well as many new aristocrats, among them George Hearst, Adolph Sutro, and John Mackay.

In addition to the mining district's cosmopolitan population, other aspects of the new community tied it to someplace else. Virginia City imported its technology, its architecture, and nearly every other element of its existence. Its mining industry produced tons of gold and silver that flooded the international marketplace. At its height, the Comstock startled visitors with the noise of its machines, the clanking metal of its stamp mills, the shrill whistles, and the constant rhythm of its engines. The cacophony of this industrial colossus echoed in the hills day and night. The entire world eventually heard the roar of the Comstock. Although the mining district was on the periphery, it was intimately connected to the core of the international system.[3] Significantly, John Mackay's silver helped lay the first transatlantic cable, thereby making the world a smaller place, translating into a broader context what the Comstock had done as a microcosm.

Still, Virginia City was sufficiently removed from the rest of the world that its rocky slopes provided a habitat for an extraordinary type of society, fashioned from a synthesis of Asia, Europe, the Americas, and Africa. The depth of its ore required a new technology. Distances to be traversed for transportation of goods, timber, and water also mandated innovation. Virginia City was stamped out of imported materials, but at the same time it created its own persona, which eventually influenced the entire mining world.

A fourth characteristic of Virginia City is that it was always in flux. It is not possible to point to any one year or span of years as giving definitive expression to what the mining town was like. Virginia City alternated between boom and decline, again and again. A single portrait cannot capture the nineteenth-century community; at the very least two snapshots are needed, to illustrate both sides of its economic cycle. Even this is insufficient, however, since each period of prosperity and depression assumed its own distinct form.

With these four observations in mind, it is possible to proceed. Virginia City was and is a remarkable place, and its riches continue to stir the imagination.

THE ROAR AND THE SILENCE

1: A GLIMMER OF OPPORTUNITY

The Setting

And they came to Ophir, and fetched from thence gold, four hundred and twenty talents.
—*1 Kings 9:28*

. . . rich and extensive deposits, the real importance of which has not yet been fully appreciated.
—Territorial Enterprise, *Genoa, Utah Territory, May 21, 1859*

In the early 1850s a small colony of would-be miners began to scour the hills of what is now called the Virginia Range, running parallel to the eastern slope of the Sierra Nevada. Flakes of California gold occurred in gravels and sands known as placers along a broad north-south swath on the western side of the Sierra. Conventional wisdom maintained that these deposits resulted from the disintegrating fringe of the mother lode, the hypothesized golden core of the mountain range. It seemed reasonable that the other side of the Sierra might also bear the eroded residue of this miraculous, mythical deposit.

As early as 1850, Mormon settlers along the eastern Sierra were reputed to have discovered placer gold, dust left in alluvial sands, but the church hierarchy of the Latter-day Saints discouraged mining.[1] Gold rushes had the potential to inspire uncontrollable human tidal waves that might dilute Mormon society. The Saints had, after all, come to the Great Basin to establish the State of Deseret, a utopia populated only by the faithful.

No church prohibitions could effectively discourage outsiders, however. Using the same technique to wash placer deposits that was employed in the California gold country, prospectors in 1851 slowly worked their way up from the Carson River. They soon found what Mormons had apparently discovered a year or two before: the sands, deposited over the millennia by occasional flooding from Sun Mountain above, contained specks of gold, or "color," as they called it.

These early prospectors were part of the backwash from the California Forty-Niner days. That rush had drawn more people than could profit for long there. Like a large colony of bees in search of pollen, the California mining community sent scouts all along the Sierra. Each time word returned of mineral wealth, the resulting feverish excitement rivaled any dance honeybees could muster, and off swarmed the miners to investigate the possibilities of new riches. The prospectors who searched the eastern Sierra were part of this movement, but throughout most of the 1850s they located only meager deposits.

A severe terrain greeted the adventurers who crossed the Sierra into the Great Basin. When visiting Virginia City, it is still possible to perceive what the early explorers encountered, although years of stripping the land have destroyed the fragile ecosystem, leaving the high, cold desert even harsher than before. During the summer, the ravines and gullies leading up to Sun Mountain (now Mount Davidson) and the site of Virginia City are hot and dusty. The air is filled with the sounds of insects clacking about. Nearly everything is brown or gray, whether soil, rock, plant, or animal. Only a desert flower or the gray-green juniper and piñon pines change the monotony. Occasionally, one encounters an overpowering and pungent floral fragrance. At this, even those who are most immune to allergies feel their lungs tighten, and breathing grows difficult.

In winter, the same place can be a scene of sudden change. Snowmelt turns sparse soil into quagmires. A fierce wind can bring a storm in a matter of minutes and send it away just as quickly. It is not uncommon for the blizzard of the morning to yield to the sun and melt away in the afternoon. Without clouds, even the winter sun can bake everything dry in a few days, but a clear sky also signals bitter cold at night. The commonly heard saying, "Nevada has two seasons, winter and summer, and they alternate daily,"[2] is hardly an exaggeration.

At the foot of ranges and ravines, where water could collect, settlers found rare oases, which raised their hopes for agriculture. It was the higher elevations, however, where neither soil nor flora hid the bones of the land, that beckoned the early prospectors. During the 1850s, these explorers searched the eastern slope of the Sierra and the ranges immediately to the east. A few promising deposits of gold-bearing sands provided some encouragement and income, but throughout most of the decade, there was no reason to favor one ravine over another. Still, the discoveries gave hope to those who believed in an eastern counterpart to the California gold country. In 1859 the *Territorial Enterprise* observed:

> A great number of very rich quartz specimens have been found in Gold Canyon and vicinity [present-day Virginia Range], gold is known to exist in considerable quantities *throughout the entire range of hills on the east of the chain of valleys skirting the Sierras* from Walker's River [south of the Virginia Range] to Pyramid Lake [north of the Virginia Range], and it is reasonable to conclude that there must exist also in those hills a vast amount of gold-bearing quartz which will at no very remote period be a source of great profit to the capitalist and the miner, and be one of the chief sources of wealth to our country.[3]

The California experience had shaped expectations in the Great Basin. Prospectors looked for placer gold and assumed that deposits were widespread, as they were in the multitude of California valleys and streambeds.[4] And as they had on the other side of the Sierra, these miners used simple methods to extract the gold, relying on the weight of the mineral to cause it to sink faster than worthless sand. Most often, they used rockers or long toms, wooden troughs with small ridges at the bottom, in which they shook and washed sand until the gold settled out and the worthless dirt flowed away. When water was scarce, miners used mercury, which attracted the gold, but that method was more costly. Groups of men thus worked just as they had in California, up and down the ravines, wherever likely-looking sands had gathered. Wealth would be cumulative, not concentrated, and strikes would emphasize the promise of the entire region, not of specific locations.

A few people realized that there were possibilities other than gold in this land. Much has been made of Hosea Ballou Grosh and Ethan Allan Grosh, brothers who identified a ledge of local silver as early as 1856. A succession of tragedies prevented them from revealing or developing their discovery. On August 19, 1857, Hosea struck his foot with a pick, and two weeks later he died of blood poisoning. In November of that year, Ethan Allan stumbled onto a mountain blizzard while crossing the Sierra and died of exposure.[5]

The brothers figure prominently in local folklore because their plight underscores the chancy, dangerous nature of early prospecting—and, of course, it makes a good story. The anecdote also suggests that some were not limited by the idea of the golden mother lode and instead sought other mineral possibilities in the austere mountains of the Great Basin. Nonetheless, for most in the 1850s, the canyons of the eastern slope of the Sierra provided an opportunity to eke out a modest living from placer gold, while continuing a search for more promising sands.

Records of the early community in Gold Canyon and the vicinity of Sun Mountain, although rare—and suspect—do offer some information. Those working in the area established a mining camp in Gold Canyon by the early 1850s, but they almost all abandoned it during the height of summer for lack of water, which was needed to work the claims. The mining itself was "monotonous and colorless," according to nineteenth-century historian Eliot Lord, who interviewed many of the participants. Miners' crude dwellings made of stones, sticks, and brush dotted the landscape near promising diggings. In winter, the miners retreated to abodes only slightly better, constructed of stone, mud plaster, canvas, and boards. Window glass was an unobtainable luxury. Chimneys were rare. Holes in the roof were the standard means of getting rid of smoke.[6]

Entrepreneurs built a crude station house at the foot of Gold Canyon during the winter of 1853–54, and soon afterward they added a combination store, saloon, and bowling alley farther up the ravine. These facilities supplied local miners with provisions, liquor, clothes, and entertainment. Fresh meat relieved the drudgery, and miners always welcomed a successful hunt or local ranchers who would occasionally "drive a cow or calf up the cañon, slaughter the animal at some convenient point and sell portions as required, or roast the whole by a barbeque." Journalist-author William Wright maintained in 1876 that "the people . . . though not numerous, were jovial. They were fond of amusements of all kinds. Nearly every Saturday night a 'grand ball' was given at 'Dutch Nick's' saloon."[7]

Many Mexicans numbered among the early miners. Only a few years before, the entire region had been part of Mexico. Treaties may have placed land under the control of the United States government, but Spanish-speaking people continued to live there. Like Euro-American prospectors, Mexicans found reason to cross the Sierra into the Great Basin. In addition, Chinese laborers and miners worked in the area of Gold Canyon. Together with American Indians—the Northern Paiutes, whose existence continued largely unchanged—these groups formed a complex, international society early in the history of the mining district.[8]

A crisis occurred in the community in 1857. Both earnings and the number of miners had declined since 1855. Nearly two hundred men had depleted the richest sands, and no one had made comparable new discoveries. Gold grew more elusive, and to make matters worse, miners now

1.1 Bullion Production from Gold Canyon, 1850–1857

Year	Number of Working Days	Number of Miners	Daily Earnings in Dollars	Total Yield in Dollars
1850	—	—	—	6,000
1851	100	120	5.00	60,000
1852	120	130	5.00	100,000
	220	20	5.00	
1853	120	90	5.00	76,000
	220	20	5.00	
1854	120	130	5.00	100,000
	220	20	5.00	
1855	140	180	4.00	118,400
	220	20	4.00	
1856	120	100	4.00	70,000
	220	25	4.00	
1857	70	80	2.00	18,000
	140	25	2.00	

Source: Lord, *Comstock Mining*, 24. Lord used contemporary newspapers from California to estimate bullion produced and the average number of days worked for two types of miners: (1) those who worked only a few days a year and (2) those who worked more. Lord's method is as flawed as the sources available to him, but his computerations show basic trends.

faced drought, endemic throughout the West. Without water, it was nearly impossible to work the gold-bearing sands. Earnings dropped dramatically (table 1.1); 1857 was a bad year, and 1858 showed no improvement. Many left for more promising possibilities.

It was clear that the placer diggings would fail without development of a more reliable supply of water. The remaining miners organized, created rules for what they called the Columbia Quartz District, and hoped that investors would find the prospects attractive enough to fund ditches to supply the region with water from the nearby Carson River. At the same time, the *Territorial Enterprise* began promoting the mineral possibilities of the region, calling for capitalists to develop the single ingredient—water—that distinguished the eastern slope from the California goldfields.[9] Again, the miners relied on the familiar; some drastic change would need to occur if the mining colony was to survive.

This was the situation in which miners and prospectors of the western Great Basin found themselves early in 1859. The Great Basin is prone to false springs, times when temperatures rise, plants blossom, and mountain snows begin to melt. Typically, blizzards and freezing weather follow, killing young buds and dashing hopes that summer has nearly arrived. A false spring settled upon the Comstock in January 1859, and the wintering, dormant miners used the opportunity to return to work. There must have been little hope of finding new, profitable placer sands at the base of local ravines, because most of the exploration appears to have been higher on the mountain. Groups heading toward Sun Mountain from Gold Canyon to the south and from Six Mile Canyon to the east focused new attention on rocky slopes, where alluvial deposits were unlikely.

It would have been natural for the prospectors to search such heights apprehensively. At the base of canyons, they had only needed to shovel gold-bearing placer sands into rockers to earn a living. Nature had already accomplished most of the milling and processing. As miners searched for the source of the gold higher on the mountainside, they feared that it would be locked in stone when they found it. To work a solid ore body, a miner had to remove large amounts of rock, crush it, and transport it to a stream at the base of the canyon—all of this necessary before the material could be treated like placer sands. It was hard work, and it was costly.

The year before, for example, the Pioneer Quartz Company had resorted to working a ledge near Devil's Gate down Gold Canyon, where workers crushed and processed about five tons of rock. It yielded only three and a half ounces of gold, worth about $42. The company evaluated the result and decided it did not justify further work.[10] With such experience in mind, the miners likely ascended Sun Mountain without enthusiasm. Still, they had little choice, aside from leaving the district. With the profitable alluvium exhausted, the remaining option was to search for the source of the placer sands in the hope that whatever ore resulted would be sufficiently rich to justify the arduous task of milling.

In all likelihood, astute prospecting and intuition had little to do with identifying the quartz gold-bearing ledge on the mountainside. Besides the Grosh brothers, James "Old Virginny" Finney, longtime local miner, stumbled into and claimed the ore as early as 1858. Local tradition failed to note others who certainly had also recognized the mineral-bearing strata. Logic had long before told the miners that higher deposits must exist, since there had to be a source for the placer gold-bearing sands at the base of the canyon.[11]

Unexpected success awaited those who explored the heights of Sun Mountain. Indeed, it would be months, arguably even years, before people fully understood the degree of success. During January 1859, Finney, John Bishop, Alexander Henderson, and John Yount returned to the steep slope of Sun Mountain. They began working with the soil of a mound at the head of Gold Canyon and easily obtained a yield that quickly proved promising. Although clearly not alluvial sand, the outcropping was sufficiently decomposed to make pulverizing of rock unnecessary. The material caught in the bottom of each rocker glittered. The men named the site Gold Hill and immediately established a camp. Prospects in the district suddenly looked better than they had since 1857, if not before. The site promised about $12 a day for each miner.[12] The only question was how long it would last.

Word spread through the district, and local miners came to declare their interests in the area. Consistent with custom in the California gold country, the original prospectors claimed only that portion of the site that they could reasonably expect to work. Promising deposits nearby remained there for the taking, the miners working on the assumption that the wealth was diffuse, not concentrated, and that it was impossible for a few men to monopolize an entire district.

Alva Gould soon joined the original four. Within days, Henry Comstock, James Rogers, Joseph Plato, Alexander "Sandy" Bowers, and William Knight posted their own claim nearby. They immediately began to sink a shaft to determine the depth of the resource. Eight feet of consistent yield justified the investment of labor to build a crude flume to carry water from a stream on the south side of Sun Mountain.[13]

By the time spring arrived, work was under way on the new prospects, and entrepreneurs were developing more secure sources of water. The return for a day's labor ascended to unprecedented values. The *Territorial Enterprise* reported in April 1859: "The diggings are in depth from 3 to 20 feet, and prospects from 5 to 25 cents to the pan, from the surface to the bedrock. Bishop & Co., two men, made during the week from $25 to $30 per day; Rogers & Co., two men, from $20 to $40 per day, and F. D. Casteel $10 per day, working alone and packing the dirt about 60 yards." The newspaper added that with sufficient water, "from $50 to $100 per day per man could be made with ease." A week later, the newspaper estimated that claims were worth "from $4,000 to $5,000 per share." Even if such a statement were outlandish exaggeration, clearly the prospects had improved dramatically. Gold fever gripped the community. The miners worked from dawn to dusk, obsessed with the quest for wealth. The normally freewheeling society turned serious. As the *Territorial Enterprise* observed in April 1859, "The miners are generally temperate and industrious, and whiskey has, therefore, become a drug in the market, with a downward tendency." They had been bitten by the gold bug, and a reversal of fortune was nowhere in sight.[14]

Miners began sinking more shafts in May 1859 to test the depth of deposits. Again they found the resource deep and consistent in value. The *Enterprise* reported that a small claim, 5 by 40 feet, had sold for the unprecedented sum of $250. It was the equivalent of about a month's earnings at the new inflated rate.[15] Gold Hill would remain a hotbed of activity for some time.

During the late 1850s, a nearby colony of miners also worked and prospected Six Mile Canyon, extending to the east from Sun Mountain. Encouraged by the success of Gold Hill and driven by a similar lack of viable alluvial sands, some from this group began to search uphill during the spring of 1859, just as their Gold Canyon counterparts had. Finally, on June 8, Patrick McLaughlin and Peter O'Riley, two Irish miners from the Six Mile Canyon community, started working a new, even higher site. It contained black crumbly rock unlike any they had seen before.

Most important to the miners at the moment was that water, a rare commodity in the district, flowed through the area. While building a crude dam, the two threw some soil into their rocker as an afterthought. Like the finds at Gold Hill, the results were astonishing and immediate. With every shovelful, gold glittered at the base of the rocker. There was no question that they had found a valuable claim, even though its pale color led the two to conclude that the gold was alloyed with some worthless metal, and the black rock weighed nearly as much as the gold, making separation difficult.[16]

That evening, after working all day to obtain handfuls of gold, McLaughlin and O'Riley began to retire their equipment. At that point, Henry T. "Pancake" Comstock of Gold Hill arrived on the scene. What ensued soon became a pivotal episode in local folklore, recounted endlessly and used to illustrate the nature of the times and the people involved. Although it has become legendary, there is no reason to doubt its accuracy. Comstock quickly realized that the miners had made a significant discovery. Large amounts of gold lay nearby, and besides, it was common knowledge that miners had sought to define the extent of the Gold Hill ledge.[17]

Whether these men were working the same body of ore or a separate one remained to be seen, but Comstock recognized opportunity. He immediately declared his right to the area and began negotiating. He benefited from two factors. First, even today it is not uncommon for prospectors to claim many sites, most of which they do not plan to develop. Although such assertion of ownership can be weak, there is always the hope that the first claimant can secure concessions from any subsequent miners who profitably work the property. It is not known whether Com-

The Great Strike of June 1859, depicting Comstock (with the horse), McLaughlin and O'Reilly to the left, and presumably Penrod to the right. Lithograph from William Wright's *The Big Bonanza.* (Courtesy of the Nevada Historical Society)

stock had actually claimed the McLaughlin-O'Riley site. Indeed, he may not have been certain himself. Still, it was a reasonable possibility. Miners confronted with such prior claims frequently conceded a portion of an ore body to avoid dispute. Second, the assumption in the 1850s that mineral wealth was diffuse and not concentrated meant that it was not much of a concession to admit Comstock to the site. Additional partners, after all, meant that they could claim more feet, according to regional mining custom. McLaughlin and O'Riley also complied with Comstock's insistence that his friend Immanuel "Manny" Penrod be included.[18]

On June 11, 1859, the growing community held a meeting to adopt rules to govern the mining district. Although this had been done before, a new compact, designed to suit the situation at Gold Hill and accepted by its newer citizens, seemed appropriate. The miners also felt a need to fill certain offices for the community, and the meeting saw the election of J. A. "Kentuck" Osborn as justice of the peace, James F. Rogers as constable, and V. A. Houseworth as recorder.[19]

Although word of the new discovery soon spread, Gold Hill remained the more promising location because it yielded purer gold and the site was untainted by what some called "the annoying blue stuff."[20] Still, the new diggings added to the excitement, and a local rush ensued during the summer of 1859, drawing on the agricultural areas at the eastern base of the Sierra. Rush after rush to mining districts in pursuit of gold far rarer than initial reports promised had left the West reluctant to respond to word of new discoveries. Nonetheless, more and more people realized that the gold of this Great Basin district was plentiful. As the *Enterprise* reported in early July 1859, "We have heretofore been somewhat skeptical, but recent developments have proven to our entire satisfaction that the resources of these mines, incredible as it may seem, have not been overestimated." With the growing attention, values of claims inflated. Some of the first to mine the area saw it as a good time to sell, reasoning that the gold would run out and the bubble burst.[21]

Slowly the community acknowledged that no matter how easily the first shovelsful of dirt surrendered their gold, the quartz ore deeper down would require milling. Again consolation came from the other side of the Sierra. The *Territorial Enterprise* reported as early as February 1859 that "quartz operations in California are every day more clearly demonstrating the fact that as an avenue for the safe and profitable investment of capital, quartz mining is destined eventually to superceed [*sic*] placer mining."[22] Only a few years before, for example, the Empire Mine of Grass Valley, California, had begun an underground operation that was to last more than a century. Profit was available for those willing to work such a mine, but that did not make it more palatable for Sun Mountain's miners, who recognized that their future work would take a different, more complicated form.

One of the easiest ways to solve the problem of dealing with obstinate ore was to sell one's claim and leave. Although later ridiculed for accepting absurdly low prices, the first claimants received as much as $12,000 for their interests.[23] Compared with the millions eventually reaped, these were paltry amounts, but they were more than most miners could hope to see in two or three years of hard labor even at the high yields the district had just begun to provide. As long as the excitement held and newcomers were willing to pay such prices, the old-time placer miner was able to profit without attempting something beyond his experience.

Excitement and wealth, however, were about to be redefined. As O'Riley, McLaughlin, Comstock, and Penrod dug down, they noticed that the black rock plaguing their claim coalesced into a seam that grew thicker the deeper they probed. While the material was a curiosity, it was also a nuisance, its heavy weight hindering the retrieval of gold. These early miners found the situation complicated even more when they reached rock that required crushing before it could be washed. Work with picks and sledgehammers proved arduous and time-consuming. Profits diminished. The miners weighed the possibilities and decided to give John D. Winters, Jr., a local rancher, and Kentuck Osborn, the newly elected justice of the peace, one sixth of their claim in exchange for the construction of two arrastras—simple mills—furnished with two horses or mules to run them. The agreement for this project, documented by Houseworth in his record book, is elaborate and illustrates the attempt to consider all possible scenarios.[24] Without the counsel of lawyers, the miners nevertheless understood the natural affinity that has always existed between gold and costly lawsuits.

A few weeks after the discovery of the new site, B. A. Harrison, a ranch hand from the nearby Truckee Meadows, took a sample of the black rock to Placerville, California. On June 27, 1859, he gave it to Melville Atwood, an assayer, who found its value per ton to be $876 in gold. The big surprise was that a ton of the ore would also yield $3,000 in silver. Together with Judge James Walsh and Joseph Woodworth, both of Grass Valley, Harrison hurried back to profit from the discovery before it became widely known and the values of claims exploded. But as William Wright pointed out in his farcical 1876 history of the mining boom, "each man had intimate friends in whom he had the utmost confidence . . . and those bosom friends soon knew that a silver-mine of wonderful richness had been discovered over in Washoe country. These again had their friends."[25] And so on. A rush of major proportions had begun.

Walsh, like all those who followed with money, began buying claims throughout the region. By fall, many of the original discoverers were out of the picture, replaced by entrepreneurs and experienced quartz miners from California. Those who came late and without significant funds, merely hoping to find gold or silver, asserted ownership of anything they could. The result was a network of overlapping claims further confused by the hundreds who arrived before winter. The rules established in the June 11 meeting were not sufficient to handle the multitudes now pressing in upon the hillside. Houseworth's book became filled with dozens of vague claims and ill-defined notices. Wishing to avoid legal disputes so that they could focus on mining, all concerned strove to untangle the confusion and eventually established customary boundaries of claims.[26]

As the miners set to work, they found wealth on an immense scale. The first who worked in the area of the McLaughlin-O'Riley strike during the early summer of 1859 had called the incipient community Mount Pleasant. By August 5, after news of the assay, those in the area had renamed it the Town of Ophir, the site of the original strike becoming the Ophir Mine. With biblical references at their fingertips, these miners knew of the Old Testament's fabled gold mine, the riches of which exceeded all others. Throughout history, many had searched in hope of finding its long-lost treasure.[27] Now the miners of the Great Basin claimed to have found comparable wealth of biblical proportions. Although local miners held a meeting in September, during which they renamed the growing town for an early claimant, James "Old Virginny" Finney, one of its richest diggings continued to take its name from ancient Ophir.[28]

Silver, together with gold, was now the game. In hindsight, writers would later point out that this should have been more apparent to those who had worked the canyon for gold. Henry DeGroot, an eyewitness at the scene in 1859, later wrote that many people had believed that silver deposits probably existed in the Great Basin. In addition, he noted that Spanish legends of silver in the region were well known. Still, silver was far from the miners' minds. Again the California experience colored their perceptions, and gold remained their only goal. Even if the miners had recognized silver earlier, they might have ignored it: Mexican mining tradition maintains that "it takes a gold mine to run a silver mine." A lack of capital to pay wages, buy material, and build mills restricted the first miners to working gold deposits.[29]

Because of capital and experience, the second wave of quartz miners in the district was able to accomplish things that had been impossible before. And in the process they were about to create an international legend. In September, Judge Walsh took 3,150 pounds of high-grade ore to San Francisco for milling. It yielded $4,871. Over the next few weeks, 38 tons followed. Although it cost $512 per ton to transport and mill, the $114,000 that it produced more than offset the expense. The resulting bullion, prominently displayed in the windows of bankers Alsop and Company, provided tangible proof that tremendous treasure had been found.[30] Before long, not just California but the entire globe answered the call of yet another rush for mineral wealth.

On November 2, 1859, one foot of snow fell in Virginia City. The storm cut off the Sierra passes, ending both the rush to the Comstock Lode and the transportation of ore to San Francisco. Virginia City settled in for the winter. By then the community consisted of several hundred people, living in crude accommodations. Everything from tents and brush hovels to the mining tunnels themselves served as abodes. As Wright pointed out, they became a community of cavemen, or "troglodytes," as he called them.[31]

Winter progressed, and the ground froze, putting a stop to most work. Many, finding the cold of their meager shelters unbearable, descended to communities in the valleys. Food was scarce everywhere, however, since many of the local farmers had forsaken their fields to try the new diggings during the summer. On the western slope of the Sierra, a good part of California waited for the first opportunity to cross and inspect the new El Dorado. For a brief time, Virginia City was not in flux. The quiet would not be repeated for thirty or so years.

By the winter of 1859, success had permanently altered the community in the Gold Canyon area. A new society of outsiders had replaced the original ragtag band. One by one, old-timers were fleeing; placer mining was a thing of the past. The Comstock Mining District now prospered with hard-rock mining. As the community moved forward in pursuit of its destiny, it occasionally looked back, assessing and reassessing its past. The more removed the 1850s Comstock became, the more incomprehensible that period seemed.[32] This circumstance created a fertile environment that spawned a rich tradition fabricated to explain the earliest history of the region.

The first miners who worked the district died for the most part in obscure poverty. Local oral tradition has been unkind to the discoverers of the Comstock Lode, depicting them as mad, lazy, drunken, and unimaginative or incredibly unlucky. From the Grosh brothers to "Old Virginny" Finney and "Pancake" Comstock, the early miners became characters in a comedy of errors, complete with the obligatory tragic moments. With celebrated or invented idiosyncrasies, their tale filled the repertoires of local writers and continues to provide tour guides with material. By tracing the evolution of Finney's and Comstock's stories in particular, it is possible to understand an aspect of western legend. The accounts dealing with the early discoverers are a marbled mixture of myth and reality, so sorting out the specifics may never be possible. Looking at the period from the point of view of folklore, it is possible to work with the material as it exists: the focus of people's image of the past then becomes as important as what actually happened.

The citizens of mining camps often revel in celebrating the eccentric nature of their origins. Perhaps many people also like to think that were they placed in similar circumstance, they would not make the same mistakes. They consequently characterize history's players who failed as inept or unfortunate. Ironically, local tradition ultimately affords these early miners the notoriety that they might have won directly by profiting extravagantly from the discovery. Finney and Comstock acquired immortality because of the ore body, in spite of doing little to develop the mines. Finney gave his nickname to the principal community of the district, and Comstock's name identifies the silver and gold deposit.[33] Virginia City and the Comstock Lode won international fame even though their namesakes' monetary gains were minimal.

The earliest sources feature Finney and Comstock as hardworking, locally prominent miners. James Finney[34] earned a reputation for having a good nose for prospecting. All sources consistently support the idea that he was one of the first to recognize the mineral-bearing potential of the future Comstock Lode. Still, he was apparently a simple man, unable to write and with limited aspirations.[35] As early as April 1860, *Hutchings' California Magazine* remarked on the incredible fact that Finney sold his claim to Comstock for "an ancient horse." Articles dating to the summer of 1861 appearing in the *San Francisco Herald* and the *Alta California* state that Finney sold his claim to Comstock for "an old horse, worth about $40, and a few dollars in cash."[36] Henry DeGroot wrote in 1862 that Finney was an "honest old pioneer" who was "ignorant but generous." He also records a story that has Finney selling Comstock a claim for "an old horse and a bottle of whiskey."[37] By 1876, DeGroot elaborated on Finney's association with the bottle, suggesting that he died of old age and alcoholism. He pointed out that Finney "received nothing for his interests" although, being first, he had deserved more. DeGroot also recounted the story of Finney's selling his claims for an Indian pony and a bottle of whiskey, adding "some supplies" to the purchase price. Whatever the cause, Finney died in June 1861.[38]

Early references to Henry Comstock are far more abundant. A rich assortment of newspaper articles reveals a gradual shift in the perception of this miner. Sources dating to 1859 described Comstock as a hard worker who took advantage of opportunities. For DeGroot, writing in 1862, he was shrewd.[39] After insinuating himself into the early success of the district, Comstock sold his interests and used his capital to establish himself in a mercantile business, providing support for the new mining community. His failure at the commercial enterprise sent him wandering throughout the West, prospecting for a new claim to equal the lode he had once owned, which was by then making international headlines.[40] In 1863 Comstock attempted to reestablish his link to the district through legal dispute.[41] Following the failure of that tactic, the *Virginia Evening Bulletin* reported that Comstock could be found "ekeing [*sic*] out a miserable existence by working a poor claim on the Powder river," a turn of fate written off as "miner's luck." One year later, the *Gold Hill Daily News* contradicted the idea that Comstock was doing poorly, suggesting that he was discovering rich ore bodies in Idaho and that "money is no object with him; he has too much energy to eke out a miserable existence anywhere."

Henry "Pancake" Comstock. Lithograph from William Wright's *The Big Bonanza.* (Courtesy of the Nevada Historical Society)

During the summer of 1865, Comstock returned to Virginia City and established a company intended to exploit a claim north of town. The *Gold Hill Daily News* reported: "Comstock is acting Superintendent of the mine, and we trust that this old pioneer of the silver fields of Nevada may reap a rich reward for his long and continuous labors."[42]

During or shortly after his 1865 visit, Comstock began to be known for an increasingly irrational mental state. In 1868 he wrote a letter from Butte, Montana, making outlandish assertions. The *Territorial Enterprise* pointed out: "Many of the statements in the letter are unquestionably true, while others, no doubt are sheer fictions. When Comstock was last here . . . it was impossible to listen to him and believe in his perfect sanity." His reputation for mental illness growing, Comstock said in late 1868 that he was hard at work on a new claim. Two years later he apparently committed suicide in Bozeman, Montana, but word of the event did not reach Virginia City until 1875. At the time, the *Territorial Enterprise* maintained that "he shuffled off the coil [and] was led to the rash act by dissipation and want." Later that year, the *Enterprise* described Comstock as "an illiterate man, being unable to read or write" and added that "the notoriety which attached to his name led him to frequent exaggerations of his personal importance—a weakness which may be readily excused in the old pioneer."[43]

Writing in 1876, DeGroot characterized Comstock as full of bluster and as having lost his money to an "artful woman." He reported that Comstock sold one of his claims for an old horse and $40, but that he received larger sums for other properties. On one occasion, Comstock boasted that he had won the better part of the deal from a so-called California mining authority. DeGroot also held that Comstock's suicide occurred in the midst of a bout of temporary insanity.[44]

The earliest evidence concerning Henry Comstock suggests that he was a hardworking, clever miner who was able to recognize opportunities and exploit them. After the local transition to quartz mining, Comstock attempted to succeed by using his profits to address the need for mercantile support of the community. Like Finney, however, he was clearly out of his league in the new environment.

Even before he died, Henry Comstock was becoming the object of a growing local legend. The idea that he was illiterate is a fiction, given the letters he wrote. In addition, the *Enterprise*, which mistakenly called him John Comstock, attributed to him the sale of his claims for "a mule, a shotgun and a bottle of whiskey,"[45] thus confusing him with Finney. The introduction of an "artful woman" as the source of Comstock's economic demise also appears to be a later, fanciful addition. In contrast to Comstock, Finney died much earlier, and so he could not make himself a current topic of conversation as Comstock had through repeated visits and letters. Except for the constant reminder of the city's name, "Old Virginny" was in danger of dropping out of popular view. After all, he, like Comstock, possessed the unremarkable traits of hard work and lack of economic success. It was left to William Wright, newspaper reporter and writer, under the name of Dan De Quille, to place the immortal, legendary stamp on both Finney and Comstock.

It is impossible to know whether Wright created or merely used the tradition surrounding Finney and Comstock. Perhaps a combination was responsible for the author's final product. For Wright, Finney was a drunk and a fool. He was capable of hard work, but tended to stop his labors as soon as he had enough money to buy whiskey. Wright maintained that he was attempting to correct the record by refuting the story of Finney selling one of his claims for an old horse, a pair of blankets, and a bottle of whiskey, but the author was equally capable of passing off legend as fact.[46]

Wright also is the earliest source for the story that Finney himself named Virginia City. To this day, local residents relish his recollection of the inebriated "Old Virginny" tripping and breaking a bottle of whiskey on the ground. Not wanting to waste the precious substance, the prospector announced that he was baptizing the new town in honor of Virginia, the state of his nativity.[47] Whether this happened or not, evidence clearly indicates that local miners decided in a meeting to name the community Virginia City. Still, Wright's story lives on.

Wright immortalized Finney's drinking habits with a tale about a lawsuit that called on the old pioneer to find his original claim notice. Perpetually drunk, Finney refused to cooperate. The lawyers consequently locked him up, away from whiskey, and waited until the next morning to discuss the issue with the sober but surly prospector. After negotiating for one drink to start the day, he was able to lead the lawyers to the notice. Wright held that alcohol proved the ultimate demise of Finney, his having been thrown from a horse while drunk.[48] The author also characterized Finney as a hunting enthusiast, recounting how he shot a skunk and then surreptitiously fed it to the camp. Only after the meal was well under way did Finney, in trickster fashion, reveal the nature of the beast.[49] None of these anecdotes appear in the original primary documents, and there is no reason to believe in their historical veracity.

Comstock fared no better under Wright's fanciful pen. For the author, Henry Comstock was a lazy, fast-talking charlatan who deserved none of the fame he acquired. He had, according to Wright, incompetently and recklessly squandered the opportunity that he had won through neither fairness nor work.[50] Wright stated that people called Comstock "Pancake" because he never baked bread, professing to be too busy. Instead, he fried the ever-simple pancake.

Unwilling to work, Comstock employed American Indians to labor at his claims. As Wright put it, Comstock swindled his way into the Ophir claim. He then "elected himself superintendent and was the man who did all of the heavy talking." Wright suggested that Comstock "in the early days, was considered by many persons to be 'a little cracked' in the 'upper story' . . . [and] was a man flighty in his imagining." According to Wright, Comstock's only positive attribute, besides sobriety, was generosity. One story has Comstock offering visitors a panful of dirt that they could wash, keeping whatever they found. Once, some visiting women received a pan that proved to be worth $300, to the delight of "Old Pancake." Wright also told of Comstock wooing a wife away from a Mormon who was passing through the area. Confronted by the jilted husband, Comstock paid a sum in compensation but demanded a bill of sale. Later, when the woman ran away from Comstock in favor of yet another man, he enforced his bill of sale and demanded her return. Eventually, however, she managed to escape, and that, like so many of Comstock's other investments, proved poorly placed.[51]

As in the case of Wright's treatment of Finney, it is possible to verify the attributes or stories regarding Comstock in the contemporary documentation. Indeed, the earliest sources indicate that Comstock was hard-working, and although he appears to have suffered from growing insanity by the late 1860s, there is no documentation of earlier mental illness.

Wright's *The Big Bonanza* is recognized as a quasi history with literary aspirations. Historians rightly treat it with suspicion. Eliot Lord's *Comstock Mining and Miners*, on the other hand, is regarded as a reliable source. For Lord, Finney was a good prospector but little else. He "remained sober when he was too poor to buy whisky and would never work longer than was necessary to obtain the means of filling his bottle." Lord repeated Wright's story of needing to deprive Finney of alcohol so that he could find his original claim notice. He also echoed Wright in asserting that Finney died in 1861 after falling from a horse. He portrayed Finney as a miner who eked out "a precarious livelihood by bartering feet in sundry claims in exchange for drink and food-money."[52]

In the case of Henry Comstock, Lord reiterated the idea that he was lazy, used American Indian labor, and was a fast talker. He employed Wright's story about the purchase of a Mormon wife and agreed that Comstock was insane and committed suicide. Lord also introduced an additional story, though without a source. In this anecdote, Comstock allegedly sold his claim to one Herman Camp for very little money. After being ridiculed for the bad deal, Comstock persuaded his friends to conduct a mock trial to judge the validity of the transaction. This court invalidated the exchange and in spite of Camp's protestations forcibly obtained and destroyed his deed. Like Wright's stories, most of Lord's assertions about Finney and Comstock cannot be confirmed by earlier documentation.

Lord focused his study upon the evolution of mining technology and the history of the claims associated with the Comstock Lode; he meticulously researched and documented these aspects of his work. He spent less time in the assessment of personalities, and although he salted his writing with characters and anecdotes, literary color was not his principal objective. Instead, it seems likely that he relied on Wright for perspective and stories in this regard. Because subsequent historians have given Lord so much credibility, these stories became codified as accepted history in a way that would not have occurred had Wright alone been the source.

Understandably, twentieth-century historians of the Comstock invariably repeat, if not the actual stories, then at least the evaluations of Finney and Comstock as presented in Wright and Lord.[53]

The first decade of mining within what was to become the Comstock Mining District played a crucial part in the discovery of wealth and in the laying of a foundation for the meteoric success of Virginia City. In spite of their importance, the earliest miners were destined to be misunderstood. Placer mining quickly became an oddity on the Comstock, making the 1850s incomprehensible to later residents of the district. In an attempt to reconcile the prospectors' behavior with the point of view of the community after the rush, Comstock oral tradition cast the earliest players as drunkards and madmen, and that transformation has become part of the history of the community.

Ironically, most if not all of the early prospectors searched for gold rather than silver, the precious metal that ultimately brought fame to the region. That the first miners ignored all but gold even while digging into one of the wealthiest silver lodes in the world is nothing short of comical. People see the environment according to their preconceptions, however, and the actions of those first on the scene are understandable when considered from their point of view. For a decade, local miners regarded placer gold-bearing sands as more economical to mine than the quartz veins high on the mountainside. Further, for nearly half a year after the first big strike, the retrieval of gold remained the objective. Miners ignored the mineral-bearing black rock and pursued the precious metal they knew best. And finally, when the original claimants to these fabulously rich mines had a chance to sell out, they did so, often to the first bidder, and then trotted away gloating over their good fortune. All this made sense from the perspective of the miners of the 1850s. With hindsight, authors such as William Wright, writing a decade or more after the earliest events, regarded the nature and value of the resource as obvious and judged the earliest prospectors accordingly.

In fact, nothing of the district's ultimate development was evident to those early miners. The prospectors who worked Gold Canyon and the surrounding area during the 1850s were largely industrious men who approached their craft seriously and methodically, but the California experience had shaped their expectations. They understood gold. They knew how to acquire it, and they preferred to work alone or in small teams. They also knew that mineral wealth concentrated in a single area was a rarity. They probably doubted it was possible. Ore bodies, after all, almost never met the expectations of the rushes they inspired. To resist a willing buyer offering thousands of dollars for a claim to a vein that might pinch out a few feet from the pit's bottom would have been reckless gambling. The possibility that a claim was worth significantly more than was being offered was slim. The miners' decisions to sell represented calculated risks that would have been regarded as shrewd maneuvers in almost any other situation. Certainly they understood their profession well enough to know that it would take considerable capital, an understanding of wages, accounting, logistics, and mining law, and a good deal of luck to bring success to the new kind of operation needed to work the district. Considering that they had neither the money nor the experience, they made the best rational choice. That these particular men happened to be dealing with the Comstock Lode made their gamble appear foolhardy in retrospect, leaving them vulnerable to later judgment. It also created a situation ripe for folklore.

The traditions generated about the original miners, and in particular those about James "Old Virginny" Finney and Henry T. "Pancake" Comstock, were well in place within fifteen years of the time they sold their claims. They had become vivid characters in a myth developed to explain their acts. No amount of historical research and source criticism is likely to overcome the persuasive nature of an oral tradition that has assumed the role of local history. Indeed, it thrives today and has all the promise of surviving into the next century in spite of anything written here. Of paramount importance is that it took less than two decades for this myth to assume its place. Hardworking, serious miners transformed into eccentric, devil-may-care drunks—the latter fit easily into the legendary Wild West during its carefree frontier period; the former would contradict that image. The explosive nature of the Comstock's growth and shifts in population allowed for a citizenry so removed from its own past that its heritage could be profoundly misunderstood and transmuted. Ultimately, the story of the discovery of the Comstock Lode illustrates the power of the myth of the Wild West and exemplifies how quickly it could claim the imagination.

2: THE FIRST BOOM

Building the Community

The discovery of the silver-mines in Nevada gave all an excellent opportunity of gratifying their migratory instincts, and miners and men of all classes and all trades and professions flocked over the Sierras, in the spring of 1860.
—*William Wright,* The Big Bonanza

[Virginia City] is expected to grow more rapidly this spring, though the entire absence of wood, and water fit for drinking, in the neighborhood, will operate as a great drawback on its prosperity.
—Hutchings' California Magazine, *April 1860*

California emigrants, each hoping for a fortune, poured over the Sierra in 1860 at the first sign of winter's cessation, long before prudent travelers would have thought the trails passable. After slogging through mountain mud and snow, the would-be millionaires descended into valleys at the eastern foot of the Sierra that were lush with the first spring runoff. Still, it was a narrow band of green, and the travelers found a more characteristic Great Basin landscape of browns and grays as they ascended the foothills of the next mountain range to the east on the way to the new diggings. From valley floor to Virginia City, they climbed nearly 2,500 feet in a little more than a dozen miles. The wagon road led through spectacular, stark scenery. After following the Carson River to Chinatown, the oldest mining camp in the area, they took the trail north through Johntown and into Silver City, founded only a few months before at the mouth of the rock-bordered pass known as Devil's Gate.[1] From there the road grew steeper until it reached Gold Hill, and with each step, the travelers saw imposing Mount Davidson grow taller as their perspectives shifted.

On an early spring evening, the sun sets far enough north for its rays to slide between Mount Davidson and Mount Butler, just to the south. Huge, ragged rock outcroppings surrounded by patches of snow rise from Davidson's southern slope, looming over Bullion Ravine below. Brooding clouds often hover over the mountain late in the day, a sign that winter's storms could still unleash their fury even in April or May. At the same time, piercing rays of sunlight, bright in the thin air of almost 8,000 feet, catch the mountain and make the rock buttresses and snow gleam against the dark sky. Even the most insensitive is awestruck. Certainly this image greeted many emigrants traversing the last few miles before journey's end as they neared Virginia City in the spring of 1860.

J. Ross Browne, writer, artist, and sometime government official, was among those who set out for the Comstock early that year, only a few months after the call of the strike had sounded in the West and echoed throughout the world. The trip he describes was full of hardship, yet he was led on by his dreams of the riches to be won. He found a community that had clung to the mountainside through the previous winter, and it had not been easy. Snow fell incessantly, making warmth and supplies rare commodities. The few crudely built, drafty saloons became town centers, where miners collected more to spend the day by a well-tended stove than for companionship or drink. The local piñon and juniper trees, twisted and full of pitch, served poorly as firewood, but there was little else. Ancient groves fell to the ax that winter, forever changing the delicate ecosystem of the high, cold desert.[2]

Construction had halted as snow blocked supply routes. With the ground frozen, miners could not open new diggings. A few continued excavation on existing adits, but in general there was "not much doing, owing to the extremely cold weather," as reported in the *Enterprise.* At the first hint of spring, the community set to work and began its remarkable metamorphosis. Wright records something of this: "At first there was not sufficient shelter for the newcomers, and they crowded to overflowing every building of whatever kind in all the towns along the Comstock range. But houses were rapidly being built in all directions, and the weather soon became warm enough to allow camping out in comfort almost everywhere; men who had rolled up their blankets and slept on the snow, high up on the frosty Sierras, did not mind much sleeping in the open air on the lower hills."[3]

It was in the midst of this excitement, occasionally dampened by a late winter storm, that Browne and a thousand or more other fortune seekers arrived. For Browne, "the great mining capital of Washoe, the far-famed Virginia City" in the spring of 1860 could be summed up in the following way: "Frame shanties, pitched together as if by accident; tents of canvas, of blankets, of brush, of potato-sacks and old shirts, with empty whiskey-barrels for chimneys; smoky hovels of mud and stone; coyote holes in the mountain side forcibly seized and held by men; pits and shafts with smoke issuing from every crevice; piles of goods and rubbish on craggy points, in the hollows, on the rocks, in the mud, in the snow, every where, scattered broadcast in pell-mell confusion."[4]

Besides describing a setting long gone, Browne's account, published in *Harper's Monthly Magazine*, is a key to understanding the mind-set of the first to arrive in Virginia City. Initially, people came, as Browne points out, because of "the promise of unlimited reward" in a land where "any man who wanted a fortune needed only to go over and pick it up."[5] The lure of wealth in the mining districts of the West enticed emigrants to leave family and friends and to endure a dangerous journey. So intoxicating was the promise of gold that even when disappointed, the same people often followed the next call of a mineral bonanza.

The rush to Washoe and its Comstock Lode was nothing more than another in a long series. Some undertook the journey filled with hope. Other, more experienced citizens of the mining world wrote it off as "humbug," to use a term common at the time. They knew that rushes most often led to failure, and although the first to reach a truly rich mining district stood to gain the most, long-term prosperity was against the odds. Only the inexperienced answered the call of a rush, particularly one involving a hard-rock district. It was a naïve traveler who believed he could arrive early enough to stake a valuable claim and assume his place among the millionaires made rich from digging for gold and silver.

There were some exceptions, however. A few capitalists and entrepreneurs arrived early with an eye to exploiting the boom one way or another. Some people with money to spare realized that regardless of the actual value of a mining district, it was possible to ride a boom and turn a profit. By purchasing claims as they were rising, an astute investor could sell before the inevitable crash, clearing the difference as pure earnings. During the period of ownership, he could also mine some of the richer ore and enjoy the proceeds. Most of the earliest capitalists had little intention of lingering long. Many never visited Virginia City, but rather conducted their Comstock transactions from California. They hoped to exploit booms and knew that the best bet was to sell a claim when intuition suggested that its price had peaked. The most successful among them were a cynical breed who invested with the assumption that the mining district was destined to fail, intending to turn that probability to their gain. Timing was the only variable.

A drop in Comstock stock prices in April 1860 fulfilled the expectations of failure. For many experienced mining investors, the crash certainly would have signaled the end of the boom: the wise capitalist would have pulled out before stocks fell; the naïve rode them to the bottom. What may not have been immediately clear in the spring of 1860 was that the viable mines had survived. The crash involved peripheral properties with little or no real potential except that imagined during the early San Francisco feeding frenzy that demanded Comstock-area stocks regardless of worth.[6]

Although damaging to the entire district's credibility, the collapse of stock prices could not alter the fact that every day, miners brought fantastically rich ore to the surface. The reality of gold and silver provided its own momentum, and mining continued despite fluctuations in the stock market. As it turned out, the crash of 1860 was only the first of many for the Comstock. Unscrupulous capitalists found it all too easy and tempting to use worthless claims to mine the wallets of imprudent investors, so eager to participate in the boom that they would blindly buy anything within the Comstock Mining District. As Lord points out, "Paper fortunes were made in days by shrewd sales or rumors of rich 'strikes' and 'assays.'"[7] While the Comstock will always be remembered for the tons of gold and silver it produced, it was also the subject of stock manipulations and scams, bonanzas of a different sort. The two "mining" operations, one into pockets of ore and the other into pockets of fools, paralleled one another for decades.

Some entrepreneurs also cynically exploited rushes. They knew that during a boom resources were scarce and money free-flowing. Merchants, lawyers, and others arrived, planning to satisfy the needs of the mining community at the highest price possible. When Eliot Lord wrote his history of the Comstock twenty years after the first rush, he interviewed John L. Moore, one of the "shrewd traders, who saw a richer prize in the fortune-hunters than in their loadstone." Moore arrived in Virginia City in March 1860, his pack mules laden with blankets, tin plates and—most important—brandy, gin, whiskey, rum, and "assorted wines and liquors of various kinds." With this merchandise and a tent, Moore was able to establish a lodging house and bar capable of making immense profits. Moore claimed that he had "refused a cash offer of $8,000, five times the cost of his whole outfit, 'for the lot' on the day before reaching the camp."[8]

J. Ross Browne records a similar incident when, on his way to Virginia City, he met a man who specialized in sharpening tools. This small-scale capitalist pointed out that grindstones were scarce in the mining district and yet those who labored there were in constant need of refitting their tools. He brought a grindstone to Virginia City and made between $20 and $30 a day before his implement wore out. Browne met him as he traveled the road to California in search of a replacement.[9] Most of these entrepreneurs did not intend to stay a lifetime, but like the more powerful capitalists of San Francisco, they saw the short-term gain as sufficient motivation to participate in the boom.

Fortunately for subsequent historians, 1860, the first year after the strike, was also the time for the U.S. census. The manuscript census that the enumerators kept provides a snapshot of the incipient boomtown and its first entrepreneurs. Predictably, single miners dominated the area, representing nearly three quarters of the district (table 2.1). Others pursued diverse occupations and gave further definition to the budding community. Some, such as engineers, ore mill operators, and assayers, served miners and the mining industry directly. Merchants and clerks sold goods along the young commercial corridor that eventually became the main street of Virginia City, extending down through Gold Hill and Silver City. In addition, a full range of construction workers—carpenters, masons, laborers, painters, tinsmiths, stonecutters, and paperhangers—turned imported lumber into mine supports, mills, and places to live and work.

Laundrymen and -women, shoemakers, blacksmiths, butchers, bakers, gunsmiths, wheelwrights, saloonkeepers, lawyers, doctors, and cabinetmakers were all part of the vicinity's increasingly complex infrastructure by 1860. Other tradespeople, such as jewelers, gardeners, confectioners, and cigar salesmen, seem to have offered unlikely luxuries in a town struggling to meet basic needs, and yet they too were there. Shortly after the census, Virginia City even boasted a pottery shop, which supplied both domestic and industrial customers.[10]

Relatively soon after the rush to Washoe was well under way, newspapers with their assortment of editors, journalists, and printers began to arrive. The *Territorial Enterprise* was the first and most notable, gaining an international profile for reporting on some of the richest mines in the world. It would also become famous as the place where young Samuel Clemens got his start. He came west in 1861 with his oldest brother, Orion, who was to serve as secretary in the new territorial government.

2.1 Occupations for Comstock Men, 1860*

Occupation	Number	Percentage of Workforce
Mining	1,984	71.4
Construction	221	8.0
Teamsters/packers	176	6.3
Service	125	4.5
Mercantile	124	4.5
Saloons	61	2.2
Manufacturing	24	0.9
Infrastructure	23	0.8
Mills	7	0.3
Other	1	—
None	32	1.4
Total	2,778	

Source: 8th U.S. Manuscript Census of 1860 for Virginia City and Gold Hill, Utah Territory.
*The adult male workforce is defined as those 15 years or older. The total population for Virginia City and Gold Hill during the census of 1860 was 3,017.

The younger Clemens hoped that he too could secure a position in civil service, but as it turned out, Nevada was ill-prepared to support more than a handful of officials. He then turned to a variety of undertakings, including mining—at which he failed, lacking both skill and an interest in hard labor. Eventually Clemens settled on reporting for the *Enterprise*, where he eventually took the pen name Mark Twain. Because of his later fame, Clemens's Comstock period became legendary. Unfortunately, little concrete information from those early days exists to add detail to the early development of Twain in Virginia City. His autobiographical *Roughing It* chronicles his western experience and is filled with valuable portraits of the early Comstock, but like so much crafted with his sly pen, fabrication often shades the details.[11]

None of the earliest Comstock development, from opening a mine to building a house to running a newspaper or a cigar store, would have been possible without the teamsters and packers who transported tons of material over Sierra passes on roads that were little better than trails.[12] Winter inhibited these couriers, but with the spring, they responded to opportunity like everyone else did. By early March, mines were extracting ore, and since work on mills had only just begun on the Comstock, most of the rock had to be hauled over the Sierra, to San Francisco or elsewhere.[13] On the return, the wagons and teams strained with the weight of much-needed supplies. Scarcity of teamsters and packers combined with a voracious market to drive prices up. The *Territorial Enterprise*, for example, complained that "the local trade of our valleys employ all the teams that can be procured. They are charging $100 per [1,000 board feet of lumber] . . . for hauling from here to Virginia City, a distance of 15 miles."[14]

As the weather improved, more teamsters and packers joined the ranks of their brethren, and prices for both material and hauling tended to drop. Still, with the first sign of snow the following winter, prices rose again, illustrating the volatile nature of the young society and its economy. Despite the seasonal decline in prices during the summer, the teamsters and packers could hardly meet the need. The growing communities constantly demanded more supplies, making those who hauled for the Comstock a prominent fixture of the prerailroad western landscape. As Mark Twain observes, "It did seem, sometimes, that the combined procession of animals stretched unbroken from Virginia to California. Its long route was traceable clear across the deserts of the territory by the writhing serpent of dust it lifted up." Lord estimates that by May 1860 more than two thousand animals hauled ore and material on the Comstock. The 1860 census, enumerated in August, records almost four hundred teamsters and packers in the western half of the Utah Territory that would become Nevada after 1861. This amounted to 6 percent of the region's population, and even those numbers apparently fell short of the need: Lord maintains that two hundred more teamsters joined their ranks the next year on the Comstock alone.[15]

Once in town, the hundreds of wagons, their mule teams decorated with sets of chimes and bells, created a deafening cacophony of snorting animals, clanging metal, and shouting teamsters spiced with the odor of sweat and excrement. The ensuing chaos inspired William Wright to comment that the wagons and animals produced the "most vexatious blockades in the streets . . . owing to some accident or to mismanagement on the part of some teamster, and, teams rolling in from each side."[16] These, however, were the necessary growing pains of an infantile behemoth. Even if the process turned Virginia City into an enormous traffic jam, none could curse the arrival of supplies or the departure of ore, because both were crucial to the existence of the community.

Hauling ore was no easy task. The gold-and-silver-bearing quartz made for heavy, awkward loads. The first shipment by Walsh and Comstock in 1859 weighed 3,151 pounds. It took half a dozen mules to pull

Teamsters were extremely important in building the Comstock. Wagons are taking on ore from ore bins. The newly constructed Fourth Ward School stands above on the hill, 1878. Photograph by Carleton Watkins. (Courtesy of the Nevada State Historic Preservation Office)

the wagon along the poorly engineered and maintained roads. By 1862 several toll road companies had constructed upgraded highways that allowed for larger teams and wagons. It cost a half million dollars just for the 101 miles from Virginia City to Placerville, but the effort allowed a dozen or so mules to haul huge wagons. On occasion teamsters hitched two, three, or even four wagons together so that extraordinary loads were possible.[17]

Among the ore wagons were dozens of stagecoaches, speeding along with their lighter loads. In this time before railroads in Nevada, the stagecoach driver put a unique stamp on the image of the West. Men like Henry James "Hank" Monk served the Comstock and drove his way into legend. The most famous of his profession, he was known for his skill and speed. Many of the West's more notable visitors rode with him.[18] One story places no less a personage than Horace Greeley in Monk's coach. According to the tale, Greeley urgently needed to reach Placerville and egged Monk on as the coach ascended the east slope of the Sierra. Unable to comply because of the grade, the famed driver rushed down the west side of the range at what Wright called "a fearful rate of speed." Greeley now pleaded with Monk to slow down, but the expert stagecoach man knew his business and refused to comply. The idea of Hank Monk, hero of the western highways, teaching a lesson to the famous writer from the East amused the residents of the Comstock and the mining West, and the story assumed legendary proportions.[19]

Nine Bactrian camels, first employed in 1861 to carry salt from the interior of the Great Basin, provide a peculiar footnote to the early history of hauling on the Comstock. Each camel could carry a remarkable 950 pounds. True to their reputation, they needed little water and could traverse sandy terrain effortlessly. They were less well equipped for some of the more rocky paths, however, and teamsters and packers regarded them as a public menace, since the sight of a camel reputedly would spook a mule or horse. Teamsters finally coerced the state legislature into banning camels, but the animals nevertheless symbolize the early period of regional development, when Comstockers creatively tested possibilities as they invented themselves and their remote hard-rock mining district.[20]

Whether by camel, pack mule, or wagon, teamsters and packers brought the materials that the area needed for building, and they exported ore so that capital would return. For these transients, capable of moving from place to place, it mattered little if the Comstock proved to be a long-term success or not. For others, it did. Many came to the Com-

Experiments in transportation on the Comstock included the use of camels. (*Harper's Weekly*, June 30, 1877; courtesy of the Nevada Historical Society)

stock believing, or at least hoping, that the district would survive. These people contrasted with those shrewd newcomers who considered it unlikely that the Comstock's fabulously rich ore could also be extensive. Those who would exploit the scarcity of resources and those who would build a community tugged the mining district in opposite directions. The former thrived on high prices for short-term gain; the latter hoped for a place to settle for a lifelong occupation; the cynic versus the optimist, both poised to gain from an economy whose fate remained undetermined. Regardless of their motives, those who arrived early began building a town as if it had a future: the cynic, because the appearance of longevity was important for profits; the naïve, because he did not know better. They also built, as Lord remarks, "with surprising rapidity [replacing] canvas tents and hovels [with] board cabins and business houses." Providing a firsthand account, an April 1860 article in *Hutchings' California Magazine* pointed out that Virginia City "at present contains about a dozen stone houses, two or three times as many built of wood, of every size and description, and with a number of tents, shanties, and other temporary abodes."[21]

Typically, as a mining district settled into its first boom, most of the naïve arrived with little capital and too late to join the ranks of the first discoverers. Even when these inexperienced would-be miners secured title to what might have been a profitable mine, they lacked the finances to develop it. With nothing to sell but the strength of their backs, they usually left or became laborers for those who held the best claims and had the funds to build an expensive silver mine. Others assumed the role of ne'er-do-wells, surviving, as Lord describes, by "borrowing money from lucky gamblers, haunting free-lunch counters, pledging 'feet' [of supposed claims] with reckless butchers and bakers, or picking the pockets of good-natured friends."[22]

Trained wage-earning miners rarely turned out for a rush because the experienced knew that attention-getting strikes usually involved little viable ore, a brief period of profitable excavation, and no real justification for relocating there. The *Territorial Enterprise* even suggested that the standard wage of $4 per day for experienced miners would not cover expenses and was "scarcely above the standard of mining wages in California." The newspaper advised the professional who might be attracted to the boom to reconsider the trip: "Better far he had remained in California, than ever lift a pick [on the Comstock] . . . at such rates." Thus, Cornish immigrants were practically nonexistent in Virginia City during 1860, even though those experienced miners were a fixture of the mining West. They certainly understood mining well enough to know that a

Virginia City in 1860 was still a place of tents, but a few substantial buildings had emerged. Lithograph by J. Ross Browne. (Courtesy of the Nevada Historical Society)

proven district was the best place to find a reliable wage at the highest price. Although common from Michigan to California, the Cornish mining experts rarely assumed the role of risk-taking prospectors or investors. These professionals arrived after prospects for employment were secure.[23]

The 1860 census also shows a community with few, if any, prostitutes. Of the 111 women who lived in Gold Hill and Virginia City in August 1860, 83 were with husbands, and they cared for more than 100 children. Most of the remaining single women were apparently not prostitutes.[24] It appears, therefore, that women involved in sexual commerce realized, like the Cornish, that the odds were against a booming community, making the cost and inconvenience of moving imprudent. Both groups apparently waited to relocate until a mining district demonstrated that it would last. Besides saying a great deal about nineteenth-century prostitution, this assertion challenges the stereotype of the raucous, devil-may-care western boomtown. The myth of the Wild West, like the story of the fools who discovered and then lost a bonanza, runs deep in the nineteenth century. Lord echoes the idea as early as 1883, when he writes of the early Comstock: "Prostitution flourished, as in all large camps, and courtesans promenaded the streets slowly, decked out in gay dresses and showy jewelry, drifting about with the restless tide which set to and fro through the city." Clearly the idea of a freewheeling society where prostitutes plied their trade openly, unrestrained by moral restrictions, has run deep in the regional myth for more than a century. That most Comstock women in 1860 were respectable by the standards of the day defies folklore.[25]

At least part of the image of the Wild West was valid, however. Many of the people who first came to the mining district contributed to an early era of violence, reinforcing the stereotype of the frontier as popularly imagined. Indeed, Mark Twain captured the gray area between image and reality regarding violence on the early Comstock when he wrote: "The first twenty-six graves in the Virginia cemetery were occupied by *murdered* men. So everybody said, so everybody believed, and so they will always say and believe. The reason why there was so much slaughtering done, was, that in a new mining district the rough element predominates, and a person is not respected until he has 'killed his man.'" Lord aptly calls the rough element "the floating scum of the California mining towns [that] drifted naturally to the new camp." Both he and Wright cite one "Fighting Sam Brown" as a particularly good example of the brutal sort who made for a savage time. Historian Sally Zanjani has shown that their depiction of Brown deviated from the actual character so that he could play the part of stereotypical villain and enhance the myth.[26]

Brown arrived in Virginia City in early 1860, and as Wright describes it, "he walked into a saloon, a side at a time, with his big Spanish spurs clanking along the floor, and his six-shooter flapping under his coat-tails." According to the stories that would later be told, Brown almost immediately established his credentials by killing a man with his bowie knife. Satisfied with his work, the murderer fell asleep on a nearby pool table as others rushed to the aid of his victim, who was struggling in his death throes. Lord and Wright claimed that Brown was merely one of the more notorious of the many violent men who swarmed over the Comstock in its early days.[27] Brown met his demise in July 1861 in Carson Valley, to the south of the Comstock. He tried to kill Henry Van Sickle, a local hotelkeeper and future Nevada assemblyman, who took umbrage at the assault. Van Sickle chased Brown and killed him with "a heavily loaded double-barreled shotgun."[28] Although Zanjani has been able to unravel the literary myth from what might have been the true nature of Sam Brown, he nevertheless became a symbol, fairly or unfairly, of the violent possibilities of the times.

Clearly ruffians like Brown could be a nuisance, and the violence they perpetuated claimed headlines, but whether they were the driving force in early Comstock society is a greater question. The primary documents from the Comstock in 1860 have little to say about violence. The *Territorial Enterprise*, for example, seems more concerned with daily shipments of supplies, ore discoveries, and stock prices than with what Twain, Lord, and Wright characterized as an epidemic of murders.

Since the publication in the 1870s and 1880s of the earliest histories of the Comstock, it has been all too easy to typify the early period of boomtown Virginia City as a time of murder and lawlessness. In reality, the people who lived there strove for order early on and did not see their community as others would who later fit it neatly into the myth of the Wild West. The Browns of the Comstock existed, but ultimately the Van Sickles could and did put them in their place. Just as the first Comstock families, though rare, far outnumbered the single women, peace and civility may have been more common than often characterized. The seeds of a stable city were sown early. A Swiss visitor who arrived on the Comstock in 1864 observed: "Regardless of the pugnacity of the Americans, one seldom hears of murder or injury and there are many false reports, assuming the form of folklore, circulating throughout Europe that should be corrected."[29] Indeed, the other early authors on the mining district could have as easily focused on the growth of the community as illustrated by the founding of churches and congregations during the first years of the Comstock, but that lacked the allure of mythic violence. The oft-quoted passage from Twain's *Roughing It* again reinforces the image of the Comstock as part of the wild West. He writes: "There were military companies, fire companies, brass-bands, banks, hotels, theaters, 'hurdy-gurdy houses,' wide-open gambling-palaces, political pow-wows, civic processions, street-fights, murders, inquests, riots, a whiskey-mill every fifteen steps, . . . a large police force, two Board of Mining Brokers, a dozen breweries, and half a dozen jails and station-houses in full operation, and some talk of building a church."[30] With the art of exaggeration, Twain chose to ignore the fact that congregations and churches in Virginia City predated his arrival late in 1861. The Reverend Jesse L. Bennett began ministering to a Methodist congregation as early as 1859. According to tradition, those who heard his sermons answered with heaps of gold and silver. By 1861, his successor had built a church. Similarly, Father Hugh Gallagher came to the Comstock in 1860 and soon built a small Catholic church. Within two years, the structure collapsed under the famed heavy winds known locally as Washoe zephyrs. Still, this house of God served, at least temporarily, as a bulwark against criticism that the community was lawless and immoral.[31] Religious life on the mining frontier, however, does not make good press, and it does not fit the image of the Wild West.

Some of the violence of the early Comstock grew out of the region's diversity and the contempt with which different ethnic groups occasionally regard one another. A growing number of people from throughout the world flocked to the mining district to claim some of its successes. These people were the foundation and early building blocks of the cosmopolitan character that was the hallmark of the Comstock. J. Ross Browne, after his visit to Virginia City in the spring of 1860, wrote of a diverse community, including Jewish merchants "setting out their goods and chattels in front of wretched-looking tenements. . . . Now and then a half-starved Pah-Ute or Washoe Indian came tottering along under a heavy press of fagots and whisky. On the main street, a jaunty fellow . . . dashed through the crowds on horseback, accoutered in genuine Mexican style, swinging his 'riata' over his head."[32]

The 8th U.S. Manuscript Census, in 1860, captures the diversity of the Comstock, portrayed less eloquently but certainly more precisely. The document records a community of people from throughout the world. Besides the majority born in North America and the American Indians, who had been there longer than all the rest, there were Irish, Germans, English, Asians, Scandinavians, Russians, Scots, Italians, Poles, Hungarians, South Sea Islanders, and a thirty-year-old miner from Portugal. And there were many others.[33] Each of these groups eventually played an important role on the Comstock, but at this early date they represented little more than an indication of things to come (table 2.2).

In 1860, 30 percent of the people in the part of Utah Territory that was to become Nevada came from other countries. This compares with 39 percent of the Californians and 27 percent of those living in Washington Territory, while contrasting with only 7 and 8 percent of the people of New Mexico and Colorado Territories, respectively. Even the states to the east that had major ports could not match these early concentrations of the foreign-born on the Pacific Coast. New York (25 percent), Massachusetts (21 percent), and Illinois (19 percent), often regarded as having

2.2 Nativity of People Living on the Comstock, 1860*

Nativity	Number	Percentage Female
USA	1,949	5
Ireland	310	4
Germanies	230	4
Britain	165	6
Canada	118	3
Hispanic	99	17
Scandinavia	41	2
France	27	4
China	14	—
Switzerland	13	8
Italy	9	—
Russia	6	—
Other**	36	6
Total	3,017	5

Source: 8th U.S. Manuscript Census of 1860 for Virginia City and Gold Hill, Utah Territory.
*Of those born in the USA, 7 were African American, all of whom were male. "Hispanic," while not a nationality, allows for the grouping of all Spanish-speaking people from diverse locations. Females include women and girls.
**19 groups.

large populations of immigrants, had more native-born citizens than the Far West did. When compared with two southern states that had major ports—Georgia with only 2 percent foreign-born and South Carolina with a little more than 1 percent—the significance of Nevada's ethnic composition is even more striking.[34]

In the first year after the strike, Hispanic immigrants from Mexico and South America were a well-represented ethnic group on the Comstock. Although Spanish-speaking residents eventually lost their prominent position to later tides of newcomers, during the early years this part of the community was one of the largest and most visible.

There were ninety-nine Hispanics listed as born in Mexico, South America, Spain, Panama, New Mexico, California, and Utah who were living in Virginia City and Gold Hill at the time of the 1860 census.[35] About 14 percent of the Spanish-speaking adults were women, a remarkably high percentage compared with that of non-Hispanic women, who accounted for only 3 percent of the non-Hispanic population on the Comstock that year. In keeping with this trend, about 12 percent of the Hispanics were children under sixteen years of age, living with their mothers. With less than 4 percent of the rest of the population fitting this demographic profile, Hispanics represented a startling deviation from the rest of the community. Clearly, the speakers of Spanish laid the strongest foundation of family life on the Comstock during its earliest development.

Most of the Spanish-speaking men were packers, and they apparently lived together, suggesting that Euro-Americans restricted people of their ethnicity and occupation to a designated area.[36] Still, others did have the latitude to explore different options. After all, in this group there were also three merchants, two saloonkeepers, a shoemaker, and a teamster.[37] In addition, fifteen Hispanic miners and mine owners provide one of the more provocative expressions of the diversity and importance of the Spanish-speaking community of the district. By the early spring of 1860, the Maldonado brothers and at least nine associates from Mexico had claimed and purchased several valuable ore bodies. Their efforts focused on land just north of the Ophir, and all indications were that it would be the site of an extremely profitable mine. Indeed, for several years it was. The Maldonados built a three-story brick manager's house by 1861. The building was one of the most substantial of its day, and they became prominent members of the community.[38]

Besides operating famously profitable mines during the early Comstock period, the Maldonados employed techniques that ran deep in the history of mining technology. After interviewing Philipp Deidesheimer, an early engineer on the district, Lord wrote: "In the Mexican Mine even hand-cars and buckets were not made use of, but the rock was conveyed to the surface in raw-hide sacks, bound with a strap over the forehead." The strategy employed at the Mexican Mine represented more than a mere novelty. It drew on centuries of Spanish mining theory and practical refinement that developed a method traditionally referred to as *el sistema del rato*, literally "the system of the moment." The phrase is more loosely and better translated as a pragmatic or empirical system. As Otis E. Young, mining historian, points out, late nineteenth-century British and American miners "scornful of all things indigenous, willfully mistranslated the phrase as 'rat-hole mining.'" In fact, the Spanish and Mexican method was efficient and well suited to the sixteenth century (when it was first developed), and it could be a profitable means to work certain ore bodies. Young proceeds to explain that "the Spanish rat-hole miners did almost nothing *but* drift or sink on ore, their workings winding in and out in labyrinthine fashion; so to this day do small-time operators. Engineers of a well-managed mine, however, attempted to work out the shape and size of its ore body so as to plan and construct access workings to the best over-all advantage." When contrasted with the other Comstock mines whose managers generally aspired to engineering according to contemporary standards, the Mexican Mine was an archaic curiosity. The contrast between the two methods, however, while resulting in visible differences, did not mean the Hispanic operation was less successful. The Maldonados' mine was one of the Comstock's best producers during the early 1860s.[39]

Through their role in the Mexican Mine, their families, and their participation in the transportation industry and other occupations, Hispanics occupied an important place in early Comstock society. There are examples of antagonism between the Euro-American and Hispanic communities, but in 1860 it appeared that the two would coexist and prosper in the mining district.[40] Eventually, however, other emigrants became far more numerous than Hispanics, relegating them to the role of colorful reminder of a past when Mexico controlled the area before 1848 and the end of the Mexican-American War. The Comstock population grew steadily in number and diversity, and the relative importance of the Hispanic community declined accordingly.

Of the people who arrived in the new mining district in the early 1860s, some saw insufficient profit in a place where mining required capital and the richest claims were few and already taken. Many left, discouraged, but even more new arrivals took their place, following a trend common in boomtowns. The population rotated as it grew, losing nearly as many as it gained.[41] The society was in continuous flux. During the federal census of 1860, the population of Virginia City and Gold Hill was 3,017. One year later the territorial census recorded 3,284, an increase of fewer than 300. The next year, the territorial project documented 4,437 residents. Still, thousands more certainly had come and gone during those dynamic early months. While the Comstock Mining District peaked some fifteen years later with about 25,000 people, far more had lived there over the twenty years of its prosperity.

In the spring of 1860, however, the numbers were still relatively low. Those who had wintered on the mountainside could be counted in the hundreds. Warmer weather unleashed the hordes waiting for an opportunity to join the boom, and with the first cessation of snow the chance to build a mining district seemed at hand with apparently nothing else to stop the inevitable growth. As supplies reached the newly founded stores, the Comstock began to assume some of the trappings of civilization.[42]

R. P. Leitch drew this image of Gold Hill during its earliest period of development. (Courtesy of the Nevada Historical Society)

Nonetheless, the spring of 1860 saw another kind of storm—yet another aspect of ethnic diversity—that halted mining, building, and day-to-day existence as effectively as a winter blizzard.

Nevada's motto is "Battle Born"—ironic words, since the state has had fewer casualties fall in battle within its borders than most others in the nation. The "battle" in question was the Civil War, thousands of miles away, but Nevadans have always been proud of their role as a pro-Union state when the country needed every friend it could find. In spite of this pride, the state's history is one of nearly unremitting peace.[43] One of the few military conflicts in the area now known as Nevada occurred during the initial American Indian response to the sudden and dramatic intrusion onto their land that followed the Comstock silver strike.

Increased emigration into the Great Basin by Euro-Americans caused growing violence between newcomers and American Indians. It is impossible to determine who initiated the first isolated conflicts, but clearly tension escalated between the two groups, and all eventually realized that peaceful coexistence was unlikely. Before winter ended, the Euro-Americans were calling for expeditions to punish local Northern Paiutes for perceived infractions. In February 1860, for example, the Honey Lake Rangers declared that they "intended to take an Indian hunt, as soon as the weather will warrant." Confident that extermination of the American Indians was possible, newspapers such as the *Territorial Enterprise* complained that Paiute transgressions were adding to the bellicose specter hanging over the young year. Isaac Roop, bearing the unofficial title of provisional governor of Nevada Territory (whose creation was a year away), issued a proclamation on February 16, 1860, stating that war with the Paiutes was inevitable because of trouble in the Honey Lake area.[44]

That same spring, a gathering of a thousand or more Paiutes and their close allies, the Bannock Shoshone, at Pyramid Lake heightened Euro-American suspicions. Although later interviews with American Indian participants in the events of 1860 provide valuable information, sources are problematic and should not be taken at face value. Nonetheless, later reports indicate that most of the Paiute and Shoshone leadership recognized the inevitability of war. According to tradition, there was only one lone voice holding out for peace: Numaga or "Young Winnemucca," a leader of the Pyramid Lake Paiutes, spoke eloquently against war. Addressing the gathering, he maintained that "the white men are like the stars over your heads. You have wrongs, great wrongs, that rise up like those mountains before you; but can you, from the mountain tops, reach and blot out the stars?"[45] Numaga recognized that the Euro-Americans would use superior arms and numbers to win any war with the Paiutes and Shoshone, and he counseled for a peaceful resolution of their differences. Tradition takes over at this point with all the drama of theater, suggesting that just as Numaga finished his speech, a messenger arrived with news that there had been more killings. Numaga bowed to the irresistible flow of history and declared that all should prepare for war.

The evening before, on May 7, several Northern Paiutes killed five Euro-Americans at Williams Station, a stage stop in the desert east of Virginia City. The men at the station had captured two young American Indian women, whom they had abused. When the Paiutes discovered the outrage, they swung into action and executed their own form of frontier justice. Still, as Lord points out, "pioneer lynch-law was very different from Pi-Ute lynch-law."[46] The Euro-American community demanded retribution. Swollen with confidence, the Comstock and neighboring areas raised a force of 105 men that behaved more like a mob than an army. They marched north and on May 12 encountered the first small band of American Indians, just south of Pyramid Lake. After the Euro-Americans shot at a Paiute carrying a white flag, his companions fired back and retreated. For the motley Comstock army, this first engagement certainly appeared to be more of a nuisance than a danger. In short order, however, several well-organized groups of Paiutes and Shoshone appeared, opened fire, and began cutting off avenues of retreat. The Euro-Americans panicked, and all military formalities evaporated.

During the ensuing rout, seventy-six of the ragtag army died. Many of the twenty-nine survivors were wounded, and had it not been for nightfall, many more would have been killed. It was a devastating defeat. A tombstone in California's Nevada City captures the agony of the engagement, eloquently if also romantically and fancifully portrayed. It memorializes Henry Meredith, a prominent lawyer, who lay wounded, unable to move from the path of the pursuing American Indians. The large obelisk quotes his final words to his friends, who maintained that they wished to help him. Instead, he insisted, "Go! Leave me here; I might put you in peril."[47]

No one underestimated the American Indians again. A hush descended upon the Comstock as the terrified district prepared itself for a siege. In Virginia City, women and children gathered into a stone hotel that a man named Peter O'Riley was constructing. Resembling a fort, it could serve as a last defense. The Silver City militia established a stronghold at the top of Devil's Gate and armed it with a homemade wooden cannon. Many people fled back to California, instantly cured of the rush fever that had drawn them to the Comstock. At the same time, volunteers from California began crossing the Sierra, fulfilling Numaga's prediction that the Indians could not blot out the Euro-Americans. One of the new arrivals was a doctor named Edmund Bryant, who came to help in the struggle, not realizing that his destiny, and that of his family, would later unfold on the Comstock. He wrote east to his father, "I met trains of people and stock on their way to California flying from the Indians."[48]

Within two weeks, more than 500 men had gathered in Virginia City. They headed north on May 24, 1860, to rendezvous with 260 regular U.S. troops south of Pyramid Lake on May 31. The numbers no longer favored the American Indians, and no amount of superior strategy could save the day for them against an organized, well-armed force of this size. The Euro-Americans moved north toward Pyramid Lake, meeting only minor resistance along the way.

This second campaign was decisive. With a few brief battles, the American Indians were able only to delay the advance of the invasion force, thus gaining enough time to secure the retreat of their women and children into the desert north and west of the lake. Otherwise, the Northern Paiutes and Shoshone were totally defeated. With U.S. troops on the scene, the volunteers were able to disband soon after the battle, returning home victorious. The Comstock welcomed its men back with a raucous party, and the district never felt the threat of American Indian hostility again. Sometime after the celebration had waned, the Silver City militia ignited their homemade cannon, which promptly exploded. Had the American Indians attacked, as feared, the piece of armament would have killed many defenders and proved to be one of the Paiutes' best weapons.

During the conflict, which came to be known as the Pyramid Lake War, all agreed to freeze the mining district as it was: no one could file new claims, and claims not worked were not forfeited. Survivors could sort out the needed alterations later. With nearly one hundred casualties among the mining communities and with many others who fled at a time when Virginia City's population certainly numbered no more than two thousand, some changes were likely. Nonetheless, the Comstock quickly returned to the business of mining. Within a week or two, a new arrival to town would have seen little indication that war had broken the stride of the district. The rumble of the mining machine resumed.

The Paiutes, exiled to the desert at the onset of summer and removed from their traditional seasonal food sources, faced starvation. They eventually sued for peace and returned to Pyramid Lake. Elsewhere, hostility ran high, however, and other American Indians in the Great Basin increased attacks on Euro-Americans. These skirmishes forced U.S. troops to remain and establish several forts in the area to protect east-west transportation routes.[49]

American Indian hostility during 1860 threatened the long lines of contact between the Comstock and the outside world. The expansiveness of the place, besides creating an obstacle to teamsters and packers, made communication and the administration of government difficult under the best of circumstances. In the days before the arrival of the transcontinental telegraph, news and all correspondence, whether vital, official, or personal, had to be relayed via hand-carried mail pouches. The size of the nation, particularly with its western additions, presented a dramatic challenge.

In April 1860, the firm of Russell, Majors, and Waddell initiated the Pony Express to bridge the gap between the eastern United States and the young West Coast settlements. Although it survived for only seventeen months, the Pony Express is permanently etched into the mythic image of the region. Young men carried mail pouches on swift horses, using a continuous relay between more than 150 stations. Repercussions from the Pyramid Lake War made the Great Basin one of the most dangerous expanses to cross. Without an expedient alternative, however, the Pony Express provided the quickest option: delivery took about twelve days from St. Joseph, Missouri, to Sacramento, California. It was a heroic attempt to defeat the distance, but the mail service failed to improve on a system of horse-carried hand delivery that had been available to the Romans and Egyptians. On October 24, 1861, a completed transcontinental telegraph line ended the need for the Pony Express.[50] The telegraph revolutionized communication for the nation's newest region and proved to be one of the most significant steps in dealing with the expansiveness of the West.

Although the telegraph was an important example of technology employed to master the environment, it only slightly relieved the local appetite for immediate access to government. There was no technology in the foreseeable future that could reduce the need of having lawmakers, enforcers, and a court system in one's own backyard. The embryonic Comstock was only slowly taking shape in 1860. Similarly, its place in the nation remained ill-defined. That Comstockers wanted a role in a larger setting is clear. In 1857 residents of the western Great Basin began seeking territorial status separate from Utah. With the silver strike and the subsequent rush to the Comstock, there was a growing population with little patience for the distant government in Salt Lake City.

As early as December 1859, Virginia City sent a substantial message to the nation's capital that it intended to gain independence. This took the form of a block of ore to be added to the Washington Monument, currently under construction. The *Territorial Enterprise* reported that the "mammoth lump of silver ore . . . is two feet six inches long, ten inches wide and six or eight inches thick . . . [weighing] 163 pounds and is valued at $600." Intended to "open the eyes of Congressmen to the importance of our mining interests," the shipment drove home the demand for sovereignty with the inscription "NEVADA," an act of anticipation more than a year before the creation of the territory. Attempts by Salt Lake City to institute local governments came too late. The Utah legislature authorized the incorporation of Virginia City on January 16, 1861, for example, but it was only a matter of months before President Buchanan signed the federal act creating Nevada Territory on March 2 of that year.[51]

When Governor Nye, a Lincoln appointee from New York, arrived as the territorial governor, he assumed his office in Carson City, in a valley below the Comstock. Nonetheless, it was clear then and for the next twenty years that the mining district was Nevada. With its large population and with its wealth that drove the region's economy, the Comstock dominated state politics. Carson City became a satellite of Virginia City, not the other way around.

Sir Richard Burton, world traveler newly arrived from his expedition to the Nile, traversed the Great Basin in 1860. This illustration of Virginia City dates to October of that year. Burton stood next to the Ophir Mine, the richest of the day, to draw this image. (Courtesy of the Nevada Historical Society)

The wealth of the Comstock commanded the attention of the president and the world. The mining district flashed into existence with a flame bright enough to be seen around the world and captured people's attention as only fabulous wealth can. In the year after the first strike, emigrants and the curious began arriving from everywhere. Among the first sightseers was Sir Richard Burton, world explorer and cosmopolitan author. After his attempt to find the source of the Nile, he had come to western North America to see this new wilderness. In October 1860, he sat at the edge of the growing open pit of the Ophir Mine, overlooking young Virginia City. He was one of the first of thousands of international travelers who included the Comstock as a "must-see" on their world tours. Presidents, Irish revolutionaries, suffragettes, opera stars, millionaires, royalty, advocates of free love, and a multitude of others would eventually visit and add their names to the history of the place, striving to touch, no matter how vicariously, the wealth that burst forth there.

In October 1860, however, no visionary could have foreseen the degree of notoriety that the Comstock would achieve. Burton sat there looking at a boomtown, but in fact he was watching an infant whose full potential, whose role as one of the Wonders of the World, remained to be realized. As a careful observer, Burton no doubt noted what was there at the time—a place of growth, diversity, energy, and most important, mining riches. After a brief stay, he left. As a contemporary, he could not benefit from historical perspective and would have found it harder to recognize the Virginia City that was attempting to lay a foundation for permanence.[52] A budding commercial district, the construction of hundreds of buildings, and the presence of scores of families hoping for a long and prosperous future were also part of the place. Two months before Burton arrived, on August 25, 1860, community leaders set aside ten acres for a Catholic cemetery near their new church.[53] This was no boot hill for the murdered ruffians of a boomtown, the Virginia City that Twain describes as initially burying only the victims of violence. This was a spacious plot of land, the last resting place for the devout members of a community. With the amount of real estate involved, the plan was clearly for growth and longevity. Nevertheless, whether the community would even survive remained a question unanswered in 1860.

3: THE FIRST BOOM

Building the Mines

Storey County is almost entirely destitute of timber. . . . It is equally destitute of agricultural lands, as well as grass and water.
—Henry DeGroot, from Kelly, First Directory of the Nevada Territory

Unfortunately, the Comstock, with a local government, international recognition, and community infrastructure, quickly encountered mining difficulties that proved even more of a challenge. Nature provided a wealth of gold and silver but little else. The place was generous on the one hand, but parsimonious on the other. Seemingly insurmountable problems faced those who would unlock the treasure chest. Subsequently, much of the early history of Comstock mining involved coming to technological grips with the situation. Indeed, the early 1860s saw several obstacles arise, each with the potential for ending prosperity.

Miners on the Comstock almost immediately experienced problems involving milling. While the first participants in the strike easily scooped ore from the rich surface outcroppings, their successors needed a cheap, effective means to unlock the silver and gold from a tightly bound quartz matrix. Even before the 1859 strike, there were attempts at milling, but retrieval of minerals combined with other rock is costly and time-consuming at best, and the ore must be valuable enough to justify the effort. That single fact perpetuated placer mining in the district during the 1850s for as long as it was possible to mine free gold and avoid the more obstinate quartz ore. Several factors created the need for milling. First, mining had depleted the easily processed gold-bearing sands. In addition, the 1859 ore strikes of gold ore ensured that mills would soon follow. The discovery of plentiful silver, chemically bonded to other materials, however, meant that milling would need to assume new forms.

After the strikes, some mine owners shipped ore to San Francisco for milling, and in some cases to England. Clearly this was a costly, inefficient method, impractical as a permanent solution. At the same time, entrepreneurs established local mills to handle the tons of material. Since the types of valuable minerals and the extent and nature of chemical bonding can all be different from district to district, nineteenth-century millers usually tailor-made their process for each location. Often lacking a profound understanding of the principles of chemistry, they usually relied on tradition, intuition, and trial and error to arrive at the best means of retrieving valuable minerals.

Milling begins with the crushing of ore into small bits; the next step typically employs water, heat, or chemicals to separate the mineral from the rock. The earliest and simplest method of milling employed on the Comstock involved the Mexican arrastra. In 1859, John P. Winters, Jr., and John "Kentuck" Osborn obtained partial rights to the original claim of O'Riley, McLaughlin, Comstock, and Penrod on condition that the new partners build two of these mills. An arrastra consisted of a circular bed of flat stones, upon which the miller placed the ore. An animal tied to a central pivot dragged a rock over the material until it was pulverized. The addition of water, salt, and copper sulfates with the heat of the sun turned the crushed ore into a thick paste. The miller then introduced mercury, which combined with the silver and gold in a heavy amalgam that settled to the bottom. By removing the material on top, the miller exposed the amalgam, which he then heated to drive off the mercury. Mining first used this "patio process" in Mexico in 1540.[1]

The cheap and simple arrastra remained in use throughout the early history of the Comstock. These mills were small and labor-intensive, however, and they were unable to handle large amounts of ore efficiently. In addition, the long, cold Nevada winters deprived miners of the sun's natural heat throughout much of the year. The Comstock demanded alternative methods.

A mill owner from California's Nevada City named Almarin B. Paul was the first to address this need on the Comstock. In March 1860, he gathered investors under the Washoe Gold and Silver Mining Company No. 1. After selecting a mill site with ample water, south of Virginia City near Gold Canyon, he began securing contracts for working ore at $25 to $30 a ton. A condition of the agreements stated that Paul would start

In 1875, *Frank Leslie's Illustrated Newspaper* published this image of two approaches to milling. In the foreground, horses tread around arrastras. In the background is the Gould and Curry Mill, which used state-of-the-art technology. (Courtesy of the Nevada Historical Society)

milling by August 11, 1860, sixty days after the signing of the first contract. Paul ordered the forged components for the mill from San Francisco and immediately began constructing the foundation and shed for the operation. Completed with amazing speed, his mill used a process that others followed on the Comstock in modified form throughout most of the nineteenth century. The mill first pulverized the ore with twenty-four stamps. These were pistonlike iron hammers that a steam engine lifted and dropped alternatively, slamming them down on a bed.[2] The moment the stamp mill began to work, the Comstock became a noisier place. After the stamps completed their work, mill hands shoveled the resulting powder into large iron pans, each holding three hundred pounds, together with warm water, forty pounds of mercury, a pint of salt, and some copper sulfate. A revolving muller, a device that looked like a pair of revolving doors, churned the pulp. This gave the mercury an opportunity to contact as much of the resulting free gold and silver as possible. The bits of gold, silver, and mercury, being heavier than the rest, dropped to the bottom, and the mill hands drew it off through a discharge hole. They then strained it through a buckskin bag, forcing the mercury out the pores and leaving the precious metals and remnants of mercury and ore behind in the form of a paste, which was rich enough in gold and silver to be retorted. They burned off the mercury, then poured the molten gold and silver away from the remaining worthless material.[3]

Paul's mill proved an instant success, largely because he first worked the ore from Gold Hill, where gold predominated. The process lost much of the silver, however, since Paul's method did little to unlock the chemical bonding that characterized that metal on the Comstock. His mill could not effectively work the silver-rich ore of the Virginia City end of the lode. Nonetheless, as long as ore had a sufficient quantity of gold, Paul could guarantee mine owners a profit, and, at least at first, the efficiency of the process was less important than the fact that the location of the mill required little transportation of ore.

Three hours after Paul's mill began operation, Charles S. Coover and Elias B. Harris began operating an eight-stamp mill nearby. The Washoe Gold and Silver Mining Company No. 1 answered the challenge by initiating work on a sixty-four-stamp mill near Gold Hill. The company completed this $150,000 project on January 4, 1861, but it could not keep pace with the competition looming on the horizon. A new rush was on as more and more entrepreneurs saw a chance for profit in milling. Unlike mining, where there was always the risk of ore pinching out, leaving effort and investment unrewarded, milling produced a guaranteed income as long as there was ore to process.[4] As more mills began operating, run by steam engines, the sound of the Comstock crescendoed to a roar.

Many of the millers attracted to the Comstock worked to improve the standard process. Paul remained prominent in the field, but others joined him in trying to increase the efficiency of milling through a cycle of trial and error combined with ingenuity. The resulting amalgamation procedure, developed over the next two years, became known as the Washoe Pan Process. This ultimately included the use of steam to heat the amalgamation pan and its charge of pulp, as well as new designs to improve the flow of material. Although the industry settled into a standard convention by 1862 that showed little variation over subsequent years, the adopted process never adequately addressed the silver content of the ore.[5]

Besides the few inventive, well-trained millers involved in the industry, many people with no experience in the reduction of ore were swept up in the excitement. They, too, began operating mills with little understanding of the subject. Consequently, their mills lost enormous amounts of gold and silver. As Lord points out, some requirement for credentials

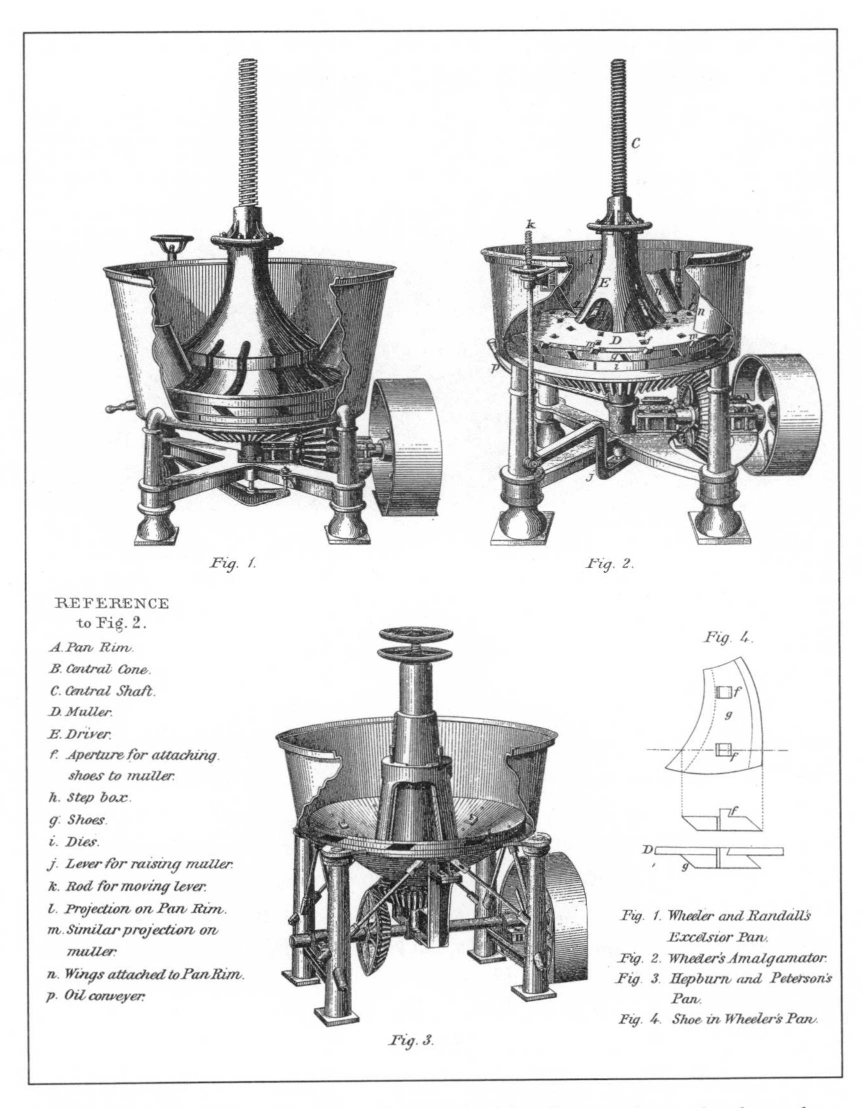

The Washoe Pan Process became the standard for Comstock ores by the early 1860s. (*King's Geological Survey* III; courtesy of the Nevada Historical Society)

for mill operators might have prevented such a scandal, but this "would have abridged the privilege of American citizens to waste the mineral resources of the public lands without hindrance."[6] Greed aggravated a situation that ignorance had fostered, creating even more waste: most mill contracts set charges for processing with no guarantee on amount of bul-

Comstock mills were often elaborately designed, captivating buildings. Carleton Watkins photographed the California Pan Mill in 1878, three years after its construction by the Bonanza Firm. (Courtesy of the Nevada State Historic Preservation Office)

lion produced. This subsequently encouraged mill owners to rush tonnage through their operation with little care about retrieving as much gold and silver as possible.

To address this problem and to monopolize the profits, several mining companies constructed their own mills. The Ophir was one of the first to build a mill, this one situated on the shore of Washoe Lake, west of the Comstock. Its complex covered an acre and claimed a capacity of one hundred tons per day, crushed by thirty-six stamps and a 100-horsepower steam engine.[7] Maldonado followed the lead in early 1861, borrowing $170,000 from Alsop and Company of San Francisco. The loan proved crippling, however, when the Comstock Lode pinched out on the north end as the Mexican's depth reached the 300-foot level. Unable to pay the lender, Maldonado settled the loan by selling the mine and the new mill to the bank for the loan's principal and some cash. With this action, Maldonado ended Hispanic control of a Comstock mine.[8]

The Gould and Curry Mine attempted to exceed all others with its mill east of Virginia City in Six Mile Canyon. Covering sixty acres, the complex had forty stamps and employed sixty workers. The mill's monumental architecture appeared to be unrelated to its industrial function. As Lord describes it, "A stranger, at sight of the stately edifice rising in the center of a group of offices, shops, stables, and laborers' cottages, would naturally have supposed it the mansion of some wealthy land-owner rather than a mill built in a barren district to crush silver ore, nor, on approaching nearer, would he have been undeceived by the presence of an oval basin of clear water, 50 feet long and 30 feet wide, in whose center three water-nymphs supported a rock shell whereon floated a white swan that with upturned head spouted a jet of water high in the air."[9]

Although it was one of the larger facilities, the Gould and Curry Mill stands as a clear statement that the mine owners both hoped to control the entire process of mining and milling and wished to project an image of permanent prosperity. Such an architectural statement could go a long way toward quelling the concerns of investors.

By the end of 1861, the Comstock and its environs could boast seventy-six mills with 1,200 stamps, working an average of one ton of ore for each stamp every day.[10] The problem posed in early 1860 of having no local means to mill ore reversed itself. It had now become a crisis of having too many mills. Mine-owned mills exacerbated the situation by providing another layer of competition to what became known as "custom" mills. Mill owners attempted to outdo each other by milling impressive amounts of ore, often with little regard for the quality of the process. Besides wasting a tremendous amount of gold and silver, this approach also left part of the mining infrastructure overinvested and underemployed, vulnerable to financial takeover and manipulation.

The Ophir Mill initiated the use of Washoe Lake for milling, thus requiring a new transportation route. While the Ophir Mining Company was constructing its complex on the lakeshore, it also built a road that left the Comstock through a pass in the Virginia Range, heading west between Virginia City and Gold Hill. Descending 1,200 feet to the lake and traversing sixteen miles, the route cost $600,000.[11] As others built more mills nearby, they also cut the Jumbo Grade, a road to the north of Ophir, using yet another pass in the Virginia Range and dropping into Washoe Valley.

These two roads augmented the earliest means of access to the Comstock. Originally, the main road extended north from Dayton, following Gold Canyon through Devil's Gate to Gold Hill and finally Virginia City. The road made a breathtaking ascent in Gold Hill, particularly at a switchback known as Greiner's Bend, giving the community the nickname of "Slippery Gulch."[12]

Another early road went east from Virginia City down Six Mile Canyon to the Carson River downstream from Dayton. It too was steep, however, and proved useful chiefly for hauling ore down to mills on the

The interior of a mill. 1878. Photograph by Carleton Watkins. (Courtesy of the Nevada State Historic Preservation Office)

river. The Gold Canyon route from Dayton was the best early means for most transportation up the mountain. In 1862 entrepreneurs built a toll road from Carson City that went more directly to Gold Canyon, bypassing Dayton and serving no mills.[13] With minor alterations, this road remains the most commonly used access from the south.

The territorial legislature granted a franchise for a northern route to Dr. D. M. Geiger and J. H. Tilton on November 29, 1861. It was to connect with Lake's Crossing, later known as Reno, which was growing into an important crossroads for the transportation of material and people. The northern road, known as Geiger Grade, consequently became one of the most-traveled on the way to the Comstock. In a modified form, it serves in that role to this day. Besides giving access to Reno and Truckee Meadows, the road also connected the Comstock with California through the pass where the ill-fated Donner Party spent a devastating winter. Although not the only access to California across the Sierra, this trail eventually became one of the most important, serving as the route for the transcontinental railroad completed in 1869.

Geiger Grade still winds its way up the valleys and ravines of the Virginia Range's western slope. The road serves as a geological field trip, introducing the novice to the mineralized wonders of the area. At the point where Geiger Grade leaves the valley, the traveler can see Steamboat Hot

Springs, so named because the plumes of vapor make it look like an approaching riverboat. The result of faults and heated water, the springs coat the nearly barren area around them with salts and minerals from far below belched forth over the millennia. The rare Steamboat buckwheat is one of the few plants to thrive in the harsh environment.

As the road climbs into the foothills, it enters a forest of plump piñon and juniper trees. Abundant sagebrush and other shrubs fill the land, at least up to a point. The traveler passes through barriers beyond which most plants do not grow. In these places, the naked soil has the color of pumpkin, the effect of ancient volcanic activity that spewed andesite here and there in the Virginia Range. The deposits decomposed, leaving dirt that inhibits the typical Great Basin mountainous plants, creating islands of rare, specialized adaptation. The slender Washoe pine, peculiar to andesite-rich soil, stands tall with sparse foliage. In the spring, a careful observer may notice the white flowers of a small plant scattered among the trees. This is Lobb buckwheat, an altered, andesite-dependent cousin of the plant found at Steamboat.

As quickly as a stagecoach climbed and wound its way into an andesite zone with its stands of Washoe pine, it returned to the piñon-juniper forest, leaving the traveler with a subliminal impression of the mineral and floral diversity of the area.[14] Geiger Grade ascends steeply, rising more than two thousand feet in only a few miles. Nonetheless, careful engineering allowed the original wagon road to have no inclines more formidable than those found on the other routes. From the summit of the pass, the road gently wound its way along hillsides until with one last turn, Virginia City, the Queen of the Comstock, suddenly appeared before the weary traveler.

While teamsters used the roads down Six Mile Canyon, and the Ophir and Jumbo Grades, to haul ore to mills, Geiger Grade and the Gold Canyon road became the principal routes for returning with the tons of material to build the mines. One of the most cumbersome cargoes was lumber, a crucial delivery since underground mining was impossible without wooden supports.

Initially, mining on the Comstock consisted of ever-widening shafts that looked like modern open-pit mines. Many of the scars from these early efforts remain as part of the Comstock landscape, but except for their respective sizes, it is virtually impossible to distinguish them from their larger counterparts of the twentieth century. Miners soon found, however, that the veins of ore took them to depths that made the side walls of the pits dangerous. Thus underground mining in the district was born.

For millennia, miners have wrestled with the problem of supporting the roofs and walls of mines. The fewer timbers used, the more economical the project, but also the greater chance of cave-ins. Traditionally, miners left pillars of the native rock here and there as a natural support system. In Cornwall, these pillars were often composed of the richest ore. The Cornish called them the "eyes of the mine" and believed that if they were never removed, they would bring luck. No matter how many rock pillars there were, however, a mine still needed wooden supports, which generally were constructed in one of two ways: miners wedged timbers individually from roof to ceiling and from wall to wall haphazardly wherever it appeared they were needed; or they built post-and-lintel structures, creating a series of inverted U's that looked like a succession of doorways. An added touch to the latter included side walls and ceilings made of planks secured to the vertical posts and lintels, designed to prevent the sloughing of dirt into passageways. Underground carpenters could employ these methods as frequently as necessary, depending on the stability of the ground.

Although nineteenth-century miners used both posts and lintels and rock pillars in the smaller drifts of the Comstock, neither proved effective for the larger mines, for two reasons.[15] The size of the Comstock Lode was one factor. Fifty feet below the surface at the Ophir, the ore was a dozen or so feet wide. By October 1860, miners had more than tripled the depth of the diggings and reached a level where the vein had widened to between forty and fifty feet. Wealth of that magnitude begged to be scooped out, but timbers long enough to bridge the resulting cavern offered little support. The slightest movement of earth could twist and snap wood that spanned such an opening.

The second factor contributing to the failure of conventional mine supports was the nature of the ground itself. Much of the quartz ore within the Comstock was crushed and water-soaked. Clays surrounded it, making the ground soft for digging. This also meant that the side walls and roofs of most mines were unstable and nearly impossible to support by conventional means. Pillars of rock offered little help, since they were made of the same malleable clays and quartz ore, and the shifting burden of the mountainside could crush them as well. The Maldonados employed pillars of native rock in the Mexican Mine and inadvertently provided a vivid example of just how disastrous this traditional engineering could be in soil like that of the Comstock. Although the procedure worked while the Maldonados managed the mine, after they left, the pillars began to disintegrate. Finally, on July 15, 1863, a cave-in occurred, wrecking the mine and damaging the neighboring Ophir. A second collapse six months later left the original workings of the Mexican Mine completely unusable. While the technology that the Maldonados employed served their temporary interests cheaply, it could not be a long-term approach in a place like the Comstock.[16]

To make matters worse, the local clays became even more unstable when exposed to air. Clarence King in his geological report of the area wrote during the heyday of the Comstock that after miners excavated a tunnel, the ground began "to swell and exert tremendous pressure, forcing itself through the interstices of rocks, bending and breaking the most carefully laid timbers and filling mine openings with extraordinary rapidity."[17] Clearly, unless the issue of adequate supports was addressed, the Comstock bonanza, no matter how rich, could potentially come to a premature end.

In 1860, W. F. Babcock, a director of the Ophir Mining Company, called on the services of a young engineer named Philipp Deidesheimer. Although he was not yet thirty years old, Deidesheimer had a great deal of experience in mining. Born in Darmstadt, Hesse, in 1832, he apparently attended the Freiburg School of Mines, the world's leading institute in the field. At nineteen, he came to California, where he worked for a decade in the gold mines of the western Sierra. He arrived in Virginia City on November 8, 1860, and studied the problem for about a month before coming up with the solution. He decided to employ a system that came to be known as square-set timbering. Milled wooden supports, each about eighteen by eighteen inches, stood from six to seven feet tall. Horizontal braces were the same size, but only four to five feet long. Carpenters fashioned the ends of all these timbers so they could be joined into blocks. These, in turn, easily fit together with other blocks. Diagonal braces provided more support where needed. Planks applied as floors, ceilings, or walls meant that mine engineers could create work platforms and protection from cave-ins.

It was an ingenious system that could fill any size underground void with supports as stable as the mountain itself. In addition, carpenters could assemble the square-sets as miners removed ore so that all were afforded maximum protection from a collapse of the mine. Once Deidesheimer revealed his approach, the Ophir Mine manager snapped into action. As soon as lumber companies could mill and deliver timbers, carpenters began assembling the square-sets in the Ophir Mine. It proved an immediate success, and miners from Gold Hill, after inspecting the system, began using it in their mines.[18]

Deidesheimer's square-set system sounded the death knell for thousands of acres of Sierra forest.[19] The trees were already easy targets for

Philipp Deidesheimer invented the square-set timbering system that opened the Comstock for exploitation. The invention was used by the rest of the mining world for several decades, and Deidesheimer gained immediate fame. He is shown here with his wife, Matilda. (Courtesy of the Nevada Historical Society)

exploitation, since mine supports and the community above placed increasing demands on the resource, but Deidesheimer's proposal called for even larger amounts of thick timbers. Later visitors to the Comstock mines described them as having the appearance of buried cathedrals. Over the years, carpenters needed tremendous amounts of lumber to build these structures, each following Deidesheimer's plan with no significant variation. It would be decades before the Tahoe Basin and much of the neighboring Sierra would recover from the Comstock's voracious appetite for wood, fueled even more by this invention. Ultimately, the entire mining West felt the effect of Deidesheimer's method, as other areas used it and even more trees fell victim.

For more than a hundred years, writers have maintained that Deidesheimer's square-set timbers represented the most significant advancement in the world of mining for several years to come. Ironically, Deidesheimer profited little from his innovation. Failing to secure a patent on the idea, he could do little more than watch as it became widely used elsewhere. Deidesheimer never wanted for employment, since mine owners found it useful to have his prestigious name associated with their companies, particularly when attempting to attract investors. In spite of this, he declared bankruptcy in 1875, the victim of a stock crash that resulted in part from his own miscalculation of the value of a body of ore. Deidesheimer's contemporaries and subsequent historians repeatedly pointed out that he had not earned the financial rewards that were his due.[20] Nevertheless, he managed to escape the harsher judgment passed on Henry Comstock, Old Virginny, and the other original discoverers of the lode, who likewise failed to realize the wealth at their fingertips. Instead, Deidesheimer occupies a distinct niche in local legend. While his story has a tragic tone, he is not condemned as a fool, perhaps because the spark of his genius in 1860 was so important to everyone's success. And after all, for most people, lightning does not strike even once.

While Deidesheimer solved a crisis facing Comstock miners, his invention accentuated yet another problem. Deeper mines increased the need to get rid of water as shafts began to function like wells. One of the ironies of the Comstock was the lack of water on the surface and the flood of it below. At the outset, the latter appeared to be the solution for the former, but the mineralization in the rocks made most underground water undrinkable and even corrosive for machinery. Rather than a boon, water was generally a nuisance, threatening to stop mining as shafts deepened.

In the spring of 1860, the Ophir Mining Company purchased a 15-horsepower steam engine to pump its shaft. It was the first of its kind in the district, a precursor of the much larger ones that would follow. At the same time, several companies—the Mexican, the California, the Central, and the Ophir—pooled their resources to pay for a single tunnel intended to drain all of their mines at the 200-foot level. Begun June 8, 1860, and finished the following October 17, the Union Tunnel, as it was called, was 1,100 feet long and cost $10,000.[21] The adit solved the problem for the four mines down to the level of contact, but the destiny of the Comstock led miners far below that point. For the next several years, workers excavated deeper, longer adits, piercing the mountain horizontally up to 3,300 feet in an attempt to intercept shafts at even lower levels. Each of these, while serving for a time, proved to be only a temporary solution. The story of flooding became all too familiar. Miners working below the most recent drain tunnel would pierce a seam of clay, releasing a torrent of water. After the miners barely escaped with their lives, whatever engines were near the disaster would work constantly, pumping against the steady stream. Sometimes it would be months before the water abated. Often, even more pumps had to be brought in before men could return to the area. Water could be a problem of herculean proportions: in 1865, for example, the Belcher Mine pumped 1,017,878 gallons a day out of its shaft, which nevertheless remained inaccessible for months.[22]

The Comstock needed a gargantuan solution for flooding of this magnitude. As soon as the excavations reached depths that made lateral tunneling unthinkable, mine managers tackled the unceasing waters with enormous engines known as Cornish pumps. The task of diverting steady streams hundreds of feet up shafts should never be taken lightly. The pumps finally installed were huge beasts made of tons of iron. The machines included a steam engine, counterweights, rods, and the ever-characteristic flywheel. The pumps followed an evolution much like the dinosaur: they became increasingly large over time and demanded more and more fuel. A pump installed at the Union shaft in 1879 had a flywheel 40 feet in diameter that weighed 110 tons. The cylinder that sucked up water hundreds of feet below ground was 5 feet 4 inches in diameter with a stroke of 6 feet 9 inches, capable of clearing a small water-filled room with one cycle of its action. With machines of this size, the casting and even the hauling of the parts from foundry to mine site assumed heroic proportions and became a matter of local legend. It took several years for Cornish pumps to attain their largest forms, but the earliest ones were still massive, noisy affairs. Pumps required even more steam engines, adding to those of the hoisting works and the mills. This, combined with the rumble of the mills, produced a twenty-four-hour cacophony that serenaded the Comstock.[23]

Together with the stamp mills scattered over the landscape, the pumps served to remind the visitor that this was an industrial community in some ways more like Pittsburgh than like a quaint outback mining town, the image that fit in better with the myth of the Wild West. The Comstock was not a place where miners, lacking machinery, relied exclusively on the pick, the drill, and an occasional stick of dynamite to pursue their profession. The might of nineteenth-century industrial technology tackled problems posed by some of the largest and deepest mines the world had ever known.

At the same time when mine owners were first buying pumps to eliminate water from the shafts, an entrepreneur named Adolph Sutro suggested the unthinkable: the ultimate drainage tunnel. Sutro, a local mill owner, secured legislative action incorporating his Sutro Tunnel Company on February 4, 1865. He proposed an adit extending 20,489 feet—more than three miles—from the Carson River valley to the Comstock mines 1,663 feet below the surface. Sutro had, in fact, proposed a drainage tunnel, though presumably of more modest proportions, as early as April 1860, in an article on the Comstock for the *Alta California*.[24]

What had been an early idea for Sutro grew both in the length of his proposed tunnel and in his obsession. Initially the project had powerful allies, and Sutro was able to secure $3 million in financial commitments from mine owners. Support for the project waned, however, when it became clear that Sutro would gain control over many aspects of the Comstock. After completion of the drainage system, the cheapest way to remove ore from underground would be to roll it down the steady incline of the tunnel. Similarly, the easiest, safest way for miners to travel to and from work would be through the tunnel. Sutro's plan was to construct a mill town and residential community—named, of course, Sutro—at the opening. He would be landlord to all who lived there, and he would own the nearest mills. Sutro could charge mine owners for hauling their ore and workers, and then he could monopolize rents and milling. Profit-hungry mine owners had no interest in surrendering control of the Comstock to anyone, and they began to choke off support for the tunnel.[25]

The Sutro Tunnel remained locked in political controversy, which wrestled the project to a standstill for several years. Regardless of the debate concerning the efficiency of tunnel versus pump, workers still faced the immediate danger of floods being unleashed when a miner unluckily picked through the wrong seam of clay. Sometimes the water released was searing hot and turned shafts into boiling cauldrons. Often the flow of water, even if not hot, was so rapid that miners had to scramble for safety or drown. In either case, water threatened injury or death, and the sudden floods could slow further excavation for months. Although Sutro portrayed his project as the panacea for many ills on the Comstock, no such scheme could reduce most of the risks that miners faced when working underground with dangerous equipment in such an extraordinary environment.

Providing potable water for the growing Comstock communities also continued to be a challenge. When J. Ross Browne, journalist and artist, arrived in the spring of 1860, he was not only disappointed at failing to win millions, he also became sick because of the water. Browne commented that "the water was certainly the worst ever used by man." The quality of water did not get better after Browne left, particularly as the population grew. The Virginia City Water Company laid pipes "through the greater part of town" by 1862, boasting a source of drinkable spring water, but complaints about its poor quality remained rampant during the early history of the Comstock.[26]

Lacking much by way of rivers, lakes, or creeks, the region's early efforts relied largely on underground sources to obtain drinking water. Since a mining district like the Comstock has no shortage of wells created as a by-product of underground drilling, the challenge became finding a source less contaminated and repugnant than others. The Virginia Water Company and the Gold Hill Water Company both formed in the early 1860s to distribute the best of the tunnel water. On May 12, 1862, they consolidated and consequently controlled most of the drinkable water, such as it was. It would be more than a decade after the founding of Virginia City before engineers arrived at an approach to obtaining a better source for the Comstock. In the first years of the community's existence, drinking water remained lacking in both quantity and quality, an endemic problem paralleling that of the floods underground.

Another issue that has always plagued the Comstock involved the difficulty of defining the boundaries of claims. The first miners on the scene in 1859 were casual about the limits of the land they worked. This nonchalance was, in part, yet one more consequence of the California experience. There, miners worked placer sands until they had exhausted that resource. They then abandoned the claim and moved on. Although there were disputes in California, the exact definition of a claim's boundary did not translate into millions of dollars, since the gold was broadly scattered rather than concentrated. As already pointed out, having an ill-defined claim also gave a prospector the latitude to co-opt a neighbor's success. Whether it was because of these factors or simply carelessness, the vagueness of early claim documents created one of the major problems in Comstock history. The failure to address this issue with precise technology and attention to detail ultimately cost many of the mine owners more than they had paid for their Cornish pumps or square-set timbers.

A claim recorded in the Virginia Mining Claim Book A, dated October 1, 1859, for example, defined the boundaries of a holding as "commencing at this notice and running north to a certain stake." Another indicates that the claim commenced "at a notice posted near the flat at the head of Six Mile Cañon opposite the Dutchmans [*sic*] cabin in Virginia District and running in a northerly direction twenty hundred feet and two hundred feet in width embracing all the Quartz ledges, with dips, spurs and angles."[27] In the same way, the Carson County Records of Plats and Surveys document a land claim that described the parcel boundaries as "beginning at a rock on the easterly side of Virginia City from which flag staff on the summit of Mt. Davidson bears W. 801 201 W. and the south westerly corner of the stone store house on D Street belonging to R. D. W. Davis, Esqu. . . ."[28] The text goes on with an attempt at precision, but with initial benchmarks as vague as "a rock" on the side of a city and the corner of a stone building, there could be little hope to reproduce the survey after people forgot the exact locations of the landmarks in the description. Each of the other claims was just as vague, similarly depending on temporary points of reference. Relying on the boundary of a neighbor's holdings was particularly problematic, since the description lacked concrete reference points. A change in the traditionally accepted location of a claim by a court could subsequently affect many other claims down the line.

The most serious legal question ultimately focused on a phrase customarily included in claims, stating that the prospector owned not only the ore in the area designated but also the vein as it left that plot of land, following "all its dips, angles, and spurs." At first, miners assumed that the Comstock Mining District consisted of a series of discrete deposits isolated from one another by masses of worthless clay. Subsequent excavation raised the possibility that the lode was actually one large ore body, each part connected to the next. The question of whether the Comstock consisted of many ledges or only one exposed a fundamental issue. If it was one large deposit linked by a complex of "dips, angles, and spurs," then the first claimants had secured ownership to much more than previously assumed. Neighbors would become illegal squatters. Millions of dollars were at stake. Either violence or court disputes were inevitable. The Comstock explored both alternatives, a process with ramifications that made and broke careers, cost millions, and ultimately affected national Civil War politics.

Armed conflict was the most direct way to handle disputed claims of ownership. Miners of rival companies occasionally broke into neighboring works, which sometimes transformed into eerie, subterranean battlegrounds. Thus in 1863 and early 1864, the Keystone and the Peerless, the Grass Valley and the Bajazette and Golden Era, the Uncle Sam and the Centerville, and the Yellow Jacket and the Gentry all tried to settle the question of title with violence. The attacks were surprisingly tame, however, usually consisting of destroying equipment, timbering, and ropes and filling shafts with debris.[29]

The combination of contested ownership, violence, and millions of dollars proved to be an irresistible magnet for the lawyers who drifted into the district. With the benefit of historical hindsight it is possible to see a sign of things to come in the form of the dozen or so lawyers who lived on or near the Comstock during the 1860 census.[30] Many more followed.

On March 16, 1863, the Ophir Mining Company brought suit against the Burning Moscow. The claim of the latter, recorded on April 19, 1860, professed ownership of mineral rights to the west of the former.[31] As digging progressed on the Ophir, however, miners discovered that their vein drifted west, and evidence accumulated that the Burning Moscow was exploiting the same ore body. This discovery inspired the owners of the Ophir to sue their neighbor. Two months later, on May 14, 1863, Ophir miners, digging across the vein toward the Burning Moscow, broke into the latter's works. A violent response repulsed the Ophir miners. Their employers sought judicial redress, and territorial judge Gordon N. Mott answered the complaint with a temporary restraining order forbidding further work at the Burning Moscow. Its stocks fell.

The case languished in court, but Mott had made it clear that he favored arguments supporting the idea of a single Comstock ledge. This view was enormously unpopular with many miners and owners at the time. They regarded a multiledge Comstock as a place of opportunity for more people. It meant that anyone could strike it rich by obtaining a claim and then stumbling onto a discrete pocket of ore—one's own private bonanza. It also left the door open for many mines, each with owners striving for riches. Most regarded the idea of a single Comstock ore body as a door slammed shut. There would be no opportunity for the common man to strike it rich, and many mine owners would lose their businesses.

In September 1863, word circulated that Judge Mott planned to resign. Rumor also suggested that his replacement endorsed the idea of many ledges within the district. The Burning Moscow stock rose. On October 23, Ophir management responded with several expeditions into the works of its neighbor, which repulsed each attack. Finally Philipp Deidesheimer, now the superintendent of the Burning Moscow, secured a warrant and entered the underground battlefield with two deputy sheriffs and the authority of the court. They arrested the Ophir's superintendent, foremen, and sixteen miners for riotous behavior. In response to the complaint, Judge John Wesley North, Mott's replacement, issued a temporary injunction forbidding any more Ophir intrusions beyond the claimed boundary of the Burning Moscow. The tables had turned. Ophir stocks slumped, but much of the Comstock rejoiced, since most viewed North's action as a blow for the common man. In December, Judge

William Stewart was a lawyer who profited from legal contests associated with mining claims. He went on to become a U.S. senator. (Courtesy of the Nevada Historical Society)

North reasserted his support for many ledges by denying an appeal of the Ophir, stating that there was no evidence that the ore of the Burning Moscow was connected with the veins of its neighbor.

In the midst of this and similar contests, a political dispute brewed. William Stewart, a local lawyer arguing the single-ledge cases in court, was also willing to bring the issue into the political arena. Representing wealthy California owners of mines such as the Ophir, he was well paid and had enormous economic resources at his disposal. Judge North, hero of the common man, became Stewart's mortal enemy. With the legal arguments on behalf of the Ophir turned aside, the crafty lawyer devised a scheme to rid himself of North, taking advantage of circumstances far beyond the limits of the Comstock Mining District.[32]

Civil War politics led President Lincoln and the Republican Party to see merit in the idea of statehood for Nevada. Removed from the controversy of slavery, and consistently backing the Union cause, the territory would be more useful to the president as a state, with senators who supported Republican amendments to the constitution and with electoral votes for Lincoln in the 1864 election. Although the territory lacked the requisite population, Lincoln and his supporters eventually pushed legislation through Congress enabling Nevada to apply for statehood. The proposal found popular support in the territory, since it would mean promoted status and a degree of political power for Nevada. A territory-wide vote in September 1863 endorsed the pursuit of a constitutional convention 6,660 to 1,502, and by December 11, 1863, the territory had a document to put to a vote of the people.[33]

At the same time, Stewart saw statehood as a unique opportunity to rid himself of Judge North, since the transition would mean a new election of officers. The lawyer managed to gain control of the political machinery of the Union Party convention delegates in Storey County. Besides working on the question of statehood, the county conventions were to nominate candidates for state offices. Stewart was able to manipulate the situation so that North failed to receive a nomination to office, and it became increasingly clear that the supporters of the single-ledge case would gain control of the new state government. Slowly, sentiment turned against the idea of statehood. Stewart had succeeded too well. The seeds of his defeat were sown in the midst of his victories. The majority of the Comstock, disliking the single-ledge premise, came to see Stewart, the monopoly he would create, and statehood as synonymous and against their interests. On January 19, 1864, the electorate defeated the 1863 constitution four to one.

More than the defeat of a constitution was to occur in the pivotal year of 1864, however. Another legal contest over boundaries reached a climax, entangling itself in the Stewart-North rivalry and the politics of statehood. The granddaddy of all lawsuits on the Comstock, the one that was to answer the question of single versus multiple ledges, was a drawn-out affair with roots predating the Ophir and Burning Moscow contest and the drafting of the 1863 constitution. As early as December 1861, the Chollar Mining Company filed suit against the neighboring Potosi over the definition of the border between their claims. Again, the debate over single versus multiple ledges was at the heart of the controversy. Much of the outcome of the Potosi-Chollar lawsuit helped define development on the Comstock and other districts, as well as the future of the Nevada Territory. It also eventually influenced federal mining legislation. The first trial in 1862 ended in a hung jury. The second, a few months later, resulted in a judgment in favor of the Chollar and the single-ledge argument. The Potosi appealed the decision to the territorial supreme court, which, following the lead of the despised Judge Mott, sustained the lower court in March 1863. The decision did little to endear the judges to the people, since most favored the idea of many discrete veins of ore.

In short order, miners in the Potosi works discovered a new ore body, which they once again maintained was separate from the vein owned and worked by the neighboring Chollar. The owners of the Chollar, however, did not see it that way, and so they launched another lawsuit. All this occurred just as Judge Mott was resigning and North was taking his place in office. The new lawsuit, consequently, promised to have a different complexion from the first one. Indeed, in Judge North's first ruling on the case, in March 1864, the Potosi and the multiple-ledge arguments came out on top. This was a repeat of the final case between the Ophir and the Burning Moscow. Once again North was the champion of the common man.

In the midst of this legal controversy, a new constitutional convention began working on a revised document. Ultimately, there were few alterations, and so its success would hang on changes in the electorate. As often happens, the economy led the way. During 1864, fortunes on the Comstock changed dramatically. Costs to exploit deeper resources mounted, and the rich ore needed to justify the expense remained elusive. Depression gripped the community, and many mines sank into *borrasca*, as the people called a time when mines ceased to profit. Following the pattern exhibited with the term *bonanza*, citizens of the mining West also borrowed a Spanish word to describe the alternative economic extreme. The hard times ruined many of the small mine owners and left workers unemployed. Popular sentiment shifted, and many came to view North's stand against the single-ledge case as hurting the common man. Popular opinion now maintained that only large corporations would be able to exploit the deep Comstock ores and bring back prosperity.

Interest in statehood returned to life. At the same time, Stewart and other supporters of statehood were able to blame the depression on a corrupt, lazy judiciary. They maintained that the failure to process mining lawsuits expeditiously and fairly had crippled the industry. For months, North and his two colleagues on the territorial supreme court withstood a blistering attack, as Stewart, the newspapers, and others accused them of all sorts of wrongdoing, from the inept to the criminal. In August 1864, North reassessed his stand on the controversy. He had received a report from a mining expert that provided scientific evidence that the various veins and pockets of ore ultimately ascended from one gigantic and single Comstock Lode in the depths of the mountain. North ruled, consequently, in favor of the single-ledge argument. It was too late, however, to save the judge. The tide of public opinion was too strong to be turned with a last-minute change of position. When North and the other two territorial judges resigned on August 22, the electorate took it as an admission of guilt, and support for statehood increased even more.[34]

Voting on September 7, 1864, the electorate overwhelmingly embraced the revised constitution, 10,375 to 1,284. After territorial officials counted the ballots, they telegraphed the constitution to Washington, D.C., where Congress needed to review it to make certain that it met the conditions of the enabling act. It was the longest telegraph ever sent, but Nevadans did not intend to do things in a small way. With the paperwork in order, Lincoln signed Nevada into statehood on October 31, 1864, just a few days before the November presidential election. Nevada sent back a thank-you in the form of three electoral votes for the president's reelection.

Although forced to resign as territorial judge, North contested the accusation that he was corrupt by suing Stewart and the newspapers that had written about his alleged wrongdoing. The new state court system found no evidence of criminal activity on the part of North, judging against Stewart and the codefendants and directing them to pay for the costs of the proceedings. The court also pointed out, however, that North's activities had been questionable. North left Nevada and went to Reconstruction Tennessee, where he continued his career, although under the shadow of the term *carpetbagger*.[35]

As for Stewart, he proceeded to take control of politics in the new state of Nevada. With the formation of the first state legislature, Stewart had himself elected senator, thus beginning a decades-long distinguished career in Washington, D.C.[36] Still, the creation of the state and the resignation of North and the other territorial judges did not completely resolve the question of title hovering over the Comstock mines. The single-ledge theory had won the day, but opportunity for legal contests remained. During another lawsuit in March 1865, the Chollar and the Potosi reached a compromise, resulting in a corporate merger on April 22.[37] And with that whimpering finale, the great age of Comstock litigation came to an end.[38]

The Comstock lawsuits were another in a long series of complications that grew out of the transition from placer to hard-rock mining. While the 1850s was a period of exploration and discovery of the Comstock Lode, the early 1860s saw the unveiling of several obstacles. If the mining district were to have a future, its developers would need to solve all the problems, no matter how costly or technically complicated. Claimants had not applied technology consistently or soon enough to address the issue of ownership as the district made the transition. While vagueness could be an asset when working placer sands, it proved disastrous for hard-rock mining. For example, the legal contest between the Chollar and the Potosi probably cost more than $1 million, while the price tag for the suits between the Burning Moscow and the Ophir was about $800,000. The latter dispute reached a final settlement when the Ophir simply bought the disputed claims from its opponent for $70,000, far less than it spent in court costs. Altogether, Stewart estimated, the lawsuits of the early 1860s cost the mine owners $10 million.[39] These were the Comstock profits that otherwise would have gone to shareholders or into more development.

Instead, the controversy served the personal careers of the politically elite. As it turned out, the lawsuits tended to create a rigid mining and professional aristocracy and little opportunity for mobility into that class. Stewart himself apparently earned $200,000 a year during the early 1860s.[40] His San Francisco employers won their case in court and maintained their dominant interest in the Comstock. With the ore body perceived as a single ledge, it would be more difficult, if not impossible, for a lucky prospector to strike it rich with a new claim on the Comstock. The time when that could happen receded into a distant, now mythical past.

Ultimately, the legal controversy affected the nature of mining, pointing the Comstock in the direction of corporate industrial development rather than small rathole mining. The latter was characteristic of the earliest Comstock and has always been in keeping with small-time exploration and family operations. Although later thought of as one of the cornerstones of the Wild West, Virginia City was actually one of the industrial giants of North America, far removed from the cliché of a mining camp. The 1969 film *Paint Your Wagon,* for example, depicted a mining town, No-Name City, that had a population listed at its entrance as "Male." This place rose from fiction to fit the stereotype better than the Comstock ever could. The factors that caused large-scale corporate mining to become the only pragmatic approach in the district ensured the fate of the Comstock by 1864.

The process of solving the problems encountered in the first years of the mining district added Paul, Geiger, Deidesheimer, and Stewart to the ranks of O'Riley, McLaughlin, Penrod, Finney, and Comstock as actors in the legend of the early district. They were the stuff of good stories, but they also functioned on another level. Virginia City, Gold Hill, and Silver City were instant communities born explosively in the midst of a rush, composed of immigrants from throughout the world. The mining industry and the desire for wealth provided the only social and cultural link among the various elements. Featuring those who played a part in the development of the Comstock in the local emerging folklore was a means of granting the place a sense of heritage and cohesion. The early 1860s added its layer to the bedrock established in the previous decade.

The stories of two of the mining district's most legendary characters from the early 1860s also serve to illustrate how the Comstock was changing. The careers of Alexander "Sandy" Bowers and Samuel "Mark Twain" Clemens typify the extreme of success that was possible when the district was young. The Comstock matured, however, and as corporate structure solidified the area, the hallmarks of the freewheeling early days vanished.

Sandy Bowers was one of the first claimants of the rich 1859 strike in Gold Hill. When he married his neighbor Eilley Orrum, who had acquired some feet of the lode as payment for a bill in her lodging house, the two had the makings of a fortune. Unlike so many of their counterparts, they kept their claims and dug hundreds of thousands of dollars out of their glory hole. Because the ore was close to the surface and "friable," or easily milled, they could continue to exploit the claim without large amounts of initial capital to build an infrastructure. The couple became two of the first Comstock millionaires and represented a degree of success in sharp contrast to the other early claimants who settled for thousands and left after selling their chance for more.

Having won the lottery, Sandy and Eilley Bowers proceeded to enjoy their newfound wealth. They built a mansion in Washoe Valley and traveled to Europe to purchase furnishings. Their expectation of meeting Queen Victoria was not realized, however, and upon their return, they found that in their extravagance they had become financially overextended. Lacking the internal psychological makeup to manage success, they failed to maintain their financial and social status. As their ore depleted, Sandy's health failed. With his death in 1868, his widow faced poverty. She abandoned her mansion to creditors and earned a meager living marketing herself as the "Washoe Seeress," capitalizing as best she could on her former legendary success.[41]

There would be other Comstock examples of fortune's fickle rise and fall, but the story of Sandy and Eilley Bowers is born of the wide-ranging alternatives of the earliest days. As the district matured, the possibilities narrowed. In the new corporate environment, luck would have less to do with success.

Mark Twain's Comstock career provides yet another example of extremes both allowed and then curtailed as the Comstock matured. His success was not of the monetary sort, like that of the Bowers family. Instead, it is remarkable for Twain's ability to market hoaxes and fabrications instead of working at honest, labor-intensive reporting. For almost two years, he wrote articles in the *Territorial Enterprise* based largely on gossip he heard in saloons. When tidbits were not forthcoming, Twain was not above manufacturing his material out of nothing. His writing was often libelous, and yet in spite of this, he maintained his position. Ultimately, however, he offended the wrong party. To avoid a fight, he fled the area in 1863.[42] The early Comstock provided an environment in which the genius of Clemens could invent Twain. Of course, without his special gift for writing, this episode would have been the end of Mark Twain, and Clemens would, no doubt, have drifted to something else. A more established Virginia City, however, might have curtailed Clemens's brand of experimentation, and Mark Twain would have needed to find other soil to yield the first blossom.

Twain, like Sandy and Eilley Bowers, is the stuff of legend. Each of these three typifies the early, freewheeling Comstock. They found their own successes during a time of chaos when anyone clever or lucky enough could do well. A third example, that of George Hearst, illustrates yet another possibility. He was one of the first to arrive on the scene after the 1859 strikes, and he was able to buy a claim that subsequently proved extravagantly profitable. Hearst parlayed this early Comstock investment into the foundation for a fortune later won elsewhere in the mining West. His impressive brick mine superintendent's house, now known as the Mackay Mansion, remains as testimony to an early achievement in Virginia City. Like so many others, the mythic quality of the Hearst family had Comstock roots.[43] The success of Hearst was a rarity, however; only a few from this early period emerged with outstanding, long-lasting wealth, and the time for such possibilities passed quickly. The new corporate Comstock would limit the extremes of variation. Stumbling into riches would become increasingly rare. Similarly, the outrageous antics of someone like Twain quickly became anachronistic. Others would follow in his footsteps through the rest of Comstock history, but a new order drew in the limits. The failures of these people serve as clear indications that the Comstock had completed its earliest period of consolidation. Both the mines and the society of the district were now more structured.

4: GRIEF, DEPRESSION, AND DISASTERS

Successes in the Midst of Failures

Like all panics the Washoe one of 1865 terminated, and people discovered that they had been worse frightened than circumstances justified. It was demonstrated that the bottom had not fallen out of the Comstock; new developments were made; the stampede ceased; the tide of emigration began to flow back again, and there was a gradual return of confidence and prosperity.
—Territorial Enterprise, *April 9, 1869*

When the telegraph returned word that Lincoln had signed legislation on October 31, 1864, admitting Nevada to the Union, the Comstock erupted in celebration. The new state had a deep commitment to the North during the Civil War, and acceptance as an equal served for many as an acknowledgment of that support. For years, Comstock newspapers carried daily columns about the campaigns and the enormous cost of preserving the nation. The names of battles—Antietam, Fredricksburg, Chancellorsville, Gettysburg, Chattanooga—leapt from the pages of the *Territorial Enterprise*, the *Gold Hill News*, and the *Virginia Evening Bulletin*. These sleepy villages turned powerful with the searing word of thousands dead and wounded during the epic struggle. Grant, Sherman, Davis, and Lee assumed their roles in the drama as the heroic, the sinister, and the tragic. For the Comstock, it was like watching a boxing match or a melodrama from afar. Unable to contribute much by way of troops, most Nevadans could only cheer when word arrived of battles won and mourn the sacrifices of defeats.

Although the thousands of miles separating the Comstock from the struggle gave birth to frustration, people followed reports of the conflict emotionally, with deeply felt attachment to the cause. One example, originating in a central Nevada mining district, serves as an expression of this commitment. Reuel Colt Gridley was a store owner and Democrat in Austin, Nevada. He had made a bet with a merchant named H. S. Herrick concerning the political election of 1864. They agreed that whoever's candidate lost, that person would carry a fifty-pound sack of flour through town to the other's store. The wager included the provision that if the Republican-unionist lost, Herrick would walk to the tune of "Dixie," and if the Democrat failed to win, Gridley, a secessionist, would carry the sack to the tune of "Old John Brown." They also agreed that the winner would donate the flour to the Sanitary Commission, a precursor of the Red Cross that assisted the wounded from the Civil War's battlefields. Gridley lost the bet, so he hoisted the burden on his back and took his walk, preceded by the Austin Brass Band. As arranged, Herrick gave the flour to the Sanitary Commission, suggesting that it be auctioned. The sack sold for $350.

After this, Gridley and his flour stepped into legend with a single act. M. J. Noyes, the man who had placed the highest bid, paid his money, then put the flour back on the auction block. In this way, the commission sold it again and again. When Austin seemed drained of gold, after raising $4,549.80 in coin, Gridley took his sack on the road. On the Comstock, the first stop, the sack garnered $25,042. Gridley then toured California, and in all, the sack sold for about $175,000.[1] For a war that cost millions, it certainly did not add up to much, but the anecdote illustrates the concern that many people of the mining West felt for the great American conflict.

On April 10, 1865, the Comstock received a telegraph describing the surrender of Lee at Appomattox the day before. The mining district erupted in celebration. In his diary, local newspaperman Alfred Doten noted: "At noon all the bells & whistles in City were sounding—I helped ring St. Mary's . . . bell myself . . . in less than 3 hours everybody were crazy drunk—such drinking never before was seen—The military were all out—Provost Guard came up with 2 pieces of cannon & fired in streets—flags flying everywhere—anvils, guns, pistols, everything that could make a noise did so." The Civil War had ended. Residents of the Comstock, having finally witnessed the conclusion of the national torment, could not imagine how their feeling of elation could stop.[2]

Nevertheless, five days later, yet another telegraph plunged the Comstock into despair. John Wilkes Booth had shot Lincoln. The assassin plucked the hero of the nation's greatest drama from the stage. Victory would never again be as sweet. The *Gold Hill Daily News* summarized the mood of the day: "The flags of our town, which so lately waved and fluttered in the breeze, in joyful honor of peace to our country, were raised in sorrow at half-mast; our church-bells tolled the solemn knell of the hour; our local gun, on Fort Homestead, fired its hourly shot in remembrance of the National calamity."[3]

Within hours of receiving word of the assassination, vigilantes searched for anyone who would speak ill of the Union or Lincoln. As Doten noted on April 15, no secessionist ". . . dared shoot off his mouth at all—if any had, they would have been killed."[4] Still, some did, and they met with harsh treatment. Lacking the assassin or even a conspirator in their midst, the mourning mob of loyalists had to settle on what they could find to satisfy their need for revenge. Posey Coxey of Gold Hill, for example, proclaimed about Lincoln's murder, "I'm damned glad of it. It's a pity he wasn't killed years ago." Vigilantes apprehended Coxey and sentenced him to thirty lashes with a bullwhip. Ten strikes apparently proved sufficient, so the crowd tied a placard to his back reading "A Traitor to his County" and marched him to Virginia City. There, the Provost Guard, under the command of General Jacob L. Van Bokkelen, placed him in prison. It is unclear whether the incarceration was for punishment or for protection, since both Doten and the *Gold Hill Daily News* noted that the mob could just as easily have lynched Coxey.[5]

On April 19, the Comstock held memorial services for the martyred president. Two thousand people, dressed in their best, turned out for a procession in Gold Hill that ended at Fort Homestead, a prominent flat-topped hill overlooking the community. Following the officials who led the memorial march were the Metropolitan Brass Band and then five hundred volunteer firemen. Next came "the Mexicans with the flags of the two Governments borne abreast," a mounted contingent of the Provost Guard, and the Emmet Guard. After that was a hearse bearing a flag-draped (though empty) coffin. Then came the National Guard, followed by the local clergy riding in carriages, representatives of the bar, elected officials, the Freemasons, the Odd Fellows, members of the Washoe Typographical Association, the Sons of Temperance, the Jewish Order of B'nai B'rith, the Eureka Benevolent Society, the Irish Fenian Brotherhood, the German Singing Society and Turnverein Society, the Swiss Association, the Virginia Board of Brokers, superintendents and employees of the Gould and Curry, Savage, Potosi, and Chollar Mines, and "a large number of citizens." At the end of the procession marched "our African residents, who bore a beautiful banner engraved with the

General Jacob Van Bokkelin commanded the National Guard in Virginia City. He was a prominent leader who imported dynamite for the mines and ran the Beer Garden below Virginia City. (Courtesy of the Nevada Historical Society)

following words: 'He was our friend—faithful and just to us; though dead, he liveth. Hail! and Farewell.'"[6] The memorial parade illustrates not only the depth of local despair but also the diversity and complexity that the mining district had achieved in the four years since the rush to Washoe had begun in earnest.

It was an enormous outpouring of grief from a community that shared its mourning with the nation. The assassination, like the surrender at Appomattox less than a week before, became one of those benchmarks that all alive at the time remembered. Pearl Harbor, the death of Roosevelt, the end of World War II, the Kennedy assassination, the *Challenger* disaster—brought to us almost instantly through today's media—these are the events that loom in the memories of modern America. News of the end of the Civil War and the death of Lincoln were remarkable expressions of lines of communication that had only recently made news bulletins quickly available across the continent. As a consequence, the Comstock could participate in the ultimate demonstrations of national joy and sorrow, and the mining district's residents could have the opportunity to join the collective experience of the country.

4.1 Production of Comstock Mines During the 1860s

Year	Dollars in Bullion	Year	Dollars in Bullion
1860	$ 1,000,000	1865	$15,184,877
1861	2,275,256	1866	14,167,071
1862	6,247,047	1867	13,738,618
1863	12,486,238	1868	8,499,769
1864	15,795,585	1869	7,528,607

Source: Lord, *Comstock Mining*, 416. Lord uses two sets of figures beginning in 1867, one from the State Tax List and the other from a source he lists as "U.S. Commissioner." The latter is employed here because it includes information on the entire decade. The only significant deviation in the 1860s between the two is in 1868 for which the tax rolls show a figure of $12,418,023.25. This is the figure Smith uses in *History of the Comstock Lode*, 99, to illustrate a contrast between 1868 and the depression year of 1869.

When the mourning ceased and the celebration over Appomattox was only a dim memory, the Comstock returned to the world of mining slump and economic depression. As early as 1864, it had been clear that unless miners could find new ore bodies the district would fail. As mentioned earlier, economic troubles encouraged people to see an advantage in corporate mining, inspiring passage of the state constitution in the fall of 1864. It was not a good year for the Comstock—and 1865 was no better. Although actual production remained high, the best ore began to pinch out as miners dug shafts past 300, 400, or 500 feet. Smaller mines closed, and mills starved for business. Panic gripped the Pacific Coast in December, and even though Comstock mines had produced more than $15 million in gold and silver in 1865, stocks reached an all-time low because of the widespread belief that the district was all but finished (table 4.1). From 1863 to 1865 the estimated total market value of the Comstock fell from $40 million to $4 million.[7] Low-grade ores and the remnants of the rich veins were all that kept the district from failure. Only a few new discoveries sparked hope. When the Mexican Mine collapsed in July 1863, for example, it exposed a vein three hundred feet long that extended into the Ophir claim and was between five and fifteen feet wide. Still, it pinched out at the 350-foot level, and though it provided short-term income, miners exhausted the profitable ore in the new vein during 1864.

Historian Grant Smith estimates that ten thousand people left the Comstock during the depression of the mid-1860s. Although it is not possible to verify such a figure, clearly many sought more-prosperous communities, having consigned the Comstock to the status of yet another has-been of the mining world. Along with people, buildings left the district, dismantled and moved to new mining camps.[8] Austin, the home of Gridley in central Nevada, had recently struck rich ore and attracted attention. It became a magnet for people and houses, ultimately capturing Virginia City's International Hotel from the town's commercial core. The building was so noteworthy that it even appeared in the border illustration of a poster depicting an 1862 bird's-eye view of Virginia City, and yet the Comstock yielded to what seemed to be the next bonanza.

Mining, however, is a speculative industry, and for every pessimist there is an optimist. It was in 1864, after all, that the Daughters of Charity arrived and established an orphanage and school in Virginia City. Beginning their mission there, at a time when others sought to leave, the sisters apparently had faith in a new day for the Comstock. Still, many found it difficult to believe in the promise of wealth. Initially, investors were willing to build an infrastructure for the Comstock on the assumption that it would be like the famous silver mines of central Europe, Mexico, and South America, where the resources were sufficiently widespread to provide income for decades, if not centuries. By 1865, it would have been prudent to regard gambling based on this premise as ill-founded. Thousands sought the next bonanza in the new mining districts of Montana, Idaho, the hinterland of Nevada, and along the Sierra Nevada range.[9]

At the same time, others remained, hoping that the pessimists were wrong. Veterans of the mining world knew that the industry, even when prosperous, rarely provided stability. Those who obtained stocks, claims, or businesses during a depression, when prices were lowest, captured the most return on investments. Similarly, workers had the best chance of good employment if they lingered until the mining camp sprang back to life: by the time people arrived in the hundreds or thousands to take advantage of such an upswing, those who remained or had arrived first had inevitably taken the more desirable jobs. Thus, the optimists hung on to the Comstock, hoping that they would look back with hindsight on their decision and see wisdom rather than foolishness. Clearly, the prospect of staying presented a difficult choice. Unemployment could wipe out life savings if the economy slumped into oblivion. The mining camp could become nothing more than a ghost town, with stocks, businesses, and claims left worthless. The workers who lingered optimistically risked a great deal. They stood the chance of missing the next bonanza elsewhere, and with little or no income, they could quickly find themselves in a desperate situation.

Preliminary archaeological surveys suggest that the poorer residents of the Comstock frequently used a depression-era strategy of living on the outskirts of the urban centers. Tradition holds, quite rightly, that the founders of Virginia City arranged their town so that elevation and class status corresponded to one another. In other words, the higher one lived on the mountainside, the richer one was. That is why the largest, most ornate houses in the district stand proudly above the town's commercial district, while modest nineteenth-century residencies are scattered downhill. Archaeology identifies the limits of this observation, however. Artifact distributions suggest that the humblest of abodes were situated even farther up the mountainside, from which it was a long, exhausting hike up and down the steep hill to places of shopping and prospective employment. People evidently fashioned shacks and cabins from scavenged building parts high on the slopes of Mount Davidson. Food and household goods were the cheapest possible. People apparently squatted there, paying no rent and merely waiting for the economy to turn around so that they would be in a position to secure the best jobs. Thus artifacts from this area often date from the mid-1860s, the 1880s, and the 1930s.[10] Each of these periods was a time when the economy and the mining industry slumped, inspiring pessimists to flee and optimists to hang on as best they could.

In the midst of this early depression, the Comstock set the stage for one of its more charming plays, this one alternating between drama and comedy. In November 1862 a group of publishers established the *Virginia Daily Union*, challenging the preeminent position that the *Territorial Enterprise* had enjoyed in the region since 1859. The following spring, others organized the *Virginia Evening Bulletin*. In October 1863, Philip Lynch and John H. Mundall began printing the *Gold Hill Daily News*. Journalists such as Joe Goodman, Alf Doten, James W. E. "Lying Jim" Townsend, and William Wright's "Dan De Quille" became notorious, humorous, much respected, and sometimes loved actors in the district's early story. Firemen competed by racing to fires; mining speculators strove to corner the best claims and highest profits; and all the while, these newspapermen fought to get the jump on the best stories, rushing into print with the latest headlines, inventing items when times were dull, and disclaiming opponents' opinions and liberalities with the facts. They continued a tradition that Mark Twain enjoyed in the earliest days of the Comstock, but they acted out the parts with the grace, wit, and finesse of people who intended to amuse but also to settle in for years. Although they never became wealthy bonanza kings, these journalists were Comstock princes as they jousted and, like troubadours, sang the song of the great mining district with documentation, elaboration, and occasional fabrication. Ironically, the depression seemed to spawn, rather than inhibit, the creation of rival newspapers. Once again, there were those who saw sufficient promise in the mining district to build rather than to dismantle everything and leave.[11]

In 1864 a new brand of optimist assumed a place at the Comstock card table, and the poker game was never the same again. William C. Ralston and Darius Ogden Mills of San Francisco opened the Bank of California on July 5 of that year, with $2 million in gold as capital. Based in and named for California, the institution and its organizations nevertheless had plans for Nevada. Ralston had served as treasurer for several Comstock mines, including the Ophir and the Gould and Curry, once they incorporated in 1860. His experience with the mining district ran deep, and he apparently recognized the depression as an opportunity for profit. Ralston had established himself as one of the more formidable businessmen of the West Coast. Besides his involvement in many profitable Comstock mines, he had been a partner in the banking firm of Donohue, Ralston, and Company for several years. More important, Ralston was daring and charismatic. His forming a partnership with Mills, a banker and businessman from Sacramento, demonstrates the nature of his character as well as the value of his instincts. Ralston ensured that the Bank of California would gain a steady hand and an impressive reputation for business acumen.[12]

The Bank of California almost immediately opened a branch in Virginia City. Ralston and Mills hired William Sharon to serve as manager of the Comstock office. Sharon was to play a prominent role in Comstock history and folklore, but in 1864, no one could have guessed his destiny. Born in 1821 to Quaker parents in Ohio, Sharon left at seventeen to work on a riverboat, but the enterprise failed. He returned home, studied to become a lawyer, and set up practice in St. Louis, Missouri. The California gold rush brought him across the continent in 1849, and upon arriving, he became a merchant in Sacramento. A flood destroyed his business, so Sharon retreated to San Francisco, where he opened a real estate office. By 1864 he had amassed the tidy sum of $150,000, which he invested in mining stock and then lost. In short, William Sharon had given the world no reason to recognize him as the potential architect of a financial empire. Lord astutely described Sharon as "a small, though compactly formed, person, quiet in manner, and reserved to the point of coldness. . . . [He] was to all outward seeming the antitype of the burly, frank-spoken, domineering lawyer," as one found in William Stewart, for example. Stewart, however, was on the path to Washington

William Sharon represented the powerful Bank of California as it monopolized the Comstock in the late 1860s. (Courtesy of the Nevada Historical Society)

to serve as a senator, and his days of dominion over the Comstock were ending. It was Sharon who stepped into the vacuum Stewart left behind. Although Sharon was an unlikely replacement, Ralston recognized promise in him. Like Ralston, the forty-three-year-old failed lawyer and entrepreneur was daring and would not shy away from a gamble. Times were changing, and the Comstock called for a new type of risk-taker.[13]

The bank's strategy apparently was to manipulate the Comstock in such a way as to acquire as much of the district as possible.[14] Bank loans on the West Coast typically had interest rates of 3 to 5 percent per month. Sharon's office loaned money at 2 percent, and he had no problem finding customers. Indeed, many mines and mills came to the new bank for funds to weather the depression. What followed was the most cynical of hostile corporate takeovers. As the economy worsened in 1865, properties began to default to the bank's ownership. Sharon's web of control grew firmer and more complex, and he was able to target his surviving customers for failure, one by one. For example, he could acquire a mill simply by having his mills undercut standard prices and by controlling where the mines under his influence sent ore for processing. He could, consequently, starve a rival mill so that it failed and fell into the bank's hands.

In this way, Sharon acquired seven mills by early 1867. About the same time, Alvinza Hayward, an investor made wealthy from the California mines, joined the group that had become known as the Bank Crowd. With his additional support, they were able to gain control of two of the best Comstock mines—the Yellow Jacket and the Chollar-Potosi. In June 1867, the Bank Crowd formed the Union Mill and Mining Company to manage the property they had and that they hoped to acquire. Within two years, all the leading mines and a total of seventeen mills had fallen to the monopoly, and Sharon controlled almost everything that mattered on the Comstock.[15]

This glowing destiny could not have been apparent during the depths of depression in 1865. Still, the optimists had hung on because of the assumption, based on the precedent set by silver mines elsewhere, that ore would be widespread and able to yield profits for a long time. This point of view sustained investors and served to apologize for losses in the early years. It was this same perspective that also motivated the Bank Crowd, but in 1865 a consensus began to build that the pessimists had been right. Ralston and his associates may have been on their way toward gaining total control after investing millions, but many wondered about the value of such a monopoly. After all, complete ownership of a ghost town is complete ownership of nothing. Mining is gambling. The Bank Crowd came to the table with high stakes, and it appeared in 1865 that they might have lost everything. Nonetheless, in 1866 new discoveries in many of the mines vindicated the Bank of California and the other optimists. The flush times lasted another three years. It was the perfectly timed combination of depression followed by economic upswing that fueled the success of the Bank Crowd. The gamble had paid off in a way that even the optimists would have been embarrassed to imagine. Fortune at last shone on Sharon.

The three years of ore production from 1866 to 1868 sustained the Comstock, but profits were not spectacular, and when miners again began to exhaust the meager ore bodies in sight, a new depression loomed. As before, economics separated the pessimists from the optimists. Coincidentally, in 1867 a rich silver strike at Treasure Hill in eastern Nevada culminated in yet another rush throughout the mining world. During the following two years, thousands left the Comstock for the new mines, assuming that they would soon eclipse Virginia City.[16] Others scattered elsewhere. Repeating the situation of 1865, the pessimists suspected that the Comstock was finished.

In spite of the depression, 1869 was also the year when optimists undertook two spectacular, extravagant engineering feats noted throughout the mining West. The more famous of these was a railroad that became one of the most important short lines in the world. The nineteenth century was the railroad century, and the industrial world sought to connect itself with ribbons of steel. In its bid to establish the transcontinental railroad, for example, the Central Pacific pushed north of Virginia City toward Reno in 1868. It was clear to the Bank Crowd, as well as to anyone who would have reflected for a moment on the subject, that a line to the Comstock was both inevitable and desirable. Consequently, the Virginia and Truckee Railroad incorporated on March 5, 1868, with William Sharon, William Ralston, and D. O. Mills, together with two others, Thomas Bell and J. D. Fry, serving as directors. Sharon immediately launched a two-pronged effort to make the railroad a reality. First, he hired prominent local engineer Isaac E. James to survey a route. Sharon then proceeded to obtain the support necessary so that he could keep the financial obligation of the Bank Crowd at a minimum. Ultimately, Sharon was able to secure $500,000 in public bonds from Storey and Ormsby Counties and several hundred thousand more in subscriptions from mining companies in exchange for future service.[17]

Finances would be crucial, since any route ascending the 2,000 feet from valley floor to Virginia City would be tortuous. James's survey took nearly seven months, ending in June 1869. His plan called for twenty-one miles of track from Carson City to Virginia City, thirteen of which would need heavy construction to establish the roadbed. The route required six tunnels, at a total length of 2,400 feet, and about 6,000 degrees of turns, the equivalent of almost seventeen complete circles. Playing on one of the line's early names—the Virginia, Carson, and Truckee River—people nicknamed it the Very Crooked and Terrible Rough Railroad.[18]

Construction on the monumental task began from Carson City in February 1869 before James had completed his research. Work progressed along the Carson River to serve the mills of Brunswick Canyon, shunning a more direct overland route. By mid-April, there were 750 men, mostly Chinese, working on the line. Within two weeks, the number had grown to 1,200. Soon 1,600 Chinese and Euro-Americans were constructing the route, scattered across thirty-eight camps. Some predicted that the number would reach 2,000 to 3,000 before the line was complete.[19]

Work progressed rapidly and uneventfully until September, when the crews began to approach Gold Hill. With the economic slump, the Comstock did not lack for unemployed miners. There had long been a concern about Asian workers in the mines, since Euro-Americans regarded them as pariahs in the mining West. Many felt that they could undercut the wages of others. There were early calls to prohibit Chinese emigration to the Comstock, to save the district from the fate of integration experienced by many California mining camps. Asians arrived early on, but union activity had maintained their exclusion from the mines. Because there had been ample bullion to pay high wages and because Comstock shafts would quickly flood if workers did not properly tend machinery, the miners' union had only to threaten a strike to secure a daily minimum wage of $4, one of the highest in the industrial world. When Chinese gandy dancers ascended the heights of the Virginia Range with their railroad, particularly to areas that required tunneling, the union drew the line. On September 29, about 350 miners marched down to Gold Hill, along the proposed right-of-way, and confronted the Chinese. When the sheriff told the union men to disband because their acts against the railroad were unlawful, his order had no effect. The miners continued, chasing the Chinese for several miles down to an area known as American Flats. A week later, Sharon conceded defeat in a speech to the union and exchanged his Chinese laborers for unemployed miners to complete the line into Virginia City.[20]

On November 29, 1869, the first passenger train arrived in Gold Hill. A crowd of celebrants cheered the event, and "General Grant," the cannon at nearby Fort Homestead, fired a salute. Free food, champagne and beer, speeches, and music from the Gold Hill Brass Band abounded. The Comstock finally had its rail link. Efficient delivery of ore to the mills and supplies to the district was now possible. Almost lazily, two months passed before the Virginia and Truckee sent its first passenger train the next few miles to Virginia City. It would be more than two years, not until August 24, 1872, before workers tied the Virginia and Truckee to the transcontinental railroad in Reno. It was an important event, but the slow progress was a vivid contrast to the speed and evident urgency that the initial tie between mine and mill had demanded in 1869. The rest was luxury.[21]

Riding along the Virginia and Truckee, with its twists and turns and its steady incline, revealed the genius and careful planning that went into its construction. Climbing the "long grade," the lengthy ascent between Gold Hill and Virginia City, the traveler would no doubt be impressed by the struggle of the engine as it slowed and heaved with each drive of the piston, puffing its complaint at the demands placed upon it. Still, it made the journey, thanks to James. Passengers could take engineering for granted, however, and were probably more easily awestruck by the remarkable views afforded from the right-of-way. From parts of the line, it was possible to see more than a hundred miles on a clear day.

One of the most striking features of the Virginia and Truckee was the trestle spanning the Crown Point Ravine near Gold Hill. Five hundred feet in length, the wooden structure stood a maximum of eighty-five feet high. It was, as one Comstock history colorfully put it, "one of the engineering wonders of the West."[22] Still, wonder can just as easily inspire fear and concern. Central Pacific authorities, uncertain of the trestle's stability, refused to allow their Silver Palace Sleeping Cars to travel to Virginia City until June 1873, when railroad entrepreneur George M. Pullman himself rode one of his famed sleeping cars across the wooden structure. The Central Pacific conceded defeat and subsequently allowed its cars to climb the heights of the Comstock.[23] The trestle, together with the remarkable vistas available from the line, caused the Virginia and Truckee to become the nineteenth-century equivalent of a tourist attraction. People had long come from throughout the world to see the famous Comstock mines, but after 1869 a ride on the railroad became a pivotal part of the trip.

Gold Hill's Crown Point trestle appears in the center of this photograph. The steeple of St. Patrick's Catholic Church stands in the foreground. (Courtesy of the Nevada Historical Society)

The Great Seal of Nevada dates to 1866 and captures the optimism of a young state. The seal highlights agriculture, mining, and industry. (Courtesy of Eugene M. Hattori)

The trestle was so impressive and figured so prominently in local legend that it eventually transcended time: people came to believe that the trestle depicted on the Great Seal of Nevada was the structure spanning the Crown Point Ravine. This assertion stood in the face of the facts: the seal's trestle was stone in contrast to Crown Point's wood, and the seal dates to 1866, three years before the construction of the Virginia and Truckee. These details mattered little, however; the facts be damned! People could not imagine the Great Seal depicting anything less than the wondrous Crown Point Trestle, and pronouncements to the contrary have done little to turn the tide of folklore.[24]

The monopoly, which the Virginia and Truckee completed, left Sharon and the Bank Crowd with all the profits from the mines to the mills. Ultimately, the control of profits extended all the way to the U.S. Mint, since in 1869 the government opened a facility in Carson City, where a spur of the Virginia and Truckee served it directly. Transportation costs plummeted. Within a year, a cord of wood dropped from $15 to $9. Shipping of ore fell from $3.50 a ton to $2. As Lord points out, "A natural result of this reduction was to bring into market a large amount of ore lying on the mine-dumps or still left in the lode as too poor to pay the charges for transportation and milling. The first ore shipped over the railroad was 7 car-loads, 60 tons in all, from the 700-foot level of the Yellow Jacket Mine, of a grade which had been considered too poor to reduce and had been used as waste rock to fill abandoned drifts."[25]

The railroad instantaneously increased the longevity of the Comstock by years. It also practically cut off a source of income for teamsters. There was no way that they could compete with the railroad. Hauling by wagon became a costly anachronism, a specialty service to and from the Virginia and Truckee line for the odd item too heavy for the average person to handle or bound for locations out of the railroad's reach. Lord goes on to say: "Competition was useless, and the files of mules, with their creaking cars, soon disappeared from the roads." The U.S. Census manuscripts tell the story dramatically. In 1860 teamsters and packers represented more than 5 percent of the Virginia City and Gold Hill population, and people complained that there were too few. In 1870 only a few months after the completion of the Virginia and Truckee, teamsters and packers had dropped to a little over 1 percent of the total population. That figure fell to 0.5 percent by 1880. Congested streets and high transportation costs were a thing of the past, but a way of life and many salaries fell victim to mechanization. Still, few shed tears, since this was progress in the midst of the industrial revolution.[26]

The Virginia and Truckee also played an important role in extinguishing Washoe City, the community that had grown up around milling in the valley west of the Virginia Range. Just as the railroad made the transportation of ore to the Carson River mills cost-effective, it caused milling in Washoe Valley to seem absurdly expensive. Washoe's facilities closed and its leading town dwindled in size. One of the few industries remaining was local government, but in 1871 Reno, to the north, managed to steal away the seat of Washoe County, leaving Washoe City to decline into oblivion. Today all that survives are the rocky, grass-covered mounds of collapsed buildings, scattered across the meadows like the landscape of some long-forgotten Middle Eastern city. A few walls stand here and there, but with every season another submits and falls to the ground. One of the most striking features in the area is a strange line of fence posts cutting diagonally across the valley and through the shadows of streets that once served the town. They mark the Virginia and Truckee right-of-way as it connected the Comstock with Washoe City and Reno. The community slumbers in its grave while the path of its railroad endures.

Industrialization, however, creates more victims than just the unemployed and the displaced. Week after week, the Comstock newspapers featured stories about those injured or killed in the mines. It became such a common event that many isolated fatalities went unreported.[27] The most famous underground disaster on the Comstock has ever since served to exemplify the dangers local miners faced. Sometime in the early-morning hours of April 7, 1869, a fire started at the 800-foot level of the Yellow Jacket Mine. People speculated that someone on the night shift may have left a candle burning, which subsequently ignited the underground timbers, but no one could be certain. The damage that the conflagration caused obliterated any evidence of origins. The fire spread unnoticed in the vacant level at a time after the night shift had left and before the day workers had arrived. Finally, by about 7:00 A.M., the fire had consumed so many of the timbers that the rock ceiling collapsed. The force of the cave-in caused the smoke-filled, oxygen-starved air to blast through other parts of the mine like a gale force wind of poison.

Because the day shift had not entirely arrived at its stations, casualties were lighter than they might have been. Nonetheless, those already there and those being lowered at the fateful moment came face-to-face with a miner's greatest fear. The foul air quickly choked out the lives of some fifteen miners working nearby at the 800-foot level. One of the few survivors, a station-man named John Murphy, was far enough away so that although he fainted because of the foul air, he survived. He recalled that he could hear one of his workers a hundred feet below crying out, "Murphy . . . send me a cage; I am suffocating to death."[28]

Over the years a maze of adits and tunnels had been built to interconnect the mines of the Comstock. This had the potential of providing many means of escape in case of cave-ins or other disasters, but it also gave the smoke and foul air access to the neighboring mines. In this way, the wind fouled the 800-foot level of the Kentuck, killing a few more who were working there. The worst incident occurred in the Crown Point Mine, where cages lowered some forty-five men to the various levels. When the miners descended to 700 feet, they met with rising billows of smoke and poisonous gases. The operator, not knowing the circumstances below, continued to send the cage lower, stopping it at the 800-foot level. There, those who were descending found a gallery filled with dying miners, who rushed to the cage to be saved. Not all could find room, however, and when the cage ascended, it left behind many who all feared were condemned to death. As soon as the survivors could leave the cage at the surface, the operator lowered it again to the 800-foot level. With the signal to hoist the cage, the operator raised it to the surface. This time it carried only two men, George and Richard Bicknell, brothers from Yorkshire. The fumes had overcome Richard, who slumped, mangling his head between the timbers of the shaft and the cage as it rose. George, too, nearly fainted, but he stood resolutely, firmly holding the lifeless body of his brother. George died later that day, joining yet another brother, James, who had succumbed below ground.[29] With that, all realistic hope of removing more survivors from the depths of the Crown Point faded.

Soon, dense black smoke billowed from the shafts of the Kentuck, the

Yellow Jacket, and the Crown Point. Steam whistles called for help. The local volunteer firefighters answered. Frantic families and neighbors followed, seeking word of loved ones and friends. Within two hours, drafts cleared the shafts enough to encourage rescue attempts, but smoke and fumes sent the rescuers back repeatedly. Finally, the Kentuck was sufficiently clear to allow some exploration, and at 10:00 A.M. crews were able to retrieve two bodies from the 700-foot level. At noon they found four more at 900 feet in the Yellow Jacket. It was extremely dangerous work, since the fire had fouled the air and caused multiple cave-ins. Still, as the *Territorial Enterprise* reported, "There is no end to the stories of narrow escapes made by the gallant miners and firemen. It seems wonderful that men could be found to risk their lives in such a place, yet the strife was to see who would be the first in, and it was a matter of more difficulty to keep men out than to get them to go in."[30]

The Crown Point shaft remained a chimney, but operators maintained the blowers, with the risk of feeding the fire, hoping that that might sustain someone deep within the mine. In order to warn those who might have survived below the fire, the mine superintendent sent a cage to the 1,000-foot level with a lantern and a note reading, "We are fast subduing the fire. It is death to attempt to come up from where you are. We will get to you soon. The gas in the shaft is terrible and produces sure and speedy death. Write a word to us and send it up on the cage, and let us know how you are."[31] When the cage returned to the surface, there was no reply on the note, and the noxious air had extinguished the lantern, together with hope itself.

Besides the six bodies retrieved, at least twenty-eight men remained missing. Hour by hour over the next few days firemen and miners made slow progress into the bowels of the disaster. They eventually returned with bodies and with word of grisly scenes below. In the Crown Point, rescuers found nine of the dead clustered around an air pipe that they had severed in a desperate attempt to find enough fresh air to stay alive. Elsewhere they found a miner with a death grip on a ladder, his head hanging backward. Here and there, strewn about the floors of the levels, were the scattered remains of the unfortunate morning shift. Those who explored the underground morgue found the mangled bodies of still others at the bottom of shafts. Having run through the dark passages in panic, those victims no doubt hoped to find a cage but, misjudging their exact location, plunged to their deaths.

Scenes of suspense and grief played out with the crowds waiting at the mouths of the shafts. Father Manogue, the parish priest from Virginia City, and many of his fellow clergymen stood with families and neighbors

A priest, perhaps Father Manogue himself, comforts families as smoke billows during the Yellow Jacket disaster of 1869. (William Wright, *The Big Bonanza;* courtesy of the Nevada Historical Society)

waiting for word of those underground, but as the *Territorial Enterprise* pointed out, "Even the reverend fathers could find little to say upon such an occasion. The poor women with their weeping children clinging about them, stood about with hands clasped, rocking themselves to and fro, yet scarcely uttering an audible sob." Every time workers brought a body to the surface, women and children would cry out, praying that it was not their loved one. Comstock journalist Wells Drury recalled how "one shift boss, simple yet kindly soul . . . went with a delegation of his fellows to a humble cottage and sought to break the news by inquiring of the woman who came to the door, 'Does the *Widow* Williams live here?'" Each wife and child with a miner below eventually, however, took a turn to grieve. There were no survivors after the first few who escaped. Though no one could be certain how many died, estimates ranged from thirty-four to forty-five. The prolonged, intense fire cremated some bodies before searchers could find them. Victims with no relatives or close friends could have been forgotten, lost in the tide of sorrow. Mine managers had an idea of who was probably underground, but records were unreliable and circumstances encouraged those in charge to minimize the numbers. Nonetheless, all recognized the profound nature of the loss. As the *Gold Hill Daily News* pointed out, "One of men who was killed . . . had been married only ten days. . . . Eleven of the men . . . were married. . . . Two of the men . . . were fathers of families containing five children."[32]

The funeral processions, occurring almost four years to the day after that of Lincoln, rivaled the commemoration of the president's death. Military units, bands, fraternal organizations, and the miners' union accompanied the dead and their families to the final resting place. The entire Comstock mourned.[33]

On April 10, four days after the disaster first struck, miners discovered that the fire was growing. The managers made the decision to seal the shafts and build walls underground to contain and strangle the fire. Over the next two months, miners repeatedly opened the tomb to see if the flames had died, but each time they found the fire smoldering and eager to spring back to life. Reports indicate that the rock walls nearest the sealed fire remained hot three years later.

The Yellow Jacket disaster profoundly affected the way people looked at many things. In San Francisco rumors spread that mine owners had set the blaze in a most cynical and deadly form of stock manipulation.[34] On the Comstock, people began to question the benevolence of the Bank Crowd's monopoly. Miners talked of exploitation rather than the need for corporate investments, as they had in 1864. Ultimately, this line of thought proved beneficial for one of the greatest works of the Comstock, a mining district with no shortage of dramatic accomplishments. As mentioned earlier, there had long been a proposal for a tunnel nearly four miles long that would connect the Virginia City mines at a depth of 1,663 feet with the Carson River valley below. Adolph Sutro, a Jewish immigrant from Prussia, had incorporated his tunnel company in 1865 by act of the state legislature, but opposition from the Bank Crowd stalled its implementation.

With the Yellow Jacket fire, Sutro found the leverage needed to sway public opinion.[35] Ironically, in 1869 the need for the tunnel greatly diminished. The initiation of construction on the Virginia and Truckee meant that an efficient means to transport ore and supplies would exist. It was clear that with the railroad, the tunnel would never be able to shift the center of milling to the proposed town of Sutro, away from established facilities on the river along the railroad right-of-way. Nonetheless, Sutro met with the Gold Hill and Virginia City miners' unions on August 25, 1869, and underscored the value the tunnel would have for safety. With only a little persuasion, the unions gave him $50,000 for his project. Sutro initiated work almost immediately, but progress would be slow. After weeks of preparation, he swung the first pick on October 19, 1869. His dream proved to be an enormous undertaking, consuming 1,000

The entrance to the Sutro Tunnel was three miles away from Virginia City. Adolph Sutro had planned a large city at the site. (Courtesy of the Nevada Historical Society)

board feet of lumber for every five feet of tunnel.[36] Salaries eventually exceeded $1,000 per day. The tunnel called for complex engineering of nearly four miles of excavation with no dips or turns, at a steady, slight incline to facilitate drainage and wide enough for ore carts traveling in two directions at once. Still, the designers overcame technical difficulties more easily than Sutro solved the habitual problem of financing. It would take almost a decade to complete the tunnel, but the people of the Comstock had decided that the district was here to stay.

The mining district had proved it was able to stumble away from the downward spiral of depression that had created many western ghost towns. The slump of 1865 yielded to ore discoveries the next year. Shifts in the economy caused changes in society. With the emergence of the Bank of California, the Comstock went through yet another metamorphosis. Like the change from placer mining to hard-rock, underground excavation from 1859 to 1860, the transition from small-scale industry to corporate monopoly created a radically new social and economic environment.

In addition, every time one of these changes occurred, people became even farther removed from the earlier periods, misunderstanding them and their players. It was an increasingly fertile ground for folklore about a past made distant not so much by years as by comprehension. At the same time, the late 1860s, like earlier periods, fed more characters and things into local legend. Thus Sharon and Ralston, the Bank of California, Sutro and his tunnel, the V&T Railroad with its magnificent Crown Point Trestle, and the Yellow Jacket disaster figured as giants in Comstock legend.

In 1869 the economy may have faltered, but the earlier depression helped develop a mind-set on the Comstock that its silver mines were immune to failure. Gold-mining camps throughout the West had come and gone, but experience seemed to substantiate the hope that the Comstock rested on deeper resources. For this reason, developers initiated some of the most remarkable undertakings on the Comstock in 1869, a year when mining proceeds declined. Thus the Bank Crowd's construction of the Virginia and Truckee Railroad, perhaps its most dramatic achievement, began when the economy slumped. At the same time, Adolph Sutro was finally able to begin his tunnel. These two projects, initiated in the midst of depression, serve as testimony to the new confidence bred by the Comstock silver mines. Although the projects may seem extravagant in the context of depression, the developers marketed them as a means to make the Comstock more efficient, granting it a longer life. The railroad immediately met this expectation. It would take nearly a decade to determine the effect of the tunnel.

Two articles in the same July 1869 issue of the *Territorial Enterprise* capture the sentiments of the time. The first proclaimed that "the White Pine excitement is over" in Virginia City. The idea that the new rush would replace the Comstock was actually short-lived. The second article addressed the Comstock depression itself, pointing out, "There is just at present much complaint about dull times, yet there is still a dollar or two about the city. Even the worst of our grumblers say when cornered down to it, that they know of no place on the Pacific coast where they could do better. Although somewhat dull at present, Virginia will yet do."[37]

5: A TIME OF BONANZA

The lid, so to speak, of that wonderful ore-casket, termed commonly the Big Bonanza, had been lifted off. . . . The plain facts are as marvellous as a Persian tale, for the young Aladdin did not see in the glittering case of the genii such fabulous riches as were lying in that dark womb of rock. The miner's pick and drill were more potent than the magician's wand.
—*Eliot Lord,* Comstock Mining and Miners

When enumerators for the 9th U.S. Census began working on the Comstock in June 1870, they found a mining district radically different from the one documented ten years earlier. Gone were the Mexican packers, the prospectors, the thrown-together buildings, and the society in which women were a scarce curiosity. The Comstock was now a place of industry and engineers, boasting almost four hundred men employed in milling, nearly three hundred in the manufacturing industries, and roughly three thousand working in the mines.

The Comstock also included, however, a wide variety of nonmining businesses and services. While the ranks of miners, like those employed in most occupations, had increased since 1860, their proportion of the overall workforce had decreased from more than 70 percent to about 43 percent in 1870 (table 5.1). Furthermore, the variety of professions had increased phenomenally. Besides the men involved in the large corporate industries, the census recorded twenty-two bakers, forty-nine butchers, five people operating confectionery stores, two oyster vendors, a coffee

5.1 Occupations for Comstock Men, 1860 and 1870*

Occupation	Number in 1860	Number in 1870	Percentage of Workforce in 1860	Percentage of Workforce in 1870	Ratio of 1860 to 1870
Mining	1,984	2,808	71.4	42.6	1:1.4
Construction	221	325	8.0	4.9	1:1.5
Teamsters/Packers	176	146	6.3	2.2	1:2.1
Service	125	499	4.5	7.6	1:4.0
Mercantile	124	766	4.5	11.6	1:6.2
Saloons	61	234	2.2	3.5	1:3.8
Manufacturing	24	623	0.9	9.4	1:26.0
Infrastructure	23	126	0.8	1.9	1:5.5
Mills	7	381	0.3	5.8	1:54.4
Railroad	—	44	—	0.7	—
Other	1	498	—	7.6	1:498
None	32	148	1.4	2.2	1:4.6
Total	2,778	6,598			1:2.4

Sources: 8th U.S. Manuscript Census of 1860 for Virginia City and Gold Hill, Utah Territory; 9th U.S. Manuscript Census of 1870 for Storey County, Nevada.
*Men are defined as 15 years or older. The total population for Virginia City and Gold Hill during the census of 1860 was 3,017; in 1870 the population of Storey County was 11,319.

vendor, and one man who ran a peanut stand. There were also nineteen people involved in the theater, including actors, actresses, managers, and a "tragedian." The census recorded fourteen musicians and two gymnasts, and there were gardeners, a librarian, photographers, milk dealers, tailors, stockbrokers, politicians, federal tax agents, doctors, lawyers, and several people in jail. Along with these pursuits, the enumerators listed dozens of other livelihoods.

Besides offering more diversity in occupation, Comstock communities had transformed in other ways. In 1860, there were almost three thousand adult men on the Comstock.[1] By 1870 that number had more than doubled. The truly spectacular examples of growth, however, occurred in other parts of society. There were now twenty times as many women and children. Of course, the nature of the nineteenth-century mining town attracted single men, and bachelors would remain a solid block in the district for decades. Still, by 1870 the Comstock was an established community with churches, schools, and hundreds of families. Wives and mothers together with their single sisters pursued a variety of occupations ranging from domestic service and dressmaking to laundry work and teaching (table 5.2). In addition, eight Daughters of Charity now cared for eighty children in their orphanage.[2] There were also seven actresses and as many teachers, as well as several saloon tenders, waitresses, milliners, cooks, an upholsterer, and a melodeon operator.

Prostitutes appeared in force by 1870 after apparently waiting to make certain that the new Comstock Mining District was worth the trip. The presence of 160 ladies of the evening working in Gold Hill and Virginia City during the census testifies to the fact that these women now regarded the Comstock as a good gamble. Virginia City had a well-defined red-light district between the main commercial development above and Chinatown downhill, a vivid expression of the social hierarchy. The brothels and cribs housed a cross section of the community, if not the world, filled with diverse women who were "serving the public," as several described their occupation to the census enumerator. Most of the states, as well as Canada, Germany, Mexico, Ireland, England, France, and Italy, gave daughters for the line. More than half of the prostitutes listed in the 1870 census came from China and worked the district from the western, upper boundary separating the Chinese from the non-Asian part of the red-light district. House after house for several blocks offered a wide variety of facilities with a spectrum of nationalities, costs, and perceived quality. There were also other brothels and individual prostitutes scat-

Virginia City, 1878. Photograph by Carleton Watkins. (Courtesy of the Nevada State Historic Preservation Office)

5.2 Occupations for Comstock Women, 1870*

Occupation	Number	Percentage of Total
Keeping House	1,604	72.9
Prostitution	160	7.3
Seamstress	58	2.6
Servant	53	2.4
Lodging/Boarding	28	1.3
Laundry	14	0.6
Milliner	13	0.6
Daughter of Charity	8	0.4
Teacher	7	0.3
Restaurant work	7	0.3
Health care	2	0.1
Other	19	0.9
At school	29	1.3
None	199	9.0
Total	2,201	

Source: 9th U.S. Manuscript Census of 1870.
*Women are defined as 15 years or older. The total population for Storey County in 1870 was 11,319.

tered throughout Virginia City and Gold Hill. Much of the community regarded these establishments and entrepreneurs as a public nuisance and repeatedly launched efforts to see them removed from blocks other than the accepted D Street red-light district. The most notorious of these areas was known as the Barbary Coast, a collection of cribs, brothels, and tough saloons situated on South C Street. The neighborhood had a reputation for danger and intrigue, and it attracted only the hardiest, most adventurous souls.[3]

Although prostitution was the most common wage-earning occupation declared for women in the 1870 census, other factors must be considered. Women more than men were apt to pursue several sources of employment at a time, and yet the convention of the census forced simplification. Thus a woman who preferred to be known as keeping house for her family may have also washed clothes or sewn dresses for money. The widow Mary McNair Mathews, who wrote her memoirs of living on the Comstock during the 1870s, worked as a teacher, nurse, seamstress, laundress, and boardinghouse operator, often simultaneously, as she pieced together a means of support for herself and her son. The pattern was certainly not uncommon, and yet the census enumerator could list only a single occupation. Women pursuing income from a variety of sources, most of which the community would have regarded as respectable, far outnumbered the prostitutes (who may have been involved in other occupations as well).[4]

By 1870, the Comstock had fulfilled its destiny to become an international community (table 5.3). For example, there had been 310 sons and daughters of Ireland on the Comstock in 1860. Ten years later that number had grown to 2,160, making the Irish one of the most important components of Virginia City. Paralleling this, the Cornish, internationally famed as miners, had yet to arrive in 1860. Satisfied that the mines would survive, immigrants from Cornwall numbered in the hundreds in 1870, claiming their place among the major ethnic groups of the district.

The Chinese community, represented by only 14 men in 1860, had developed into a diverse element of the Comstock by 1870, when the Asian inhabitants of Storey County numbered 647 men and 103 women. These people lived in several segregated Chinatowns in Gold Hill and

5.3 Ethnicities on the Comstock, 1860 and 1870

Nativity	Number in 1860	Number in 1870	Number of Females in 1870	Percentage of Females in 1870	Ratio of 1860 to 1870
USA*	1,949	5,560	2,077	37.4	1:2.9
Ireland	310	2,160	735	34.0	1:7.0
Britain	165	1,150	220	19.1	1:7.0
China	14	744	102	13.7	1:53.1
Germanies	230	578	148	25.6	1:2.5
Canada	118	488	87	17.8	1:4.1
Hispanic	99	116	48	41.4	1:1.2
France	27	111	29	26.1	1:4.1
Switzerland	13	82	13	15.9	1:6.3
Italy	9	53	8	15.1	1:5.8
Scandinavia	41	82	8	9.8	1:2.0
Portugal	1	43	3	7.0	1:43.0
Other**	41	152	27	17.8	1:3.7
Total	3,017	11,319	3,505	31.0	1:3.8

Sources: 8th U.S. Manuscript Census of 1860 for Virginia City and Gold Hill, Utah Territory; 9th U.S. Manuscript Census of 1870 for Storey County, Nevada.

*Of those born in the USA, 7 were African American in 1860 and 71 were African American in 1870. Eleven African Americans came from other countries in 1870.

**19 designations for 1860; 26 for 1870.

Virginia City. In addition, Asian laundry workers operated wash houses throughout both communities, and servants also lived in Euro-American neighborhoods, typically staying with their employers.[5] Besides these stereotypical occupations, the Chinese worked as merchants, cooks, peddlers, laborers, druggists, doctors, and carpenters. There were also several professional gamblers and woodcutters as well as a cigar maker and a jeweler among their ranks.

The enumerator listed all but nine of the Chinese women as prostitutes, but that designation may have had more to do with racist assumptions and cultural misunderstandings than reality. Some of the alleged prostitutes appear in the handwritten manuscript census of the enumerator as the wives of merchants and others with social and economic standing within their own community. It is improbable that these men also played the role of pimps for their wives. Historian Sue Fawn Chung points out that there may have been confusion about the role of second wives, adding to misunderstandings of this society. These women often accompanied their immigrating husbands while the first wife remained at home to attend to the extended family. The first wife would have been a peer of roughly the same age as her husband, their union the result of an arranged marriage. The man had subsequently taken a younger woman as a second spouse either for economic or other convenient reasons or for love, and it would be this second wife who would travel with him. When the census enumerator asked what the woman's role was in the household, a response of "second wife" may have caused the enumerator to assume that she was a concubine and therefore a prostitute.[6]

According to the census, a few Asian women joined their male colleagues in the laundry while others appear in the census as "keeping house." There may have been more women than those whom the enumerator identified, hidden away in the mysterious Chinatown. The official may have assumed the ethnic enclave to be too difficult or dangerous to probe, and so the census is of little use in this regard.[7]

The Hispanics of the Comstock were one of the few groups to show almost no growth during the first decade. As other ethnic enclaves grew explosively, the Spanish-speaking population remained around one hundred, but it nonetheless changed in several ways. For example, many more Spanish-speaking nations were now represented. In 1860 the Comstock Hispanics were noteworthy for their numerous women and families. By 1870 Spanish-speaking women accounted for 48 percent of the Hispanic community, up from the 14 percent of 1860.[8] Far from signaling an increase in family orientation, however, the change was part of a dramatic shift in the nature of the group. The ranks of Spanish-speaking prostitutes had swelled to fourteen, representing roughly a third of the Hispanic women. Again the prejudice of the enumerator may have been a factor in identifying too many women with this occupation, but clearly Hispanics involved in sexual commerce had become an aspect of the Comstock, whereas ten years earlier that had been either an anomaly or nonexistent. Though the packers were gone, two thirds of the Hispanic men in 1870 were miners, giving the ethnic group a homogeneity unknown ten years before. Some of these may have been scavengers who collected scraps of ore for processing in their own hand mills.[9] Others probably worked for the mining companies. In short, the Spanish-speaking community, with no growth, fewer families, and an increased number of miners, ran opposite to the statistical trends of increased numbers, more families, and greater diversity of occupations that were typical of the community at large.

The story of the African Americans on the Comstock during the 1860s demonstrates an intriguing pattern of integration, marginal survival, and occasional success. J. Wells Kelly's *Second Directory of the Nevada Territory*, from 1863, documents thirty-five African American men, working as barbers, cooks, barkeepers, and laborers. A few were washing clothes. There were also a carpenter, a blacksmith, and a miner. The story of a man named John H. Jones appears to have been quite impressive. He was a teamster who arrived on the Comstock by at least 1862. He, and perhaps his wife, operated a boardinghouse on A Street, above the commercial corridor of Virginia City. Several of the African Americans in the 1863 directory appear as living in his household. He apparently left the Comstock by the mid-1860s, but his example demonstrates that African Americans could achieve financial success in the early days of Virginia City.[10]

Unfortunately, the convention of the 1863 directory documents no children or wives, but there were almost certainly at least a few of each. The men lived scattered above C Street in areas generally thought to have been reserved for the wealthier members of the community (map 1). Virginia City of the early 1860s, however, was a much different place from what established itself later. The town in the early 1860s was sprawling uphill, following the ore body as it slanted west into Mount Davidson. B Street rivaled C Street as the most important commercial corridor, and many African Americans lived and worked at the corner of Union and B Streets, at the Piper building and the International Hotel.[11]

A close examination of the federal 1870 census and an 1875 state census together with the 1873 Virginia and Truckee Directory yields addresses that allow for yet another map. This one documents the distribution of African Americans ten years after the first. The 1870 census

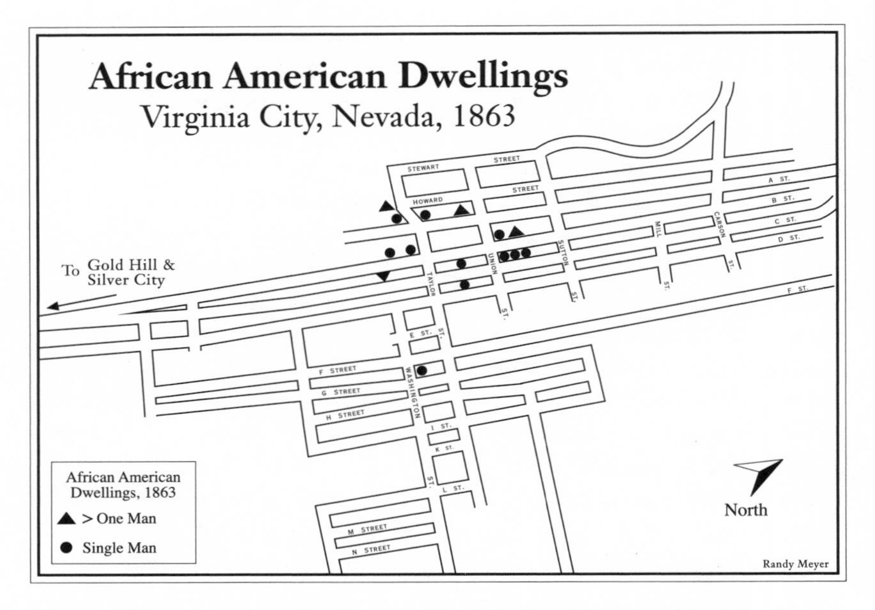

Map 1. African American Dwellings in Virginia City, Nevada, 1863. Map by Randy Meyer. (Source: J. Wells Kelly, *Second Directory of Nevada*, 1863. Not to scale.)

includes eighty-two African Americans, while the 1875 state census lists ninety-eight.[12] Only six of the men from the 1863 directory appeared in the 1870 census, which documents a larger, more diverse community.[13] Laborers, servants, cooks, and barbers still account for most of the occupations, but there were also a physician, a saloon owner, a tailor, and two milliners, women who worked at the most prestigious of the needle trades. Once again, the African Americans appear to be evenly scattered throughout the community, as demonstrated by the 1873 directory (map 2). Single men and women often lived in the commercial corridor of C Street. These included bootblacks, porters, and other menial workers, but also a lodging house operator and a saloonkeeper. The cluster of single individuals on South C Street corresponds with the notorious Barbary Coast, a neighborhood of crime, sordid saloons, and a lower class of people. The African Americans living in this area, however, appear to have been attempting to eke out a respectable living, taking advantage of lower real estate costs and neighbors who could not afford to be exclusive when it came to race. All but one, a porter, worked in hair-dressing salons.

African American families in 1873 were scattered in the eastern, downhill part of town, below the Virginia and Truckee yards. These families, apparently wishing to live in their own houses rather than in apartments above commercial establishments, found accommodations on the edge of Chinatown and the mine dumps of the Consolidated Virginia. Again, these were areas of lower real estate values.[14]

Besides the ethnic groups already discussed, there were dozens of others. The 1870 manuscript census is a veritable geography lesson, with representatives from every continent. Complementing this complexity was an increasing diversity in architecture. Photographs of Virginia City's main street show a degree of development and permanence that is surprising since it was only a decade after the community's inception. Brick and wooden buildings exhibit the latest in Italianate architecture, the stylish design choice of the time. Two- and three-story buildings were the rule, and in at least one case, a building loomed four stories over C Street, each floor with its own porch.

In the midst of the community's sophisticated architecture stood the scattered remnants of buildings hurriedly thrown up in the first year of

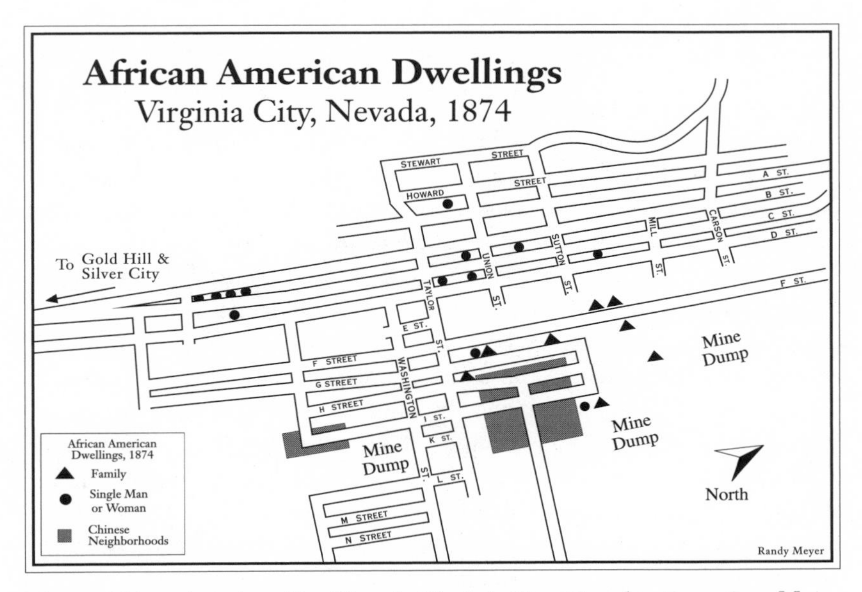

Map 2. African American Dwellings in Virginia City, Nevada, 1873–1874. Map by Randy Meyer. (Sources: Addresses from the 1873–74 V&T Directory for Nevada, compared with the 1890 Sanborn–Perris Fire Insurance Maps; ethnicity from the 9th and 10th U.S. Manuscript Censuses of 1870 and 1880 and the 1875 state census. Not to scale.)

This street scene dates from 1865, five years before the 1870 federal census, and already Virginia City is congested and highly urbanized. (Courtesy of the Nevada Historical Society)

the rush. The tents were gone long ago, as were the flimsiest of the wood shacks, having fallen victim to the Washoe zephyrs and the other elements. Here and there, however, were the survivors, primitive structures fulfilling the immediate needs of a bygone era. In the mining West, time became telescoped, and the first building efforts acquired historical significance after only a few years. Thus the old flume, the old mill, or the old market became antiques within a decade because of their utilitarian design and because few in the district could remember when they were new. Explosive change, both in the creation of buildings and in the arrival of thousands of newcomers, created a sense of history different from that of the East, with its long-established roots. The torrent of people coming and going shaped the mentality of the mining district. The Comstock of 1870 was about to see still more change, which would again render the district of a few years before a bygone era.

Among the multitude who had come to Virginia City were two miners of Irish origin who, through hard work, skill, and luck, had managed to rise slowly above the rest. Initially, little distinguished John W. Mackay and James G. Fair from the thousands of their fellow Irishmen who settled on the Comstock. Both were born during the 1830s, immigrating to the United States as young men. Gold fever captured the imaginations of both after the 1849 California strike, and they separately found their roads heading west. Eventually, each settled in the western foothills of the Sierra, the famed California gold country. Fair worked as superintendent of quartz mines in the region. Mackay, who was several years younger, gained experience as a placer miner. After the Comstock strike, both men, now veteran miners, moved to the new district, where they eventually met.

Mackay arrived with his partner, Jack O'Brien, and maintained, "All I want is $30,000; with that I can make my old mother comfortable." An oft-quoted story about the two men has O'Brien asking Mackay, as they finished the last leg of their trip to the Ophir diggings, "John, have you got any money?" "Not a cent," answered Mackay. O'Brien then said, "Well, I've only got a half a dollar and here it goes." With that, O'Brien hurled his coin into the sagebrush.[15] Although the story may be fanciful, Mackay apparently confirmed its validity. More than a part of actual history, however, the story illustrates the outlook that prevailed among many who first arrived at the strike. It appears that egalitarianism and a cavalier attitude, combined with the idea that small stakes mattered little, formed the mind-set of Mackay and others.

Mackay worked as a miner, earning the standard $4 a day, but he was frugal and invested his profits where his growing understanding of mining indicated it would be wise. By the early 1860s, he became superintendent of the Caledonia Tunnel and Mining Company and then of the Milton, a small mine that participated in the property disputes with the Chollar. Mackay slowly acquired holdings and became known in the district, as D. O. Mills maintained, for "his ability, his uprightness, and his steady judgment."[16] In 1863, a large group of investors incorporated the Bullion Mine on the Divide, the area defining the boundary between Gold Hill and Virginia City. J. M. Walker, who led the effort, invited the twenty-eight-year-old Irish immigrant to participate as one of the five trustees. Mackay sold the stock he had collected, realizing a profit in the wild market of the time, and then invested in the new mining venture. For several years the Bullion shaft, with all of its exploratory adits, probed the area, costing investors as much as $1 million. The mine was dry, which was both good and unusual; it was hot, which was bad; and it was absolutely devoid of profitable ore, which was worse and eventually intolerable.

Walker and Mackay quietly turned their attention to other prospects. In 1865, in the midst of the mining slump, one property looked particu-

John Mackay was an Irish immigrant who became one of the richest men in the world. (Courtesy of the Nevada Historical Society)

larly promising. The Kentuck, situated between the Crown Point and the Yellow Jacket, held less than 100 feet of the lode and had never produced profitable ore. Recent discoveries in both of the neighboring mines, however, suggested that the Kentuck might possess its own pocket of riches. Walker and Mackay were able to acquire the property at depression prices. They decided to offer no stocks, but instead launched a limited, crude operation with a horse-drawn whim for a hoist. It was the sort of small-time excavation that has been common throughout Comstock history, leaving the district pockmarked with what the locals call coyote holes. These rathole mines were the front line of underground exploration, but they were unusual along the core of the lode. The vast majority of them, especially those on the fringe of the district, never produced profits, leaving the proponents poorer for the effort. Still, the dream of riches to be found after the next blast kept these sideshows drilling.[17]

The Kentuck was one of those rare cases in which the gamble paid off. Walker and Mackay found a sizable ore body, making their mine one of the most profitable during the late 1860s. From 1866 to 1869, it produced $3,641,062 in ore.[18] William Sharon, in virtual control of the Comstock at the time, could not tolerate such a success under his nose but out of his control. He offered Walker $600,000 for his share, inspiring the miner into an early European retirement and ending his partnership with Mackay. When Walker left in 1867, Mackay became superintendent of the Bullion, but clearly the honor of managing a losing proposition had its drawbacks. Mackay's most important venture remained the Kentuck. Still, his lack of assets required him to seek the help of new partners in mining, as well as in his personal life.

It was during this time that Mackay met the widow Marie Louise Bryant. Her husband, the doctor who had come to the Comstock in 1860 to assist in the Pyramid Lake War, brought his family to the dynamic, growing mining district once the conflict ended. Unfortunately, he became addicted to his own medicines and eventually died destitute in 1866. His widow took in sewing and did whatever she could to acquire a means of support, but it was a continual struggle. In 1867 she met the successful young miner, and the two began a quiet courtship culminating in a wedding officiated by Father Manogue on November 25, 1867. One of the most eligible bachelors on the Comstock had taken yet another step toward respectability.[19]

Fair's professional experience propelled him into management shortly after his arrival on the Comstock. He quickly became the superintendent of the Ophir Mine, and then early in 1867 he served as assistant superintendent of the promising Hale and Norcross. For some unknown reason, management dismissed Fair in November of that year.[20]

At some time during Fair's tenure at the Hale and Norcross, he met Mackay, and the two came to realize the benefits of an alliance. Mackay was amassing a small fortune as a result of his success with the Kentuck. Fair was gaining a reputation for being a successful mine manager and a shrewd observer of ore bodies. As Irish immigrants with deep western roots, having worked at the lowest levels of the mining industry and having fought their way up the ladder, the two had much in common.

With Fair's dismissal from the Hale and Norcross, he and Mackay solidified their relationship with a joint venture in Idaho in 1868. Fair remained in that state to serve as superintendent of their jointly owned Rising Star Mine. It proved to be a losing proposition. The Comstock remained the best gamble, and Fair never lost sight of his former employer.[21]

James Fair was John Mackay's partner. Well thought of as a miner, Fair was nevertheless regarded negatively when compared to the honorable Mackay. (Courtesy of the Nevada Historical Society)

The Hale and Norcross soon fell victim to an absurd effort on the part of Sharon and the Bank of California to seize control, resulting in a rise in stock prices during the first two months of 1868 from $1,260 to $7,100 a share. Although successful, the "farcical struggle," as Lord describes it, left the new owners financially bruised.[22] When the mine produced only 16,536 tons of low-grade ore during 1868, there was no reward in sight. Stock prices of $2,900 in March fell to $41.50 in September. It was one of those opportunities that the Comstock afforded the cautious, quiet observer. Just as the stocks cost more than their value during the rush to seize control, they now reached an artificial low as public opinion turned against the Hale and Norcross. At less than $50 a share, the mine provided a chance to profit.

The Hale and Norcross now represented an opportunity for yet another common venture between Fair and Mackay. Fair was in a position to suspect, or even to know, that the Hale and Norcross could become profitable. He and Mackay entered into a partnership with William S.

O'Brien and James C. Flood, former saloonkeepers turned San Francisco stockbrokers. Like Mackay and Fair, the two were of Irish descent and had arrived in California during the Gold Rush.

When the Hale and Norcross called for the election of officers in March 1869, Mackay, Fair, O'Brien, and Flood suddenly appeared as contenders, and indeed, they won control. Mackay had himself and some friends elected as trustees. Flood became president, and the new leadership appointed Fair as superintendent. The Hale and Norcross was never the same after that.[23] Under new management the mine grew efficient and prosperous. During 1869, the business paid unprecedented dividends totaling $192,000 to its stockholders. To ensure that the Bank Crowd could not control the profits, the four Irishmen bought mills for ore reduction. Increased capital investment caused their proceeds and power to grow proportionately. The following year, after Fair discovered a new ore body, dividends rose to over $500,000. Mining eventually depleted the richest ore and payments to stockholders ceased, but not before making the four Irishmen richer for their partnership. They were now ready for still further ventures, and again circumstance provided opportunity.

At the same time, the Bank Crowd faced huge fiscal burdens and little profit, weakening the financial cooperative in the face of a challenge to its Comstock monopoly. Stock prices fell below what the group had often paid to secure a monopoly. The California corporation had participated in a gamble of extraordinary proportions: a well-timed series of discoveries would leave the Bank Crowd in complete control of huge profits; *borrasca* would prove disastrous. The year 1870 showed little promise, and the Bank Crowd's costly exploration of its Crown Point Mine yielded few positive results. Bank officials debated cutting off funds for more work and prepared to acknowledge defeat and accept their losses. It would have meant the end of pumping on the Comstock. The lower levels would flood, preventing deeper exploration and mining. They were within days of writing the last chapter of the Comstock story when miners happened upon a small bonanza in the Crown Point.[24]

Perhaps smelling the blood of corporate enemies, other capitalists began to examine possibilities. Although the Bank Crowd controlled the Crown Point, it did not own a clear majority of the stocks. One of its investors, Alvinza Hayward, a California speculator, formed a secret alliance with his brother-in-law, John P. Jones, the Welsh-born superintendent of the Crown Point. The two decided to wrest control of the mine away from the Bank Crowd. Jones slowed the operation to make it appear less profitable while Hayward worked the San Francisco stock market, quietly collecting shares. The deception took Sharon by surprise, and the Bank Crowd lost control of the new bonanza. Sharon offered his competitors a deal to end their awkward coexistence: he would sell the bank's Crown Point stock to the new owners at market value if they would sell all their stock in the neighboring Belcher. Since conventional wisdom maintained that the new ore body would extend into the Belcher, the trade seemed reasonable. All parties agreed, and on June 7, 1871, Hayward paid Sharon $1.4 million for 4,100 shares in what was probably the largest single stock exchange in Comstock history. The Belcher turned out to be the more profitable of the two, so the Bank Crowd did not suffer, but it was clear that Sharon's empire was eroding.

When Sharon faced off against Jones in a senatorial contest two years later, the limits of his power became even more obvious. Not even Sharon's absurd accusation that Jones had set the Yellow Jacket fire as part of a stock speculation scheme could stop the Welshman's campaign. The charge did, however, precipitate a crash in a stock market that had teetered for a time at unrealistically high prices. In keeping with the federal Constitution at the time, Nevada's law directed the legislature to elect senators to the nation's capital. Jones managed to secure sufficient votes in the elite club to overwhelm the political maneuvers of Sharon. Defeated, all the bank manager could do was stand on the sidelines as others found means to exploit the district he had hoped to control absolutely.[25]

In the midst of these machinations, Mackay and Fair were not inactive. They used their financial successes as a platform from which to test other possibilities. Mackay had himself elected superintendent of the Bullion. That fruitless shaft represented one of the best gambles on the Comstock. It was in the center of a mining district noted for an uneven distribution of ore. It stood to reason that a well-placed shaft that seemed barren might not be far from profitable material. With this in mind, Mackay pursued the possibilities. Fair became superintendent of the Savage Mining Company, a property closer to the original Virginia City successes. Mackay's efforts yielded nothing but mounting bills. Fair was able to pay for exploration in the Savage with low-grade ores, but a bonanza remained elusive and there were no profits. By 1871 the two veteran miners acknowledged defeat and were ready to test other areas in their trial-and-error probing of the district.

Yet another attractive gamble existed in claims between the Best and Belcher and the Ophir. This part of the lode had yielded profitable ore near the surface in the early days of the Comstock, but careful underground exploration had occurred only on a limited scale. The Consolidated Virginia Mining Company expended $161,349.41 in exploration during 1869 and 1870 without positive results. Stocks fell to $1.625 a share, giving the entire mine a market value of $18,850. Again a promising prospect was available at bargain prices, and the four Irishmen moved in.[26]

Assuming formal control of the Best and Belcher on January 11, 1872, the new owners quickly developed a two-pronged assault on the claim. Fair directed the excavation of a shaft. At the same time the new company started an adit from their property to the south, probing north at the 1,200-foot level. Ample barren rock existed to discourage the effort, but Mackay and Fair were committed to the systematic exploration of the area: their success depended on continuing the effort past the point where others had surrendered and retired. While digging through the Best and Belcher claim, miners found a vein of low-grade ore. It would yield little profit in itself, and others might have written it off as an unimportant anomaly, but there was the possibility that it could serve as the trail to something better. Fair ordered the miners to follow the seam of ore as it meandered north.[27] For a hundred feet the vein yielded nothing, and many believed Fair's instincts had gone awry. The company had spent $200,000 in fruitless exploration. It seemed possible that this experiment might prove the undoing of the four Irishmen. They could have easily become yet another in a long series of footnotes to Comstock history, which featured many miners who rose above the rest for a moment, only to tumble down again.

Fair became sick and left work for a month. His miners pursued the vein as it drifted to the east, but still there was no hint of reward. In February 1873, shortly after Fair returned, the ore became richer, assaying at $60 per ton. Continued excavation uncovered two additional small ore bodies, and miners reached a point where the primary vein widened to 12 feet. It appeared that Fair and Mackay had once again happened upon a bonanza. Miners worked in an oppressive environment, however, since their adit was long and wound into an ever-hotter area that made the air stifling. Fair directed those working on the shaft from above to pick up the pace to make a connection and improve ventilation. Miners linked shaft and adit in October, and Fair was now able to engineer the removal of ore.

Nothing he or Mackay did, however, could prepare them for the magnitude of their discovery. Miners were excavating a cavern of gold and silver ore that paid from $93 to $632 per ton. When they had created a room 20 feet high, some 50 feet wide, and 140 feet long, they found they could dig in any direction and still excavate valuable ore. Lord compared the treasure-filled ore body to "the glittering case" of Aladdin's genie, suggesting that "the miner's pick and drill are more potent than the magician's wand."[28] Under any other circumstances, such a description would be hyperbole, but this was the Big Bonanza. No other mine in the millennia-old history of the industry could easily boast a discovery of this magnitude. Teams of miners removing the ore seemed incapable of exhausting the treasure. Going far beyond realistic expectations in a mining district known for fabulous riches, this was the stuff of legend.

Because of their success, the partnership of the four Irishmen became known as the Bonanza Firm, a name synonymous in the mining world with meteoric success. Of course, the discovery radically changed their personal finances, making them kings in a mining district known for its wealth. It also profoundly affected the community. The mid-1870s became the renowned flush times. Virginia City had known prosperity before, but nothing to equal this. The roar of this bonanza resounded throughout the world. Thousands flooded into the Comstock, hoping, again, to capture some of the suddenly won riches for themselves. Many thrived and some failed, but when a mining district made such a discovery, money flowed more readily in all directions.

The strike inaugurated six years of spectacular success. During the period of abundance in the mid-1860s the Comstock reached a peak production of $16 million in gross yield. In 1873 the Comstock broke the $20 million mark for the first time. For two years, 1876 and 1877, annual production reached almost $40 million. The ore processed climbed from a previous high of 480,000 tons in 1866 to nearly 600,000 tons ten years later. The contrast, however, is even greater when the equation includes the wealth of the bonanza ore. All the tons processed in 1866 were worth less than $12 million. In 1876 the value of production was well over three times that amount. During the six years of astonishing prosperity from 1873 to 1878, the Comstock yielded approximately $166 million—this during a time when industrial workers regarded $4 a day as an excellent wage. Although other mines produced ore during these years as well, most of the profits belonged to the Bonanza Firm.[29]

Unfortunately, wealth is easily spent, often leaving little trace that it existed. One of the lasting symbols of the big bonanza is the Mackay silver, now owned by the University of Nevada, Reno. After striking it rich with the big bonanza, Mackay sent a half ton of Comstock silver to Tiffany and Company in New York to be transformed into a setting of 1,300 pieces. The elaborately carved pieces include ornate bowls, humidors, decanters, carving sets, and flatware, for example. More than 500 ounces of silver went into each of the magnificent candelabras, which can hold dozens of candles. The collection, in all its nineteenth-century opulence, rivals any of the riches at the disposal of the British sovereign and on display in the Tower of London. Mackay's table setting provides a vivid vi-

A Virginia City mine headframe building with tracks leading away for the disposition of debris, 1878. Photograph by Carleton Watkins. (Courtesy of the Nevada State Historic Preservation Office)

sual reminder of a time when the region gave birth to its own monarchy.[30]

Realizing that the big bonanza had thoroughly changed the Comstock—and as a consequence Nevada—the state government initiated a census in 1875. It captures the mining district at the height of prosperity. Like the federal censuses, this state effort depicts a complex and diverse community. Unfortunately, the state record of 1875 is superficial in the categories recorded. When comparing it with directories from the same period, it appears that the enumerators missed many people. For the Chinese it is even worse, since the census lists hundreds of individuals only as men or women, typically with no name, occupation, or age. In spite of all the flaws, however, the document can be useful, and its recording of nearly 20,000 people in Storey County is provocative. Since the state initiated the census near the Comstock's height of prosperity and growth, and since the document missed many people, it appears that the combined populations of Gold Hill and Virginia City may have reached 25,000, making this one of the larger communities west of the Mississippi. It was a glorious mining district during a fabulous time when its mines produced millions of dollars and made the name Comstock a household word for much of the world.

Even though mining yielded outstanding rewards, the achievements of the Bonanza Firm were not limited to that industry. One of the more remarkable accomplishments of these men, their mining wealth aside, occurred even before they were known as the Bonanza Firm. Lack of drinkable water proved a constant hindrance to the development of the Comstock, and engineers frequently debated the most efficient, practical means to remedy the situation. In 1871 Sharon sold his Virginia and Gold Hill Water Company to a new firm that had as directors Walter S. Dean, W. S. Hobart, and John Skae, as well as the familiar John W. Mackay, James G. Fair, James C. Flood, and William S. O'Brien. During a meeting in August 1871, the directors decided to construct a water delivery system from the Sierra to the Comstock. Although the two points are only some thirty miles apart, the principal obstacle was a valley 1,200 to 1,500 feet below Virginia City that separated the mining town from the water source. That valley was nearly 2,000 feet below a proposed reservoir site in the Sierra. The solution to the problem was bold even by Comstock standards. The directors hired Herman Schussler, a German-born and -educated engineer who had worked on a similar project in Butte County, California. That water system, however, had been less than half the size of the one that Nevada needed.

Schussler designed a series of diversion dams and flumes to transport water along the Sierra to a reservoir properly placed for the second stage of the project. His plan to cross the distance between the two mountain ranges called for a pipe that would carry the water down to the valley floor and then deliver it with the force of a siphon back up to Virginia City, a total distance of seven miles. The bottom of the pipe, consequently, would need to withstand the pressure of 1,850 feet of water. Schussler designed a system made of English wrought iron, constructed at the Risdon Iron Works in San Francisco. The company manufactured the pipe to survive more than 800 pounds per square inch of pressure. Each section had overlapping edges so that workers could joint and rivet them together. Ultimately the plan called for 700 tons of iron and 1,524 joints of pipe. Because of the problem of leaks, the pipe demanded 35 tons of lead to caulk the joints. Schussler's system also required an intricate system of valves to release pressure and provide for cleaning. To prevent rusting, the manufacturers cooked the pipe in a mixture of asphaltum and coal tar.

Workers began laying pipe on June 11, 1873, and completed the task six weeks later. When water first poured through the line, the Comstock received the treasure with fireworks, bonfires, and the firing of General Grant, the cannon at Fort Homestead. Although it would be several weeks before the pipe worked reliably and the major leaks were sealed, the event signaled emancipation for a community that had been restricted to meager supplies of water of inferior quality. Here was a source that could furnish two million gallons a day.[31]

It seemed that the Comstock would be a success story without end. Still, there were setbacks, and no story of wealth and prosperity is without its counterpoint. The morning of October 26, 1875, the wind blew. That in itself was not unusual for the Comstock, although zephyrs usually pick up in the afternoon. Nonetheless, this wind was to play a pivotal role in a news story that would flash across the nation. Early that morning, while the gale howled, a fire started in the basement of "Crazy Kate" Shea's boardinghouse on North A Street. In an age when people relied on open flames for lighting and heat, wooden buildings were vulnerable. Like other communities of the nineteenth century, Virginia City prided itself on its volunteer fire departments, which raced to fires in contests to see who would arrive first. Buildings might be lost, but the fire departments of the Comstock had a good reputation for keeping the destruction contained. In its fifteen years, Virginia City had not known the kind of urban wildfires that had destroyed many towns of the mining West.[32]

The morning of October 26, 1875, would be different. Fire quickly consumed the lodging house. It had not rained for months, and the wood ignited. Wind set the disaster in motion when it carried the flames to roofs downhill. The ensuing firestorm mixed with gales to spawn flame-filled whirlwinds ascending far above the city that it was destroying. The *Territorial Enterprise* succinctly described the scene: "A breath of hell melted the main portion of the town to ruins."[33] Accounts speak in amazement not only of the explosive combustion of the wood but also of the fact that even brick buildings seemed to wither into heaps of rubble at the diabolical onslaught. The newspaper pointed out that those who ascended Mount Davidson could look down on "a sea of flames, from which vast columns of inky smoke rose hundreds of feet into the air." It went on to portray the scenes within the city: "On all sides was heard the roar of the fire, the crash of falling roofs and walls, and every few minutes tremendous explosions of black and giant powder, as buildings were blown up in various parts of the town. . . . In all directions and on all the streets the people were seen lugging along trunks, articles of furniture and bundles of bedding and clothing. No sooner had they deposited their loads in what was supposed to be a place of safety than the advancing flames compelled them to make another retreat."[34]

Everywhere refugees fled the densely populated core of Virginia City as it succumbed to flames one block after the next. They carried the few valuables they could manage. Some hauled pianos to the street only to abandon them after the fire and heat consumed the entire neighborhood.

One woman, known to command thousands of dollars, managed to save only a mirror and a washtub filled with puppies. While fire destroyed all her other possessions, leveling economic stratification as much as it did the city itself, this woman of means discovered what her instincts told her was most valuable. Another woman, Alice McCully Crane, recounted the tragedy a week later in a letter to her daughter. She had more time to pack her things, having included her best clothes, dishes, and glasses with the help of family, friends, and her Chinese servant. "We got all the carpets up and most of the furniture at the door where it surely would have burned had we not secured a dray at $20.00 a load. We piled on as much as we could and then . . . started toward the northern portion of the city. That was the last look I had of the cozy little home . . . for in a little while it was in ashes."[35]

The firefighters worked heroically, attempting to stop the storm. Nonetheless, the odds were against them. The new water supply, while abundant, did not flow through pipes large enough to meet the challenge, and the pressure failed when the firemen tried to quench the flames. Instead, they resorted to water buckets and wet blankets. Firefighters dynamited building after building, attempting to smother the fire and construct barriers. It was all to no avail: sparks jumped from the roofs of tall buildings over fire lines and continued to progress downhill. At the Ophir Mine 1,000 cords of wood and 400,000 feet of lumber burst into flame. When the fire reached the Consolidated Virginia, it consumed another million feet of lumber. The entire city assumed the form of a stifling-hot furnace. Firefighters continued their struggle. As the *Territorial Enterprise* pointed out, "The firemen labored faithfully from first to last and by night were wellnigh [*sic*] exhausted. The Gold Hill firemen came up to the city and did good work, until the boiler of their steamer exploded."[36]

The climax of the fire survives in legend but is disputed in fact. As the fire roared downhill to the line of engine houses serving the deep shafts of Virginia City's principal mines, panic gripped the community. The Yellow Jacket disaster had clearly demonstrated that when fire took hold underground, it could be years before it was extinguished. Even the loss of hoist works and pumps meant months of rebuilding and replacing and would leave hundreds unemployed. There is an oft-repeated story, taking several forms, featuring John Mackay and the fate of St. Mary's church. The building stood in the path between the fire and the hoist works, and it became apparent to many that they might need to sacrifice the building that was the spiritual center for the thousands of Catholics who lived in Virginia City.

One account has Mackay dynamiting the church to save his mine, but the bonanza king promised Father Manogue he would build a better one in its place. Another story has Mackay desperately working to save his shafts when an elderly Irish woman confronted him, complaining that the church was on fire. The mining baron responded, "Damn the church, we can build another if we can keep the fire from going down the shafts."[37] How much of the church the fire destroyed and whether or not firefighters dynamited the building remain points of dispute. Whatever the action taken, the fire reached Mackay's Consolidated and Virginia, where it destroyed all the buildings. Firsthand accounts suggest that firefighters dynamited the roof of the church and only the wooden parts actually burned. Workers discovered another piece of evidence during a recent rehabilitation of the church: what appears to be a drill hole for a dynamite charge remained in the granite wall on the north façade. Whether firefighters never used it or the detonation was a failure, the blemish in the building's granite foundation appears to be the remains of a strategy that called for the complete destruction of the church.[38] Perhaps it matters little. The roof and steeple, from either fire or dynamite, caved in and the surviving brick walls provided a much-needed firebreak. The fire ended shortly after reaching that point, and Mackay, regardless of his actual role in the decisions of the day, joined hundreds of others in providing large sums of money to help the homeless and to rebuild the church dedicated to the Virgin Mary.[39]

Eliza Buckland, who lived near Fort Churchill, twenty miles to the east, wrote: "About all the business part of the town including mills, hoisting works, and churches was burned. We seen the fier [*sic*] from our house and there was peices [*sic*] of paper and burnt silk came clear to our house. Peaces [*sic*] of hymn books etc. At 5 o'clock the evening of the fire there was over 8 thousand homeless people in the streets."[40]

Lists of the losses filled columns of the *Territorial Enterprise* for days. The fire had swept away lodging houses, a mortuary, saloons, livery stables, markets, the union hall, furniture stores, the International Hotel, restaurants, drugstores, a cigar store, fire engine companies, three newspaper offices, two breweries, lumberyards, Piper's Opera House, and hundreds of homes. The Methodist and Episcopal churches were gone. The brick walls of St. Mary in the Mountains Catholic Church stood in ruins. The Virginia and Truckee yards and depots in the center of the city were rubble. Flames had devoured the hoist works and mill for the Consolidated Virginia and the mill of the California Company. Insurance companies valued the total losses in the city at as much as $10 million. The fire had crept 400 feet down the shaft of the Ophir works, but fortunately miners throughout Virginia City had evacuated and no one remained underground. Most of the superintendents had time to lower cages, lock them in place, and backfill the shafts with dirt for fireproofing. The mines remained intact, but many engine houses and pump stations were heaps of ash. The Ophir and the Consolidated Virginia together claimed $1,311,000 lost in buildings and equipment. The situation required quick action to prevent the lower levels from flooding and the consequences of the fire from becoming even more severe.[41]

To protect what remained of property, the National Guard assumed control of the town during the night, patrolling the ruins to stop looters. In the community where all standards and laws, like property itself, seemed to have melted away, the military imposed order until the signs of civilization returned.[42]

While the fire represented widespread personal and financial tragedy, it apparently caused only three fatalities. A falling wall of the Carson Brewery killed one man as he hurried along the street. Another man, who appeared to be drunk, lingered in a burning structure that housed the Ash Book and Toy Store. Ignoring the pleas of people begging him to leave, the man remained inside, flinging toys to those outside. Finally the building collapsed, killing him. Demolition crews later found the skeleton of yet another victim, but it was impossible to determine the identity.[43] There may have been others incinerated beyond retrieval during the firestorm, but given the extent of the disaster, casualties were remarkably low.

Word of the holocaust flashed over the Sierra, and the San Francisco stock market responded instantaneously. The *Territorial Enterprise* noted that the price of shares "tumbled out of sight, and a panicky atmosphere seemed to fill the streets of the Earthquake City."[44] Stock prices fluctuated dramatically, and thousands of shares exchanged hands. Those who wished to see a new Virginia City rise from the ashes knew that swift action would be needed to make certain that the investors who remained would not lose their nerve.

The night of the fire, the zephyr returned in full force, knocking down the remains of buildings, which were only standing walls with little supporting them. Before midnight, snow started to fall heavily. People remembered that Eilley Bowers, the Bonanza-queen-turned-pauper who had resorted to fortune-telling, had recently foretold that a fire would destroy the city and a big storm would follow. Other visionaries had predicted the fire in the months that preceded the disaster, but people gave the most credence to Bowers, the Washoe Seeress.[45]

It was a cold night, and hundreds, perhaps thousands, were camping out, many having retreated to Cedar Hill, which overlooked the charred remains of Virginia City from the north. Again, as Alice McCully Crane recounted, she and her children, with "most of our worldly goods [were] out on the side of Mt. Davidson. I was not alone for there were hundreds of homeless people with their 'little all' keeping me company." The following day State Controller W. W. Hobart arrived with a railroad car filled with "boiled hams, small mountains of bread, and all manner of substantials and luxuries." The community set up soup kitchens at the first and third ward schools. Sam Wagner, an African American who served as town crier, called out the word that help had arrived. Turnout was less than many expected, however, since most people had friends who could spare a meal or two.[46]

That day, the first after the fire, flames suddenly roared up from the Best and Belcher shaft. The timbers had been ignited when the firestorm passed and had been smoldering during the more than thirty hours since. Like an aftershock following a major earthquake, it threw a weary community into a panic. There was clearly a danger that it would touch off yet more devastation. People nearby began pouring water and shoveling dirt down the shaft, and the flames subsided, but not before stirring up painful memories of the day before. Ultimately, the fire was of little importance other than for the terror it inspired.[47]

The real crisis in the aftermath of the fire was the shortage of clothing, blankets, and furnishings and the scarcity of lumber for rebuilding. Within days, cash donations from the surrounding valleys and from California began arriving, together with supplies needed by the homeless multitude. The hysteria prevalent the first day after the fire began to subside.[48]

Fire could not stop Virginia City. Hope and optimism, like the ore below ground, remained. And what's more, the Comstock did not intend to lose its sense of humor: two days after the fire, Judge Wright of Gold Hill heard the case of a man whom the county charged with having a defective stovepipe and fined him $25. He then said, "The next man will get a still heavier fine, and the next man dies."[49] The day following the conflagration, people set to work rebuilding. Few could imagine this as an opportunity to call it quits and move on. Fortune remained on the Comstock, and its principal town would need to rise again from the ashes. Indeed, the day after the fire, workers unloaded lumber here and there for the rebuilding. Lord summarizes the reconstruction effort succinctly: "Sixty days after the fire the principal streets running through the burnt district were lined with business houses, the majority of which were a better class than those destroyed, and habitable dwellings covered the intervening blocks." The frenzied construction resulted in a homogeneity in architecture. The railroad shipped in windows, completely assembled with glass, sash, and frame, that had been mass-produced in California. Local and regional foundries were manufacturing hundreds of cornices, pilasters, and other structural components. The resulting buildings, though diverse at first glance, share imprints of those first few months. Regardless of design choice and style, the scale of construction after the fire was extraordinary. By December 15, the Ophir Mine could boast $317,811.57 worth of new buildings and machinery. Two weeks later, the Consolidated Virginia reported roughly the same value of new construction and installations.[50]

Those with an accurate understanding of the status of Comstock mining realized that below the tons of ore and bullion produced in the mid-1870s lay the possibility of a new cycle of depression. Miners had discovered few new ore bodies, and shafts probed ever deeper, where temperatures soared and water flooded more abundantly than before. The Great Fire of 1875 represented far more than a crisis of housing and supply. There was a danger that, should the leaders of Virginia City flinch and not rebuild as well or better than before, investors could take it as a signal that the Comstock was on the edge of decline.[51]

One construction project in particular serves as a symbol of the decision-making process that led to grandiose construction. Storey County government had for years occupied the old Odd Fellows hall on North B Street. That structure was one of the first to fall victim to the fire, and the county commissioners faced the need to build anew. Rather than selecting a modest course, the commissioners asked Kenitzer and Raun, one of San Francisco's most prestigious architectural firms, to present them with design options for their new edifice. The architects gave Storey County three choices. Local leaders in turn selected the most expensive of the options, declaring that they would rebuild on the grandest of all possible scales. County commissioners dedicated the resulting courthouse in early 1877, with a commemorative plaque prominently displayed at the top of the facade celebrating the nation's centennial. No Nevada county constructed a grander, more costly building throughout the nineteenth century or well into the next.[52]

At the same time, the Catholics reconstructed their church dedicated to the Virgin Mary, making it far better than its previous incarnation. The International Hotel provided yet another example. Entrepreneurs had hauled the first building away during the depression of the early 1860s. The Great Fire of 1875 destroyed its replacement. Developers set their sights on a building that was more extravagant than before. The resulting six-story hotel boasted the first public elevator in Nevada, as well as a degree of elegance difficult to exceed anywhere in the nation.[53] Besides serving as an expression of the sophistication of Virginia City, the International Hotel reinforced the assertion that the Comstock would not fade away easily.

These grand buildings, and the sudden rebuilding in general, provide an opportunity for insight into the big bonanza and the community it cultivated. Brick and mortar survive to this day as testimony to the wealth that could support such extensive construction and to the optimism of a community so thoroughly devastated. The Great Fire was a challenge as well as a disaster. Virginia City passed the test, demonstrating that it could not be stopped by anything that happened above the ground.

William Wright pointed out that when the Great Fire was over, "the people for a time seemed stupefied, or rather drunk, and it was almost night before many of them remembered that they were without houses."[54] Well over a century later, the Great Fire of 1875 continues to have its effect.[55] From the story of Mackay and the Catholic church to many other anecdotes, the trauma of October 26, 1875, remains etched in the folklore of the community. To this day residents speak of "the fire," and all understand that the reference is to this event. When people dig to plant a tree or lay a foundation and find ash, they refer to the fire. Virginia City residents invariably speak of buildings as pre-fire or post-fire. Not surprisingly, the former are rare, largely because the disaster was so widespread. Together with the rebuilding, the fire serves as the most important benchmark in Comstock history. It represents both the scar of destruction and the symbol of the magnitude of mining wealth that facilitated immediate reconstruction of an entire community of thousands of people.

In 1876, William Wright, as Dan De Quille, published *The Big Bonanza*, providing a remarkable snapshot of the Comstock reborn from the ashes of the Great Fire. He was a friend of Samuel Clemens's, the two having worked together on the Comstock newspapers during the earliest days of the mining district. Wright was a literary giant of modest proportions, capable of crafting fine prose but never stumbling into the success his friend had known. Clemens assisted his old Comstock acquaintance with putting his book together and seeing it through to the publisher, but it never gained the appeal of a Twain novel. Nonetheless, *The Big Bonanza* is an exceptional portrait of the mining district, capturing the place when it was at its peak of success and fame. When Twain wrote *Roughing It*, he depicted the Comstock in the first exciting years of its existence, when growth was explosive and everyone hoped for millions. De Quille's Comstock no longer possessed the energy of youth. Instead, the complexity of an international, urbanized, industrial town filled his society with the strength of maturity. The contrast between the decades is dramatic.

De Quille, naturally, wrote from the perspective of his time. Placer mining had influenced choices in the 1850s and early 1860s. In the same way, the assumption that the Comstock had longevity influenced the mentality of the 1870s. The author's passively held perspective colored his depiction of the Comstock. That mentality encouraged the idea that a bonanza would follow every *borrasca*. De Quille shared this assumption with his community, and it inspired the rebuilding of Virginia City after the fire. On this positive note, the Comstock waited for the next big strike.

6: THE WORKERS

Labor in an Industrialized Community

The underground miner as he goes about the street is a well-dressed, clean person, who takes a daily bath and changes his clothing twice a day—once when his shift goes on, and once when it comes off. He is calmly proud of his occupation. . . . His life has made him a sane, thoughtful, responsible person. . . . He knows himself responsible for the lives of his fellow-workmen; his own life hangs upon the honesty of another's work, and that other's life hangs upon the honesty of his own work.
—*Charles Howard Shinn,* The Story of the Mine

Mining the earth for its treasures is an extraordinary occupation. People have perhaps never worked in a more unnatural place than underground. A mine cuts off all sunlight. Fresh air cannot penetrate its depths without mechanical pumps. All sense of the outside is lost: it is impossible to perceive the transition from day to night or to feel the chill of winter or the heat of summer. In addition, few occupations can claim to be as segregated: traditionally only men went underground.[1] A life at sea offers one of the best comparisons to mining, with its customarily all-male workforce removed from the rest of society and situated in a dangerous environment. Nonetheless, not even the ocean can match the surreal aspects of a mine.

First, a mine is a dangerous place to work. The Yellow Jacket Fire of 1869 remains the most famous and most lethal disaster on the Comstock, but hardly a week passed without a fatality or serious injury resulting from an underground accident. The Storey County Register of Death records a litany of mining mishaps. "Killed in Imperial Mine," "Killed in Kentuck Mine," "Killed in Belcher Mine," and similar entries fill the pages of the document, interspersed among the record of the passing of others, typically from natural causes. One page of the document, for example, lists forty-seven deaths in Gold Hill from January through July 1876. Of these, thirteen, or more than a quarter, were miners, nine of whom were killed on the job. One of the others died of typhoid fever and the rest succumbed to pneumonia. Those who were not miners died of a variety of causes, including scarlet fever, heart disease, and suicide. Most of the eighteen children who died in early 1876 fell victim to croup or scarlet fever. Only the miners on that page, however, died of accidents. Variations occur throughout the document, but the trend remains the same: the Comstock's industrial workplaces claimed most of the accidental deaths for the district. Indeed, Eliot Lord lists 295 fatalities and 606 mining-related injuries from 1863 to 1880 on the Comstock. Disease took the weak, the old, and the young, as always happens, but the healthy working man found the mines particularly hazardous.[2]

Death underground could come from a variety of sources. Miners met with cave-ins, fires, floods of scalding water, bad air, falling timbers, and misfired explosives. Many miners were killed during the treacherous ascent and descent on open cages that offered little protection against falling. Cables occasionally broke, sending the occupants plummeting to the bottom of shafts. Tools hoisted on cages sometimes jarred loose and fell hundreds of feet, striking anyone unfortunate enough to be below.[3] The suffocating heat of some of the deepest mines threatened the lives of many who worked there. In winter, miners emerged sweating from their underground labors, only to face the cold. The sudden contrast caused many to contract often-fatal lung infections, and yet the mining-related statistics did not reflect these deaths.[4]

All this is not to say that the Comstock was more dangerous than other mining areas. Many of the district's innovations diminished the hazards of mining. The cages in which men rode up and down the shafts had specially engineered latches that gripped the wooden guides should the cable break. Roofs on these industrial elevator cars prevented injury due to objects falling from above. Unfortunately, cages used for hauling timbers lacked this feature. In 1876, W. F. Wilson, a shift boss, died when the cable for his cage broke. The safety held his cage in place, but since the elevator lacked a roof, hundreds of feet of cable fell on him, breaking legs and ribs. He died several hours later.[5] The reckless nature of humanity itself was a factor in some deaths. As the *Territorial Enterprise* noted in 1868, "We have safety cages, safety hooks, safety gates, and all sorts of contrivances to insure safety to the miners, and all are ingenious and valuable inventions, but the miner's own carelessness . . . seems to increase in the same ratio that provisions for his security are multiplied." Indeed, the *Enterprise* continued, "A safety cage is of no use as a guard against the recklessness of some men—they will climb on top of it at the first opportunity and make a jump for a cable on top of the cage as it comes up or goes down, just to show what they dare venture."[6]

Despite precautions, accidents continued to be part of the day-to-day life in the mines. Still, the Comstock probably had a safety record at least as good as other mines of the period. Fortunately for the miners, the local rock contained little of the silica that caused silicosis for many who worked underground elsewhere.[7] Black lung, prevalent in coal-mining areas, was unknown in Virginia City. Although most people passively accepted accidents as a given in the dangerous underground environment, a management prosperous enough to mandate cautious practices probably saved many lives. Nonetheless, nothing could be done to change the fact that the Comstock miners faced the same dangers known for millennia to those who ventured underground, and this shaped the way miners viewed their workplace.

The hazards of the mine, combined with its surrealistic nature, offered opportunity for stories of the supernatural. The Comstock mines, like others throughout the world, inspired talk of ghosts, the spirits of those who died at work and were unable to find rest. As Wright pointed out in 1876, "So many men have been killed in all of the principal mines that there is hardly a mine on the lead that does not contain ghosts, if we are to believe what the miners say." Along with these beliefs, the mining West shared the international tradition of peopling the underground with mining elves. Like those of most nineteenth-century districts, the Comstock supernatural beings descended from the tommyknockers of Cornwall, named for their habit of rapping on support timbers. In their native homeland the knockers traditionally toiled below, occasionally interacting with their human neighbors. They warned miners of danger, punished greed, and when it suited them, they led those of good heart to rich ore. After crossing the Atlantic, the knockers less frequently punished greed, perhaps because such an idea conflicted with American capitalism. The distinction also became blurred, particularly on the Comstock, between ghosts and mining elves. Nonetheless, enough of the tradition survived the crossing, and the Cornish established an active folklore in their new home. Indeed, miners other than the Cornish soon adopted the tommyknockers as their own.[8]

It would be all too easy to dwell on the sensational, dangerous, or harmful aspects of mining, but for most workers, employment consisted of uneventful monotony. Hundreds of miners left their homes every day and walked to the hoisting works of their employers. Most tried to find quarters nearby so the walk would be as short as possible, but many felt that living in areas with people of similar ethnicity was more important. Thus many Irish miners lived in one of the Irish neighborhoods in Virginia City, even if they worked in a mine removed from those places. In the same way, many Cornish immigrants preferred to live in Gold Hill.

Regardless of the neighborhood, most miners lived in houses or rooms, the choice determined by whether they had families or were bachelors. A surviving lodging house on South C Street serves as a striking example of the nature of everyday life for the single Comstock miner. Known as the Werrin Building after one of its owners, it is a two-story brick-and-stone structure. The bottom floor of such buildings often housed a store; this one sold groceries and liquor. The second story featured a series of apartments for rent, most of them consisting of two small rooms with a doorway to a central hall. Woodwork was crudely grained, a process of painting, to make it look like oak. Wallpaper clothed the walls. The rooms, which typically measured roughly nine feet square, each had a window and a cast-iron stove. A sitting room included a table and a chair. A second room had a bed, and some had another chair. The stove served for cooking and heat, but there was no running water. An indoor privy and faucet were both down the hall. Documentary evidence indicates that the place generally served miners who lived alone, but clearly other workers lived there, and on occasion a wife and child may also have called the place home. Still, these were stark, small abodes. The framed samplers that proclaimed "In God We Trust," "Remember Me," and "Welcome Home" may have been attempts to make the rooms more pleasing, and the wallpaper here and there added a domestic touch, but the cramped quarters were a dramatic contrast to the spacious mansions uphill.[9]

Whether near or far from home, miners reported to their places of employment six days a week. Once at the hoisting works, the men changed into their work clothes: the hottest mines required only trousers, a felt cap, and work shoes; other shafts called for a shirt, but cold temperatures were not normally a problem. The miners then climbed onto a cage that ascended and descended rhythmically during the changes of shifts. The larger mines had two or three of these cages operating side by side. Some even had multistoried cages to maximize the carrying capacity of each trip.

In 1877, *Harper's Weekly* used this image, drawn by P. Frenzeny, of Comstock miners reporting for duty at the C&C shaft. Notice the multistoried elevators and the flat wire cable. (Courtesy of the Nevada Historical Society)

Because of the depths of nineteenth-century hard-rock mining, engineers needed to rethink a standard piece of equipment that had been used in excavations for thousands of years. The hoist rope, made of hemp, was vulnerable to rotting and breaking. Even in shallower, preindustrial mines, hoist ropes had a bad reputation. Cornish miners often refused to "hang by a rope," as they called being hoisted up and down. Instead, most preferred to climb down ladders to their work stations.[10] The deeper the shaft, the longer the rope and the greater the strain placed on it. Com-

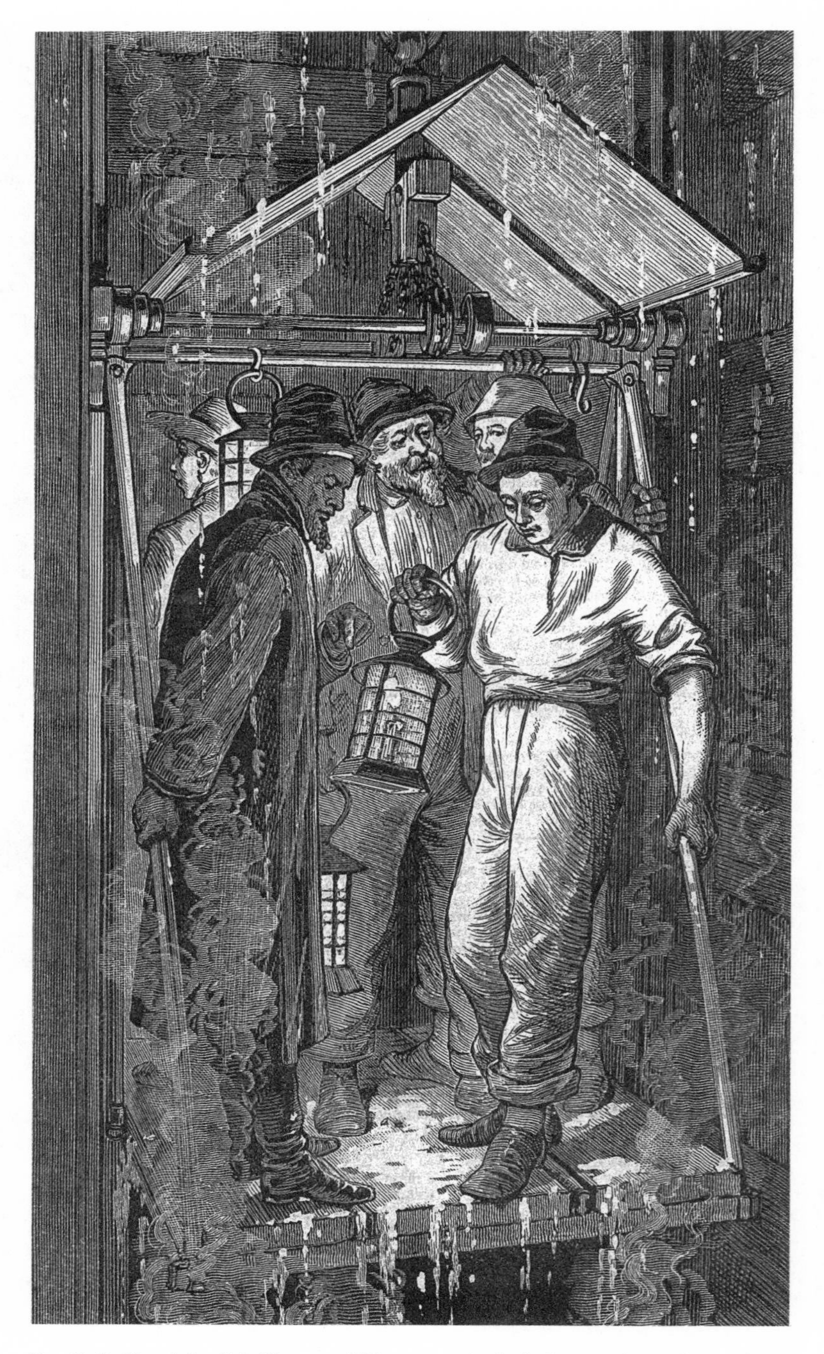

In 1878, *Frank Leslie's Illustrated Newspaper* included several articles in successive issues depicting the Comstock mines. This lithograph shows a touring party from the newspaper descending into the Consolidated Virginia shaft in a cage. (Courtesy of the Nevada Historical Society)

stock mines clearly required alternatives, particularly when an operation reached depths that made climbing a ladder impractical.

An innovation that addressed this problem came from another field of engineering. John August Roebling, later the designer of the Brooklyn Bridge, began his career by experimenting with the technology needed to build suspension bridges. As early as 1839, he was working with woven iron-wire cables, which proved far stronger and more durable than rope. Miners quickly saw the benefit of substituting iron for hemp, but they soon found that round cables proved problematic. An English-born engineer by the name of Andrew Smith Hallidie, a major producer of iron rope in California, devised a solution. He adapted Roebling's invention into a braided wire belt, the flatter cross section of which worked better on a hoisting drum as it turned to raise or lower the cage. Not surprisingly, the first use of the innovation was in Virginia City, at the Sierra Nevada Mine in 1863. As would so often be the case, the Comstock was at the forefront, and the technology first tested there became invaluable elsewhere. Hallidie subsequently used his invention to develop San Francisco's famous system of cable cars.[11]

The mining district also became home to the most recent innovation in the design of cages. The danger that the hoist cable might break remained a concern even after the introduction of braided wire, since such a calamity could result in a fall of 1,000 feet or more. Inventors consequently focused on ways to secure the cage in place if the hoist cable snapped. As mentioned above, cages soon included teeth that would sink into the wooden guides of the shaft should the tension of the hoist cable ease, and Comstock mines were some of the first to exhibit these remarkable breakthroughs of nineteenth-century technology.[12] Predictably, the innovations had ramifications for the development of high-rise elevators later in the century.

The men who took the daily trip down and back up a shaft benefited greatly from these inventions. Still, even with the security afforded by wire cables and a safety cage, the terrifying descent was awe-inspiring for many who visited the mines. In a society largely inexperienced with elevators, a trip in a mining cage represented a thrilling experience. The descent down a 1,500-foot shaft occurred with what one nineteenth-century reporter noted was "a startling and decidedly unpleasant rapidity."[13] Today it would probably provoke more than a casual reaction from even the most jaded of modern elevator riders. Charles Howard Shinn, who chronicled mining history in the nineteenth century, aptly described the descent of a mining cage:

> In an instant we are dropping noiselessly into the darkness, lit only by the flickering rays of a lantern which shows timbers seemingly leaping upward.
>
> Pretty soon a station appears, but we pass without pausing. . . . Every hundred feet a station flashes past, and the immensity of the work begins to grow upon the traveller. . . . As we reach a depth of a thousand feet or so the cable sometimes begins to "spring" with a peculiarly disagreeable bobbing motion, which gives the novice a new sensation, as if hung in an abyss by a rubber strap.[14]

Once the hoist operator stopped the cage, the miners continued to the adit or drift where needed.

Labor underground could assume a variety of forms. Sometimes the ore was loose enough that progress along the vein was possible with pick alone. More often, excavating mines entailed drilling several holes in a rock face. Miners pounded drills singly, called single jacking, or in pairs or trios, known as double or triple jacking. The miner would begin with a short drill bit, striking it himself with a single-handed sledgehammer called a jack, or he would hold it with both hands so that one or two other miners could strike the head with double-handed sledgehammers. The fists of the man holding the drill bit were vulnerable, and so this operation required immense trust and skill: the slightest error could crush the hands of a miner. After drilling a hole deep enough so that the miner's fist was flush with the rock face, he would select a longer drill bit. The miners continued until the hole was 3 to 4 feet deep.[15] They repeated this process, completing a sequence of holes in the outline of a doorway, the portal to the direction in which they wished to head.

In many places, the heat was so intense that men could work for only a few minutes at a time. Where this occurred, the company established cooling-off rooms, with barrels of ice water for lowering body temperatures that had ascended to dangerous levels.[16] Once miners had finished drilling, one of them scooped out the debris in the holes with a long metal implement known as a spoon. He then gently slid black powder and fuses into the holes. Those assigned this task used a long wooden stick (so as not to create sparks) to push the material in as far as possible. "Tap her gently" became the standard advice among the workers, cautioning each other not to jostle the unstable explosive and set off a discharge prematurely. The miners ignited the fuses, then ran for cover and hoped that all the powder would detonate, since lingering charges were responsible for many injuries and deaths. When the dust settled, they

This image of miners, laboring as much as 3,000 feet below the surface, appeared in *Frank Leslie's Illustrated Newspaper* in 1878. Notice the candlestick holder stuck into a timber and the ever-present bucket of water for cooling the men. High temperatures at the lower levels inspired miners to work with very little clothing. (Courtesy of the Nevada Historical Society)

Miners recovering from hot work deep underground. *Frank Leslie's Illustrated Newspaper*, 1878. (Courtesy of the Nevada Historical Society)

returned and began mucking out the rock into ore carts, clearing the four feet or so that now extended their adit.[17]

If the ground was unstable, miners erected wooden supports as soon as possible. Often laborers who had less skill than the true miners shoveled or "mucked" the rock into carts, then wheeled them to a hoisting station, where they were hauled to the surface. If the debris included profitable ore, the foreman directed it to a mill. If not, workers on the surface rolled the cart along rails to the edge of the mine dump and added the worthless material unceremoniously to the ever-growing mound. There, children, American Indians, and the poor watched carefully for anything of value, since some carts contained broken timbers or an odd piece of iron, items worth money as firewood or scrap. The scavengers played a dangerous game—if they were too close they risked being crushed by tumbling rocks, but to be too far away could mean surrendering a prize to someone more daring.[18]

As miners underground prepared to drill the next set of holes, others worked to extend the infrastructure. Rail lines, air conduits, wooden supports, telephone lines (after 1877), and electrical lines (after 1887) followed the miners as they excavated.[19] Assayers continually tested the mineral content of the newly exposed rock. Engineers added this information to detailed three-dimensional models of the claim. Many such models survive, consisting of dozens of glass panes that show the relative positions of adits and shafts, unexplored areas, and the nature and relative values of ore and rock. With as much data as could be gathered, mining experts worked with the superintendent to determine the most expeditious approach to finding ore and extracting it.

Although this was generally the approach taken, Comstock mining was hardly stagnant. Throughout the first two decades of prosperity, inventions constantly changed and upgraded the technology. In 1872, the Savage Mine experimented with mechanical drills, and that year saw their introduction at the Yellow Jacket. Two years later, on April 25, 1874, the Sutro Tunnel began using the new drills, with spectacular results, since the drills roughly tripled progress. Once again the Comstock had secured a first in western mining.[20]

In the early 1860s, Swedish chemist Alfred B. Nobel invented nitroglycerine, a discovery that would also affect the Comstock. Initially, nitroglycerine proved problematic because of its high volatility. Its impressive power was incredibly effective when the detonation occurred in holes drilled for that purpose. All too frequently, however, the blasting oil, as it was called, exploded spontaneously, and the resulting injuries and deaths won the invention a bad reputation. Nobel discovered that mixing his liquid with an inert substance such as sawdust rendered it stable until detonated. He also found that he could wrap the material in paper, forming sticks for safe, easy transportation.

The introduction of this explosive, which became known as dynamite, inspired labor problems elsewhere when conservative miners viewed the material as life-threatening. In California's Grass Valley and Nevada City, for example, management's insistence on using dynamite inspired a strike. Miners on the Comstock were aware of the reluctance elsewhere to use dynamite: the *Territorial Enterprise* carried testimony from San Francisco on California's controversy. Still, Virginia City's miners appear to have been most concerned about whether Giant Powder, as they called dynamite, was effective. The Gould and Curry Mine tried the new explosive in April 1868. Initially, there was some skepticism about whether dynamite was as powerful as nitroglycerine in its liquid form. As late as August 1868, miners of the Ophir asserted, for example, that its hard rock required use of the more volatile liquid in preference to the sticks. Within two years, however, the innovation had won credibility on the Comstock.[21]

Dynamite required a sudden jolt for it to explode, and so blasting caps became an integral part of mining, introducing yet another danger. Made of fulminate of mercury, these caps could safely detonate the dynamite, but miners often carelessly left them lying around. Curious children occasionally suffered lost fingers or eyes as they played with the deadly tools or attempted to pry them open. As William Wright pointed out in his overview of the Comstock: "About one boy per week, on average, tries this experiment, and always with the same result. In [Virginia City and Gold Hill] there must be scores of boys who lack the ends of the thumb and first and middle fingers. . . . Miners very frequently carry these caps loose in their pockets, often mixed with their tobacco, and thus, occasionally, get them into their pipes. Several favorite meerschaums have been lost in this way, and the ends of a few noses."[22]

Although the invention was clearly dangerous, with suitable caution it proved invaluable to the mining industry in general and the Comstock in particular. Dynamite increased efficiency, and as miners became familiar with the explosive, it proved no more dangerous than the older black powder was. Like similar technological innovations, dynamite had the potential of extending the life of a mining district as it decreased costs and made the working of lower-grade ores practical.

The invention precipitated, however, one of the worst mining-related disasters in Comstock history, an accident that neither occurred underground nor involved miners exclusively. While work underground was clearly dangerous, the ramifications of mining could be far-reaching for the entire community. General Jacob L. Van Bokkelen, the provost marshal for Virginia City who had established order after the assassination of Lincoln, was now an entrepreneur who worked with breweries as well as mining. He imported nitroglycerine for the mines and, convinced of its harmlessness, often stored his inventory in his bedroom in a B Street lodging house. He became so confident of the substance that he frequently slept on the boxes that crowded his room when he was overstocked. Near midnight on June 30, 1873, Van Bokkelen's inventory exploded. Later estimates placed the amount of explosives at 100 pounds of dynamite and 200 pounds of black powder. Many believed his pet monkey may have jostled some of the dynamite, causing the disaster. The explosion and subsequent fire destroyed several buildings and killed at least ten people, including Van Bokkelen, a merchant, a mining superintendent, a lawyer, a clerk, a housewife, and her eight-year-old daughter. One man died while working across the street at a stable, killed by the hail of bricks and steel doors that flew away from the detonation. There may have been other victims lost in the explosion and fire, and there were many wounded. Rescue workers found no trace of the monkey.[23]

At the same time that miners toiled underground, large teams of men and an occasional woman tended to the industry on the surface. The superintendent's office typically included a bookkeeper, a timekeeper, a secretary, and a mine manager who supervised a workforce of assayers, engineers, firemen, watchmen, and laborers, as well as the miners and their foremen. Hoist operators ran the cages up and down, responding to signals from below. Mechanics tended the huge pumps that drew water out of the depths. Carpenters and other smiths built and repaired equipment as needed.

Most miners on the Comstock were wage earners working for corporations in highly structured operations. For every industrial mine, however, there were hundreds of smaller affairs scattered throughout the district. People called these rat-hole mines because they were not engineered but rather seemed more akin to the older-style Mexican mines that employed the fifteenth-century *sistema del rato.* Consisting of adits and shafts of limited length and depth, these excavations drifted along the turns and dips of minor veins. Often people working for wages elsewhere owned and developed the peripheral claims, perpetually hoping to discover a rich ore body and the next big bonanza. They followed unprofitable veins, probing the depths of the hillsides, often working in small teams during off hours for no wages and using as few supplies as possible.

Claimants only slowly incorporated technological innovations in these endeavors. Hand drilling, for example, offered a cheap alternative to the more efficient but also more costly mechanical drills. The latter would require an initial outlay of capital that was unavailable to these small operations. With hundreds of pits and tunnels, small-time operators added significantly to the wealth of information about the Comstock beyond the large operations' vision at the core of the lode.[24] Miners as well as others participated in this poor man's stock market, which required only sweat and a little capital to take a chance at striking it rich. This sort of gamble also attracted some women: there is evidence that wives may have helped with the work, and in at least one instance, women owned and operated a small adit on their own. Such was the excitement generated by the possibility of discovering gold and silver.[25]

These small-time operations rarely produced rich enough ore for profitable processing, but the chance of a bonanza kept and continues to keep the effort going. Owners also hoped that they could sell their claims to people more optimistic or more foolish than themselves. Occasionally, peripheral, unprofitable operations went public and sold stock, often to investors far removed from Nevada, who were willing to assume a risk for

Hopeful freelance prospectors and miners turned the Comstock into a cratered landscape. Here a Virginia and Truckee train leaving Gold Hill passes a work site. *Frank Leslie's Illustrated Newspaper*, 1878. (Courtesy of the Nevada Historical Society)

the prospect of gleaning some wealth from the internationally famous Comstock Lode. Such scams damaged the reputation of the district, but they often proved the only means to profit from mining in an area with little or no ore. Whether the rat-hole mines were worked with honest intentions or with the idea of creating a poor investment for the ill-informed, they did provide an alternative to the corporate, industrial wage-paying businesses that made the Comstock famous. There can be little doubt, however, that especially during the first two prosperous decades of the Comstock, almost all the profitable ore came from the large industrial corporations.

The hundreds of small excavations created lifeless dumps that dot the landscape of the Comstock. Today, some have on their crests sizable piñon trees that have grown for decades, indicating that miners worked there during the nineteenth century. It takes many years for the desert to reclaim land that has been disturbed. Indeed, the vegetation on other mine dumps dates those excavations as much younger, their soils supporting only a little sagebrush; still other sites have but a few weeds, evidence that exploration with limited, inexpensive operations continues to the present.

Except in the case of a few businesses that had an on-site mill, mine operators of both the small and the large undertakings needed to take ore away for reduction. Teamsters—and then, after 1869, the railroad—transported ore valuable enough to justify processing. They took the rock to one of the many large Comstock milling operations, most of which were miles away, on the valley floors where water was more plentiful. After initiating full service, the railroad apparently never employed many more than one hundred people living on the Comstock, making them a small part of the thousands of others who labored for the industrial complex. There were also employees in Carson City at the main shops and at Moundhouse and Reno. Railroad workers included engineers, brakemen, porters, timekeepers, stationmasters, yardmasters, foremen, clerks, baggage masters, watchmen, switchmen, and others. Historians Ted Wurm and Harry Demoro use an interview conducted in 1936 with Jim Savage, a railroad employee, to piece together the experiences of someone on the V&T payroll. Savage started work as a youngster in 1883 for $15 a month cleaning the Carson City station. By the time he was sixteen he became a brakeman. He ended his career in 1937 after fifty-four years, having made the railroad his life. With the variety of occupations needed to keep a railroad operating efficiently, there was no doubt a full spectrum of possible salaries ranging from the lowest, earned, for example, by young Savage, to the highest, the thousands that the owners received from their investment. Between these extremes were the moderate but nonetheless respectable wages commanded by the various technical trades needed to keep large steam engines operating.[26]

Mill operators generally earned slightly less than most miners. Although entrepreneurs in 1859 and 1860 were slow to establish Comstock mills during the first months of explosive growth, it soon became clear that milling would be integral to the success of the district. These facilities and the people who worked in them lacked some of the romantic image that visitors typically ascribed to the mines. Still, a Comstock mill was an impressive expression of nineteenth-century industrial technology. In 1870 milling employed more than three hundred men in Storey County.[27] Five years later, their numbers had increased to almost four hundred. By 1880, however, only a little more than one hundred were left, a reflection of the failing fortunes of the district.

Like the mills, foundries were relatively slow to appear in the district, but as the future of mining grew more secure, businessmen established these facilities as well. A half dozen or so foundries supplied the mining complex with forged iron equipment. After the establishment of the first foundry in 1863, the number of workers grew rapidly. A foundry employed highly skilled artisans, many of whom received more pay than the average miner. These men had to be able to work with metals, fashioning parts and machines out of molten iron and brass. Although much equipment continued to come from California, the Comstock foundries were able to effect repairs and replace parts. Eventually they took a larger share of the contracts for iron and brass works. At least one foundry owner, John McCone, created an impressive legacy with his Fulton Foundry. McCone Steel established a California branch of the business, which later became an important operation on the West Coast.[28]

While some characteristics of the underground workforce contrasted with those of the employees of the industrial complex on the surface, other aspects of their employment were similar. Historians often dwell on the dangers in the mines, but those in the mills, on the railroad, and in the foundries had hazards to contend with as well. Anyone involved with heavy equipment and industrial materials is vulnerable to injuries, and industrial accidents also occurred above the ground. Thus James Hutchings, a carpenter, died on March 2, 1867, when a boiler exploded. J. H. Tucker, a blacksmith working in a mill, died in an explosion on July 15, 1872. Martin Whalen, a molder working at the Gold Hill Foundry, died in an accident in 1869. Four years later William Hamlin, a teamster, died when a timber fell on him. In 1873 both John Greiner and J. A. Strout died in separate accidents while working on the railroad. The Storey County burial records list dozens of additional instances of fatal industrial accidents on the surface.[29]

In addition, the mere fact that the Comstock was an industrial environment made it dangerous for those nearby, regardless of employment. As noted above, blasting caps caused injuries among the innocents who found them, and Van Bokkelen's explosion killed many bystanders. One other anecdote underscores this point. In 1869, diarist Alfred Doten recorded how a Mrs. Bridget Curran went to the Sunderland Mill in lower Gold Hill to get some condensed water from a barrel, since she and her neighbors preferred it for washing and cooking. While she was near the machinery, "her hoops caught on the end of a shaft . . . and wound round so tight that she was carried around 20 or 30 times, her head beating the floor every time—They had to stop the mill so as to get her loose, had to chop the hoops with a hatchet—She was insensible & remained so till she died, about 2 hours afterward."[30] She left a husband and several sons to deal with the effects of living in a place where heavy machinery posed a constant threat.

Besides these obvious hazards of industry, mill operators worked with mercury and other poisonous chemicals, giving an additional deadly dimension to their careers. Mercury played an important part in the milling process, and workers used tons of the liquid metal. The process of ore reduction required that workers add mercury to the pulverized ore, then draw it off together with the gold and silver. In the next step, mill opera-

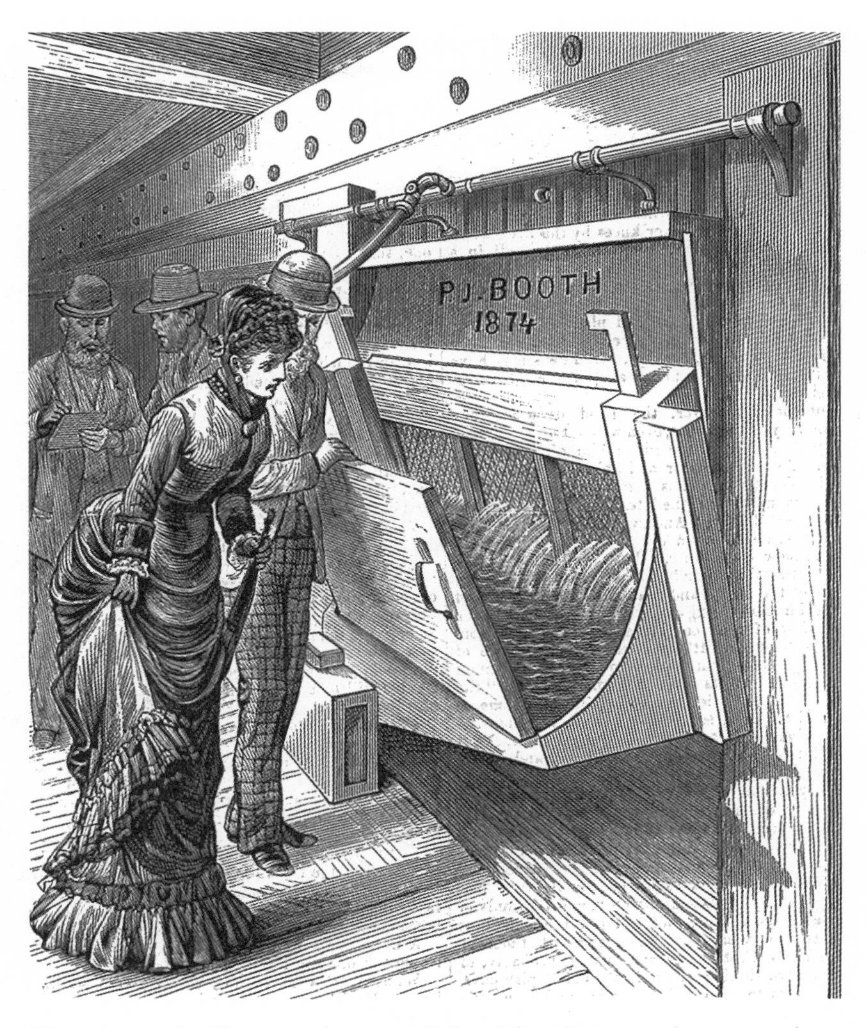

The mines and mills were a dangerous industrial environment that attracted many visitors and coexisted with thousands of residents. *Frank Leslie's Illustrated Newspaper*, 1878. (Courtesy of the Nevada Historical Society)

tors cooked away the mercury, which vaporized. Although millers made a limited effort to recapture the material for reuse, they lost most of it. Wright noted in his *Big Bonanza:* "The amount of quicksilver [i.e., mercury] used by mills working the Comstock ores alone averages 800 flasks, of 76 ½ pounds each; or 61,200 pounds per month. This in one year would amount to 734,400 pounds of quicksilver that go somewhere, and counting backwards for ten years shows 7,344,000 pounds that have *gone* somewhere—either up the flue or down the flume."[31] What mill operators did not understand about mercury was that it is poisonous and that it is most deadly when inhaled. Mercury accumulates in the bodies of those exposed, and when it reaches toxic levels, it can cause hair loss, tremors, hallucinations, insanity, and death. Early on, people first noticed the occupational disorders associated with mercury in the hat industry, where the liquid metal played a role in the making of felt. Mercury was the reason that Lewis Carroll's Mad Hatter in *Alice's Adventures in Wonderland* was mad. In spite of the intuitive recognition of a hazard, it took decades before action to remedy the situation was taken, and the milling industry followed the hatters in carelessly working with the hazardous metal. Historians may never determine the magnitude of the problem on the Comstock in the nineteenth century, but workers almost certainly suffered ill effects from the exposure.

The support services for the mines went far beyond the foundries, the railroad, and the mills of the Comstock. Workers throughout the West Coast were also part of an industrial web that served the mining district. Although local foundries produced much of the material needed for the equipment of the mines and mills, facilities elsewhere also contributed to the district's voracious appetite for raw materials and manufactured goods. Walking along the main street of Virginia City, one can read the menu of available foundries at the base of the pilasters, the iron half-columns that decorate the fronts of buildings. The Union Iron Works, the Fulton Foundry, the Virginia Foundry, the Gold Hill Foundry—all local institutions—are present, but the Vulcan Foundry and the California Foundry, both of the Bay Area, also make occasional appearances. In addition, builders imported many of the window surrounds and the tin and zinc cornices from elsewhere, playing to the Victorian tastes of a community that preferred to see itself at the vanguard of fashion rather than merely making do with local products.

The Comstock industrial machine also drew on an enormous and intricate supply network for raw materials. Ranchers and farmers from throughout the region grew food for the community, but the imports were not only from the West.[32] The Virginia City dump once included mountains of oyster shells, and the species present indicate that merchants shipped the delicacies live from both the Seattle area and the Atlantic Coast. Besides the many people involved in producing and transporting food for the Comstock, the district also relied on thousands to furnish everything from lumber to gas and charcoal. The New Almaden Mine of California manufactured tons of mercury, and the Comstock mills were some of its largest buyers. The milling process required enormous quantities of salt, as did the tables of the Comstock's many restaurants.

One of the most widespread and visible effects of the district's consumption was in the region's forests. The community and its mines used millions of board feet of lumber annually. The logging industry nearly denuded large parts of the Lake Tahoe Basin, but lumberjacks did not limit their harvesting to that area. Forests far to the north of Tahoe, all along the Sierra, fell victim to the axman's blade to serve the needs of mines that took on the appearance of gigantic wooden skyscrapers plunging into the earth. The lumber industry also attempted to meet the needs of people on the surface who required houses and other structures. The demand for firewood in stoves, furnaces, and steam engines increased consumption. Everywhere that there was a logging operation, wood mills and other facilities were not far away. In the case of Lake Tahoe, for example, the lumber industry supported a short line railroad and a steamship.[33]

After 1869 the V&T Railroad shipped bullion to the newly opened U.S. Mint in Carson City. In the quarter century of its operation, that facility produced more than 56 million gold and silver coins. The mint employed dozens of people, including assayers, weighers, refiners, coiners, mechanics, watchmen, firemen, engineers, coal men, and even seamstresses. Like other operations outside the Comstock Mining District, the U.S. Mint owed its existence to the mines.[34]

Each of these endeavors—the ranching and farming, the foundries, the auxiliary mines, the mills, the railroad, the U.S. Mint, and the logging industry—employed workers who toiled on behalf of the Comstock mines as much as the local miners themselves did. Nonetheless, many of these workers were far removed from Virginia City's society, and it was the local wage-earning labor that ultimately shaped the community.

Like the technology in the mills and mines, the makeup of the workforce changed dramatically throughout the history of the Comstock. Figure 6.1 illustrates the transformation of the district's male workforce. Not surprisingly, the first to arrive in Virginia City were in their twenties or thirties. The rigors of reaching the remote area discouraged others, creating a relatively homogeneous population for that early period. By 1870, the male population had grown older and more diverse, a trend that continued into 1880. This profile remained the same, with little variation, for all the various industrial occupation groups of the Comstock. During the 1860s, however, miners represented an increasingly smaller part of the labor force. Although there were more than six hundred additional miners in Storey County in 1870 than there had been a decade earlier, the percentage of miners in relation to men as a whole dropped from greater than 70 percent to less than 40 percent. This was partly due to the growth of the milling industry after 1860 and the introduction of the railroad after 1869. In addition, a wide variety of businesses and trades not present in 1860 now flourished on the Comstock. Still, mining continued to dominate the community (table 6.1).

The dominance of mining was particularly prevalent in 1875, shortly after the strike of the big bonanza. The number of men working directly for the mining companies increased from 2,868 in 1870 to nearly 5,000 five years later. The depression over the subsequent five years cut the mining workforce in half. The mining boom resulted in the same sort of fluctuation for the construction trades.[35]

By 1870 there also appears to have been a shift in the use of the term "miner." A decade before, virtually anyone holding a shovel and digging for gold and silver could claim the title, regardless of training or position. During the 1870 census, not all men working underground or for the mines felt it appropriate to call themselves miners. Instead, Storey County had hundreds of men who told enumerators that they were laborers, an occupation that had been almost nonexistent in 1860. These

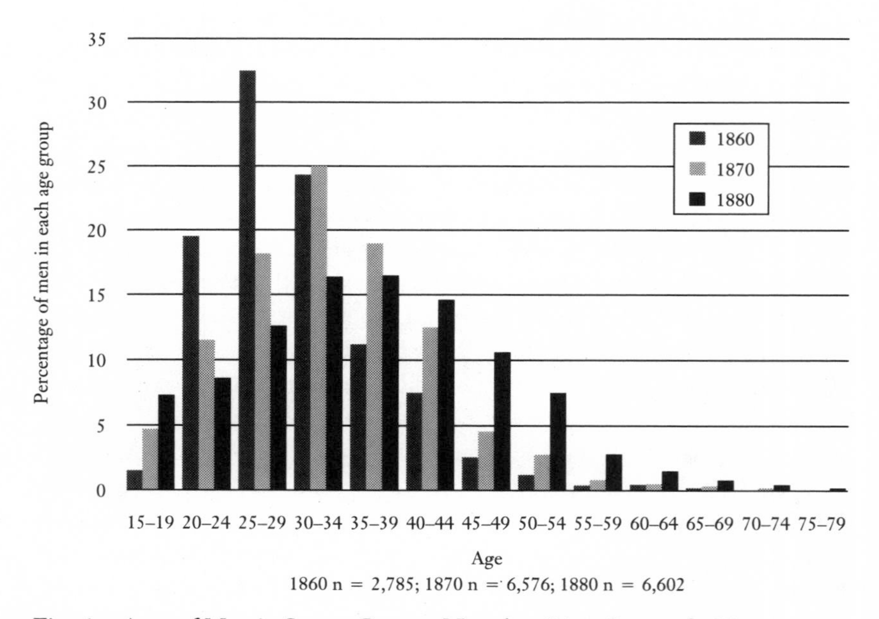

Fig. 6.1. Ages of Men in Storey County, Nevada, 1860, 1870, and 1880. (Sources: 8th, 9th, and 10th U.S. Manuscript Censuses of 1860, 1870, 1880, respectively.)

6.1 Industrial Workers of Storey County, Nevada, 1860–1880

Occupation	1860	1870	1875	1880
Mining	1,956	2,686	4,937	2,466
Milling	4	351	378	105
Foundries	5	146	153	171
Construction	123	324	683	400
Railroad	—	44	43	69
Other*	3	432	430	698
Total	2,091	3,983	6,624	3,909

Sources: 8th, 9th, and 10th U.S. Censuses of 1860, 1870, and 1880, respectively, and the 1875 state census. The 1875 state census figures are the result of hand counting: the census has not been computerized because it is often inaccurate, and it does not include categories present in the federal manuscript census. Hand counting can result in inaccuracies not present for the other census figures, based as they are on computerized tabulations. Many categories in all of the censuses are difficult to determine because of ambiguities in the meanings of specific terms.
*"Other" includes men identified as "laborer," probably those involved in one of the industrial trades most of the time as a large untrained, unskilled, and mobile workforce.

men were employed in a wide variety of industries, ranging from mining and milling to construction. Perhaps many worked in a capacity that would have allowed them to call themselves miners in 1860, but the Comstock a decade later was defining titles more rigorously as it carefully categorized talent, training, and position. The mining district had transformed from a boomtown to a structured society.

One aspect of mining remained a constant throughout the prosperous decades of the Comstock. Because a single industry provided the sole reason for existence for Virginia City, mining affected everyone's life in some way. The largest industrial mines employed three shifts of workers, and so every eight hours a large group of men left their places of employment looking for after-hours entertainment and services.[36] The nature of the industry caused the community to behave like a twenty-four-hour town: saloons, restaurants, gambling houses, and prostitutes could expect customers at virtually any time of the night or day. Certainly much of the town functioned only during the daytime, as one would expect in a "normal" community, but at least some of the businesses remained open to meet the special needs of the unique workforce of the mines.

Besides representing a large part of the population, miners also formed organizations that contributed to the character of the Comstock. Often a great deal of attention is given to the development of the miners' unions in the district, particularly because they were the first labor organizations in the mining West. These initially took the form of a Miner's Protective Association, founded on May 30, 1863, with the intent of preserving a $4-a-day wage.

While the Comstock was in bonanza, mine owners could meet this standard, but with the depression of the mid-1860s, some companies sought to reduce wages to $3.50. The Miner's Protective Association might have floundered without the incentive to organize formally. On July 31, 1864, miners marched in procession, shouting "$4 a day." The following day, delegations of miners visited the district's mines and mills with their wage-related demands, and within a week, they had organized the Miner's League of Storey County. They drafted a constitution, elected officers, and pledged themselves to the minimum wage as the most important plank of their platform. Realizing that they needed to negotiate with the owners from a position of strength, they further asserted that the mining and milling industry should not employ nonunion workers after September 27, 1864. These actions startled and deeply disturbed mine owners and those invested in the community's power structure. To combat this effort at collective bargaining, territorial governor James W. Nye called for troops from Fort Churchill to suppress labor agitation. The union's strong stand proved a disastrous mistake, since too few miners belonged to the Miner's League to support their position against such opposition. With federal troops on hand, owners ignored the miners' demands, forcing the league to withdraw from its assertion that the mines and mills should be closed union shops. Management of the businesses subsequently dismissed labor organizers and replaced them with nonunion workers, leaving the league to die a slow death. Within a year, many of the Comstock industrial facilities were paying a minimum wage of $3.50 an hour, in keeping with the depressed local economy.[37]

Union activity on the Comstock did not end there, however. Subsequent efforts were better organized and included a greater cross section of the labor force. On December 8, 1866, workers formed the Gold Hill Miner's Union and on July 4, 1867, the Virginia City Miner's Union. Critical to the success of these developments toward unionizing the Comstock mines was the strategy of controlling local elected offices. After 1864, pro-union miners won the election of fellow workers as sheriff, county commissioners, and other positions. Labor organizers made certain that the incumbents in these offices would no longer side with mine owners over workers. The two groups were then able to force employers to pay a $4 minimum wage for underground work. Eventually the union achieved such a position of strength that only a few workers were not members. Many mines actually cooperated with the two unions by deducting the $2 monthly dues from wages paid.

As previously mentioned, the unions felt the need to flex their muscle when they perceived a threat in the form of Chinese labor ascending the Virginia Range as they laid track for the Virginia and Truckee Railroad. The miners' march on the Chinese camps and their defiance of William Sharon brought the Bank of California to the negotiating table, where the unions won a guarantee that the railroad and the mines would not employ the Chinese within the Comstock. Other than this oft-cited example of union activity, organized labor needed to do little in the form of struggling with management.

Twentieth-century historian Russell Elliott summarizes the Comstock miners' labor organizations as follows: "In trying to explain this success one must consider the richness of the ore, the difficulty of obtaining skilled miners, the need for a stable working force (which high wages guaranteed) to exploit the mines properly, the high living costs when the $4 wage was first introduced, and the moderation of the miners in their demands. Then, too, it seems apparent that both operators and miners took pride in the fact that the Comstock was the model for western mining. It was the richest mineral area in the United States, and it contained the best machinery and the best engineers—why not the best and highest paid miners?"[38]

Elliott goes on to point out that the unions performed a convenient service for the mines by taking care of injured workers and the widows of miners, thus relieving the companies of a moral obligation.[39] While his observations are sound, it is equally clear that the miners who sought to deal with management through collective bargaining learned important lessons from their earlier failures. The heritage of conflict that is at the heart of their success is less visible from a historical perspective because the Comstock unions were able to assert such complete control that they rarely had to take to the streets. Mine owners may not always have wished to concede points at the negotiating table, but the union realized its goal by divorcing the capitalists from the sources of power needed to enforce oppressive positions. The lesson that the Comstock taught miners throughout the West for years to come was to control local offices to ensure that they not be used against organized labor.[40]

As a consequence of the control the miners exerted over the mining district through their union, they in turn had considerable influence over the entire state. For twenty years, the Comstock was the largest population center of Nevada, itself dominated by the single industry of mining. Since the unions and the mine owners both sought an environment favorable to their business, they could generally join forces without conflict in providing direction for the running of the state. Union representation, however, would always protect workers' rights, whether it was on the Comstock or in Carson City.

During the nineteenth century, the mining world faced a transition from preindustrial, miner-run operations to industrial, corporate enterprises. Cornwall, for example, with a centuries-old tradition of mineral processing, often found the change painfully disruptive to local culture and society. The Comstock saw something of this evolution when it moved from placer mining in the 1850s to hard-rock excavations after 1859, but the community lacked deep roots. The placer miners simply drifted to other promising fields, leaving only the stories about their contribution and the foundation they had created. In places such as Cornwall, people bemoaned a lost lifestyle and the independence it once conferred upon a miner.[41] The Comstock workforce, however, took pride in participating in a modern, technologically sophisticated industry. They represented the best of a maturing enterprise. Ironically, the industrialization of mining created new hazards for the endeavor. Miners explored record-breaking depths and used unprecedented heavy machinery, increasing the danger. While mining was transformed at the forefront of technology, safety concerns were slow to catch up. Comstock miners may have taken pride in the modernization of their industry, but some also fell victim to the changes. Many died or were hurt. In an international context, this North American machine of the Far West was leading the way for the mining world. Although there may have been problems with progress, if any of the miners ever looked back with regret at the destruction of the older ways, they failed to record it.

7: THE INTERNATIONAL COMMUNITY

Ethnicity Celebrated

The Irish . . . with the Knights of the Red Branch, the Ancient Order of Hibernians, and such societies galore . . . ; the braw Caledonians, who exulted in the games, where their men of might put the stone and tossed the caber; the French and French-Canadians, the Italians and Slavonians and Mexicans, all celebrating their own holidays in glorious care-free fashion—with many of these I was so fortunate as to foregather, welcomed into their midst, to partake of their national viands and refreshments.
—*Wells Drury,* An Editor on the Comstock Lode

Comstock wealth attracted an international array of immigrants who enriched the district with their diversity. The oft-cited litany of representatives includes large numbers of Irish, Cornish, Chinese, Germans, English, Scots, Welsh, Canadians, Mexicans, Chileans and other South and Central Americans, Italians, Scandinavians, French, and Swiss. There were also a few Russians, Poles, Greeks, Japanese, Spaniards, Hungarians, Portuguese, Turks, Pacific Islanders, Moroccans, and Caribbeans, as well as others. All these people provided contrast for the hundreds born in the United States, but even the North Americans included many with parents from Europe, as well as dozens of African Americans. In addition, there was an important Jewish population, itself with diverse nativities, and, of course, American Indians, the original inhabitants of the area, provided an ethnic bedrock for the Comstock.[1]

These groups expressed their heritages to varying degrees. The ways in which people identified themselves and interacted with one another reveal a complex society where few generalizations apply. Some residents of the Comstock were immediately recognizable as belonging to an ethnic group, while others had the luxury of deciding whether or not to declare their origin. The Chinese, Hispanics, African Americans, and American Indians could not escape notice as being distinct from the Euro-American majority. In contrast, the Irish and Cornish, as native speakers of English from northern Europe, could have blended in easily, and yet they chose not to.[2] The many other groups completed a kaleidoscope of diversity, providing an opportunity to understand an important aspect of the Comstock and the mining West.

Irish immigrants were by far the most numerous ethnic group in the mining district. In particular, they dominated Virginia City, where fully a third of the population claimed nativity or at least one parent from the Emerald Isle. The Irish came to North America by the millions, fleeing the oppression and starvation of their homeland. These exiles typically found prejudice and ill treatment by the Protestant-dominated hierarchy of the East Coast. Most remained there and learned that by standing together, and especially by forming political alliances, they could overcome many of the obstacles put in their path on the way to economic success. A few Irish immigrants traveled west, where they rarely came across established societies that were prepared to discriminate against immigrants or Catholics, as occurred in the East. In many cases the Irish arrived in numbers that made them, if not a majority, at least a significant minority. Hundreds also came as skilled miners, having worked in some of Ireland's only underground operations, in County Cork. The experience of the Irish who came to the West consequently contrasted with that of their brethren on the Atlantic Coast. The Comstock, as one of the first western hard-rock mining districts, set the stage for Irish successes throughout the region. Thousands of Irish and Irish Americans eventually made Virginia City their home, clinging to the community more tenaciously than other groups when depression loomed.[3]

On May 15, 1864, the Comstock Irish formed the Emmet Guard, the first of their military units and an organization that served as a cornerstone for the state's National Guard. The Sarsfield, Montgomery, and Sweeney Guards followed. Each of these, like the Emmet before them, had an intimate association with the Fenian Brotherhood, a nineteenth-century precursor to the Irish Republican Army. The Comstock boasted four Fenian Circles, as the local units were known, each rivaling the next for the best Saint Patrick's Day celebration and for the most money gathered on behalf of the revolutionary cause.[4] After the Fenian movement fell into disfavor in the early 1870s, the Irish joined alternative organizations, first the Ancient Order of the Hibernians and the Knights of the Red Hand, and then the Irish Land League.[5] These groups helped maintain a sense of unity and common purpose, and they reinforced the Irish identity for the immigrants and their children.

A peculiarity of Irish immigration, noted elsewhere, is that the women came to North America in numbers roughly equal to those of the men. This international phenomenon resulted in thousands of Irish and Irish American women calling the Comstock home at one time or another, making them the largest group of their gender throughout most of the mining district's history. That so many of Ireland's daughters settled there put a unique stamp on the Comstock, giving the Irish community a sense of ethnic neighborhoods and solidarity unmatched by most other groups.[6]

Like the majority of women, most of Erin's daughters list their occupation in the manuscript census records as keeping house. Diaries and other documents clearly indicate that many earned money through various pursuits. Nonetheless, professing housekeeping as the principal occupation was a way of maintaining dignity, since it fit into the conventional definition of what a married woman ought to be doing. Most Irish women who declared a wage-earning occupation in the census were servants, and indeed, the Irish represented roughly half of the domestics on the Comstock. The cliché of Bridget the parlormaid found a clear manifestation in the mining district. Other occupations that Irish women pursued included running lodging houses, working in restaurants, nursing, and sewing. A group of Daughters of Charity lived and worked in Virginia City, and most of them were Irish.

Yet another stereotypical occupation for Irish women, that of washing clothes, also occurred on the Comstock. Almost all the women who did laundry work in the mining district were of Irish nativity or descent. Many in the non-Irish community regarded laundry work, like domestic service, as a degrading pursuit and a mark of desperation. Irish women, however, tended to see both as avenues toward economic independence, and the idea of cleaning another's house or clothes was not as offensive for them as for others. Whereas servitude was best suited for single women, laundry work was one of the options available to widows with children.[7]

At least initially, there were far more Irish men than women on the Comstock. Although they would eventually become statistically important, the women only slowly followed. In the 1860 census, three quarters of the Irish men on the Comstock, like most, worked in the mines. A few

Irish men found employment in a variety of other fields, including construction and restaurant work. Others served as teamsters, laborers, and merchants. Over the subsequent years, the ranks of Irish immigrants and their children on the Comstock swelled to nearly six thousand, but the number of miners among the men was only about half. Many others worked for mills or served merely as day laborers, but there were also clergy, law enforcement officers, engineers, saloonkeepers, and railroad workers. Throughout the 1860s and 1870s, the Irish pursued increasingly diverse occupations, making it progressively difficult to stereotype them as following a single line of work. Mining, which dominated so much of the local economy and played a pivotal role in the employment of most ethnic groups, naturally remained extremely important to the Irish.

Although there were many immigrants from Cornwall on the Comstock, they were far fewer than the Irish. In addition, the Cornish had less motivation than the Irish to think of themselves as an ethnic group: these Protestant speakers of English had not known the sort of discrimination that the Irish had endured for centuries. Nonetheless, the Cornish represented something of an ethnic foil for their Irish neighbors. The Cornish had a reputation for hard work and incomparable expertise in the mines that often ensured their employment. These immigrants were commonly known as Cousin Jacks, presumably because when an employment opportunity existed in a mine, a Cornishman invariably put forward a relative (often named Jack) as a candidate. In the mining West, Cousin Jack typically won the job. The Irish, however, proved competitive through their numbers and their solidarity. There is good reason to conclude, in fact, that the Cornish expressed their ethnicity largely in response to rivalry with the Irish.[8]

The problem the Cornish faced on the Comstock appears to have centered on being out-competed on the surface rather than underground. Many Irish excelled at politics, and they frequently made certain that their fellow countrymen had ample opportunities for employment. In response, the Cornish were forced to underscore their ethnicity and to demonstrate that they too represented a unified presence.

The formation of the Cornish-dominated Washington Guard provides an excellent example of this dynamic. In early 1873, the Comstock remained in a mining depression, and many people would rightly have felt that their employment was endangered. An expression of ethnic unity, the commemoration of Saint Patrick's Day was, as Alfred Doten described, "the most celebrated I ever have seen." Doten continued by pointing out, "The Emmet, and Montgomery Guards of [Virginia City] and Sarsfield of [Gold Hill] were out in full uniform. Also Divisions No 1 & 2 of the Ancient Order of the Hibernians, and the Irish Confederation. . . . About 300 of them in all." During a time of a threatened job market, such a display could encourage preferential treatment of the Irish. Ten days later, on March 27, 1873, the Cornish miners of the Comstock formed their own guard, which they patriotically named after the first president of the United States, apparently indicating that they were not preoccupied with looking back across the Atlantic Ocean. The Washington Guard was never as successful as its Irish counterparts, and indeed, the Cornish disbanded the military unit in 1883 shortly after its treasurer left with all the organization's funds. Nonetheless, the Washington Guard had played an important role: it openly declared that the Cornish, like the Irish, could form something of an ethnic union and that they did not intend to lose jobs just because others won the political upper hand.[9]

Occasionally public posturing did not suffice and Cornish-Irish antagonism became violent. A challenge to a fistfight between Irish and Cornish combatants in August 1870 precipitated a battle that inspired one newspaper reporter to recast Gold Hill's nickname "Slippery Gulch" into "Bloody Gulch." It started as a quarrel between two men, which inspired representatives of the rival groups to ascend to Gold Hill's Fort Homestead to settle the matter with a boxing match. Before it was over, several friends and spectators joined the fray. The newspaper reported: "The fight lasted till all hands were tired out, and it is claimed that while all who participated were badly whipped, no one was victorious." With such an unsatisfactory conclusion, trouble continued to brew in the heat of summer. That evening an argument between the Cornish and the Irish started in a saloon, resulting in an even grander battle. Again, the *Territorial Enterprise* reported, "by the time it was over not a door was left upon its hinges nor a breakable thing about the premise remained intact. There were any amount of broken heads, black eyes and bloody noses, but no one was fatally injured, though there was one man who was shot in the thigh." If there was a clear winner, the newspaper failed to mention it.[10]

In August 1878, the Washington and Irish Montgomery Guards agreed to a shooting contest, with a keg of beer as the prize. The coin toss, to determine who would shoot first, resulted in a disagreement when the contenders disputed which side was heads, the eagle or Liberty. Anger mounted, and the opponents launched once again into fighting, using their rifles as clubs and drawing blood. When a few men began loading their weapons and taking aim, the sheriff made several arrests and ended the confrontation. Ultimately, the degree of friction between the Irish

Gold Hill with Fort Homestead in the foreground, 1878. Photograph by Carleton Watkins. Many from the Cornish community preferred to live in Gold Hill. (Courtesy of the Nevada State Historic Preservation Office)

and the Cornish may never be fully understood. When two policemen who were apparently Irish arrested two miners with Cornish names, one of those apprehended used the spike of his candle holder as a weapon, inflicting serious injury. This may have been ethnic hostility, or it may merely have been an objection to being detained.[11]

Much of Cornish ethnicity was probably an economic strategy. It appears that the immigrants found it useful to emphasize their origin, since it implied expertise in mining and frequently gave them preferential treatment. The Irish could not command such automatic respect, and so they worked to make certain that they constituted a formidable, unified political presence. Of course, the Irish frequently demonstrated underground expertise. It was four Irishmen, after all, who skillfully wrested control of some of the Comstock's most valuable mining properties from the Bank Crowd, and then went on to discover the Big Bonanza. Further muddying ethnic rivalries, John Mackay, the richest of the Irish Bonanza Four, hired Frank Osbiston, a Cornish immigrant, to serve as superintendent of his mine. Clearly, ethnic antagonism was sometimes not practical.

In contrast to the Cornish and Irish who had to make an effort not to blend in with those born in North America of European descent, several groups on the Comstock were visibly distinct. Of these, the Chinese were the most numerous and arguably went the farthest in defining the cosmopolitan character of the Comstock. They lived in segregated communities as a consequence of choice and of other people's prejudice. The largest of these was Virginia City's famed Chinatown, home to hundreds of Asians and a rich center of immigrant heritage.

The Chinese represented both an exotic diversion and a despised minority. Anti-Chinese leagues commonly protested the Asian presence and warned of their alleged insidious attempts to undermine wages and employment opportunities for others. The mid-1870s saw animosity toward the Chinese reach a peak. Many apparently left after the Great Fire of 1875, reacting to growing antagonism or possibly anticipating economic decline.[12] In spite of the rhetoric that permeates newspapers and diaries, Euro-Americans came to rely on the Chinese a great deal. Chinese servants had a reputation for industry, and even Mary McNair Mathews, a diarist who frequently spewed racist venom at these immigrants, had to concede that they worked hard, would take on tasks one would not wish to assign to Euro-Americans, and did not gossip about the households they served.[13] Many boardinghouses and some affluent homes employed both an Irish parlormaid and a Chinese cook or servant. While anti-Chinese sentiment abounded, about fifty Comstock households at any given time enjoyed the services of an Asian domestic who had a room in the house, living side by side with his employers.

Similarly, Chinese laundries thrived on the Comstock and provided yet another circumstance in which Asians lived in otherwise solidly Euro-American neighborhoods. More than two hundred Chinese men lived and worked in laundries scattered throughout the Comstock, cleaning the clothes of people who maintained that the non-Asian community should not employ the immigrants and should drive them from the mining district. Ironically, Euro-Americans took great pride in telling and retelling how they had taken a tough stand against Chinese labor, ensuring that Asians would not work in the mines or on the railroad within the bounds of the district. Contrary to this rhetoric, Euro-American society allowed direct competition with the Chinese. As it turned out, however, it was the women, particularly the Irish women, who faced the wage-busting efficiency of the Asian men.[14]

Although there were Euro-American men involved in the laundry industry, they were mainly the owners and operators of large industrial steam-run systems capable of processing thousands of pieces a week. Serving the large hotels, boardinghouses, and restaurants, these corporations relied on a market distinct from the one that Irish women and the Chinese sought to exploit. Instead, Asians and women washed the clothes of bachelors, families, and roomers in smaller boardinghouses, and it was in that narrowly defined market that the reputation for cheap efficiency gave the Chinese the ability to claim many of the customers. Irish racism, however, made coexistence possible: Irish women, most often widows or those separated from their husbands, were able to make a living by washing clothes because their community restricted its trade to these sentimental favorites. The Irish neighborhoods did this presumably as a means of supporting women in need, even though their laundresses were more costly. Without the ethnic trade barrier, the women whom circumstance had forced into the most menial of occupations could not have survived. Convention did not allow the Chinese to live in the Irish neighborhoods, even though elsewhere they inhabited and worked in houses side by side with Euro-Americans. In this way, a relatively small number of Irish laundresses were able to support their orphaned children in the face of fierce competition in the industry.[15]

This same principle apparently played a role in the market for domestic help. Irish parlormaids were able to exploit the prestige of their ethnicity to command a part of the job market. Again, this occurred even though Chinese servants had a reputation for working harder and costing less.

Besides servitude and the laundry industry, there were other avenues for economic relationships between Euro-Americans and Chinese. Asian men provided labor for a wide variety of undertakings. This often took the form of day-to-day employment, but Chinese also served as skilled workers for the Euro-American community: carpenters and clerks, for example, appear among their numbers. In addition, many non-Asians ventured into Chinatown to enjoy its various diversions. The Chinese community held the allure of the exotic and the dangerous, and many found its prostitutes and opium dens too seductive to resist. Not surprisingly, opium addiction among Euro-Americans became a growing concern on the Comstock.[16]

Although Chinatown had a notorious reputation, the neighborhood actually provided a wide variety of commonplace services and products. The Chinese intended most of these businesses for their own community, which included hundreds of bachelors as well as dozens of families. Doctors' and dentists' offices, restaurants, bakeries, and even some of the laundries served fellow Asians. A joss house, complete with priest and assistants, attended to the spiritual needs of the immigrants and provided something of an administrative function for them. Chinese merchants imported Asian ceramics and food for a people who kept apart and found little inspiration to assimilate. Still, some Euro-Americans came to Chinatown to buy souvenirs or to partake of exotic foods. Others came hoping for medical alternatives to address diseases that Euro-American doc-

Harper's Weekly published this multifaceted depiction of Virginia City's Chinese in the December 29, 1877, issue. Some images, such as the laundry man spitting on the clothes, the opium smoking, and the sausage making, were based on misunderstood cultural practices turned into inflammatory, unfair stereotypes meant to shock and frighten readers. Other images, however, are relatively positive. (Courtesy of the Nevada Historical Society)

tors had not been able to cure. In all, the Comstock Chinese, though they were the victims of racism and isolation, were integral to the community as a whole, giving it a distinct flair and enriching it with diversity.[17]

The relatively few African Americans on the Comstock were both the victims of bigotry and the recipients of some sympathy in the wake of the Civil War and the abolition of slavery. Hoping to gain a fierce defender of the Union, Lincoln had advocated the admittance of Nevada as a state well before it attained the required minimum population. Largely because of the influence of the patriotic Comstock, Nevada did not fail the president or his successor. The state's senators consistently supported Reconstruction-era civil rights legislation.[18] In keeping with this attitude, there were many in the district who treated African Americans with compassion and courtesy.

There were never many more than one hundred African Americans on the Comstock, and so they were not able to establish much of an identity as a separate community. A few, like artist Grafton T. Brown, arrived just after the boom and were able to find success in the early Comstock economy.[19] The 1870 census, which lacks addresses, lists nearly forty African Americans together, presumably indicating that they lived in a distinct neighborhood, apparently one of the larger D Street boardinghouses or hotels. An 1873 directory, however, shows these same people to be widely scattered in the community, indicating that the enclave of African Americans was only temporary. Other African Americans appear to be more evenly distributed throughout the community (see map 2).[20]

The 1880 census, which includes addresses, confirms this pattern, exhibiting no well-defined neighborhood: it shows African Americans scattered over a broad part of Gold Hill and Virginia City. They tended to live in apartments and boardinghouses along the commercial corridor, but this was an area that many Euro-Americans favored as well (map 3). The pattern exhibited in 1873 and discussed in chapter 5 changed, however, in 1880. While African American families had found accommodations in 1873 near Chinatown and in the area of mine dumps, by 1880 they had been able to migrate further uphill into working-class neighborhoods. The community had attempted to improve the image of the Barbary Coast on South C Street, which had provided a place to stay for a few single African Americans in 1873. By 1880, this once-notorious block served as home to more African Americans, including some families.

It would be inappropriate to overemphasize the degree of integration in 1880, but there are many examples of former slaves renting rooms from Euro-Americans, their families living side by side.[21] African American children attended public schools, and many of their parents became

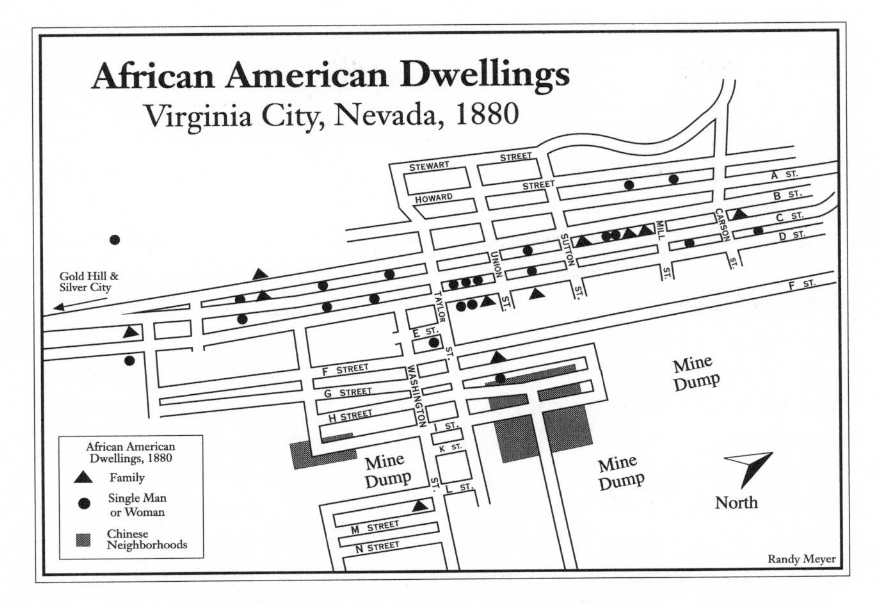

Map 3. African American Dwellings in Virginia City, Nevada, 1880. Map by Randy Meyer. (Sources: 1890 Sanborn–Perris Fire Insurance Maps and the 10th U.S. Manuscript Census of 1880. Not to scale.)

respected members of the community. Clearly, they also met with prejudice and ill treatment, but there was apparently a balance of the two. An article from the 1870 *Territorial Enterprise* illustrates the diverse ways in which Euro-Americans regarded African Americans: "Dr. Stephenson, a well known colored citizen of this city . . . registered [to vote], in order that he might exercise that right guaranteed him by the Fifteenth Amendment at the next election. We understand that a person of lighter skin but darker heart refused to register because he would not place his name under the Doctor's. We have no doubt the Doctor would have allowed this man to have had his name written first if he had appeared sooner at the registry. Dr. Stephenson has intelligence enough to see that it would not detract from him to have his name follow that of an inferior."[22] The generosity of the newspaper, reflecting feelings that others certainly shared, contrasted with the bigotry of those with narrower minds.

Besides Dr. W. H. C. Stephenson, who served as a physician, African Americans worked as laborers, servants and cooks, tailors and seamstresses, bricklayers, porters, bootblacks, barbers, and milliners. There were also a few prostitutes. African American miners were rare, but some consistently appear in census records and directories. As is always true of any aspect of the Comstock, situations often changed with the times. William Wright noted that during the mining craze of the big bonanza in 1873, "even the colored population, who seldom trouble themselves about mines, caught the infection and went out prospecting and locating mines."[23] Although prejudice often relegated African Americans to the most menial of tasks, some did well.

Dr. Stephenson and two other African Americans, William A. G. Brown and Amanda Payne, serve as examples of how it was possible to find success on the Comstock. Stephenson had arrived in Virginia City by 1863, appearing in a directory at that time as a laundry worker. Born in Washington, D.C., about 1835, he had a wife named Jane, who was born in Virginia about 1850. Presumably she came to the Comstock with her husband. Rising from the role of laundryman, he became something of a spokesman for the African American community, occasionally serving as preacher as well as doctor. In December 1873, a man almost died because Dr. Stephenson had given him the wrong prescription. Whether for this or some other reason, the doctor disappeared from the Comstock record after that time. Jane Stephenson continued to live in Virginia City, at least until 1875. She appears alone in the state census of that year as a hairdresser working at one of the salons on the Barbary Coast.[24]

William A. G. Brown was also living in Virginia City by 1863. That year's directory lists him as a bootblack. In 1864 he was involved in the killing of another African American, but he was not convicted of any crime. By that time he was running a saloon on North B Street. Two years later, his business had moved to South D Street, and he continued to operate it, and occasionally other saloons, for at least the next eight years. Like Dr. Stephenson, Brown was able to rise above humble beginnings in Virginia City to some success.[25]

Amanda Payne was the most successful of these three examples. Born about 1841, she arrived in Virginia City by 1870. She established a boardinghouse on North C Street, and eventually her fortune also included a restaurant and saloon. She was able to secure a permanence that many others did not achieve. Her obituary in 1886 declared that she was "well known to the old timers" and that she had "considerable property."[26]

Finally, the Sands family provides an expression of how the door of success, both financial and in the sense of community, could open to African Americans on the Comstock. David and Laura Sands arrived in Virginia City at least by 1865. David Sands worked as both a porter and cook. In 1883, Clarence Sands, one of five children, graduated from the senior class of the Fourth Ward School. As a member of a graduating class of seven, he and his family were certainly proud of his achievement,

Clarence Sands, an African American, was part of the Fourth Ward School's graduating class of 1883. (Courtesy of the Nevada Historical Society)

but they may have taken equal pride that they lived in a place that allowed him to stand with honor among his fellow Euro-American students to have his class picture taken.[27]

Though few in number, the African Americans were nevertheless able to raise families and form something of a community. Like many others, they were attempting to work hard and improve their lot. Mary McNair Mathews records a poignant example of this when she describes an African American woman who asked her for lessons in reading.[28] It was the African Americans' religious institutions that more than anything provided a sense of unity and common purpose. A minister established a Southern Baptist church in Virginia City as early as 1863. Angel's *History* notes that "with the exception of one person, all the members were colored people." The church dissolved in 1866, but there were several successors as the congregation repeatedly sought to establish itself. The Baptists met everywhere from the courthouse and the Miner's Union Hall to the Washington Guard Hall. Efforts also began in 1863 to establish an African Methodist Episcopal church in Virginia City, and within a short time the congregation had built a house of worship on F Street. The church existed in one form or another from 1863 to 1879.[29]

The Euro-American majority also afforded a degree of sympathy for the Hispanic community. Local newspapers usually described Hispanic organizations and celebrations in favorable terms.[30] Discussions of the Mexican Mine and its Spanish-speaking owners usually remarked on the different mining practices, but writers did not question their right to own valuable property in the district. Although, as previously mentioned, the number of Comstock Mexicans did not grow with the mining district, they nonetheless established ethnic organizations. The Liberal Mexican Club, closely following events at home, publicly mourned the death of one of its leading generals in 1865 with a display at the organization's headquarters in a building on North B Street.[31] In December 1867, the newly formed Mexican Benevolent Society held its first ball. The calendar included at least two celebrations in the Mexican community: on May 5 the 1862 victory over the French at Puebla was commemorated, and festivities on May 29 to note the anniversary of the capture of the despised Emperor Maximilian often followed. In September, the Mexicans celebrated their independence day, dating back to 1810. The event included midnight torch processions led by local bands. Doten noted that the parade in 1865 involved about sixty people and that "their head quarters on North B Street was illuminated—guns firing, speeches, etc. . . . kept it up late." The following year, the Chileans, who would normally celebrate their independence later in the month, joined forces with the Mexicans, to total about a hundred people, for a mid-September joint holiday.[32]

As noted earlier, the Mexican population, who had pursued diverse occupations in 1860, became increasingly homogeneous in the following ten years: by 1870 most of the men worked as miners, and many of the women were prostitutes. This trend continued into 1880, when the census recorded almost all the men as either miners or laborers.

The American Indians won a great amount of sympathy from the Euro-American majority. Although nineteenth-century rhetoric included a nationwide antagonism toward American Indians, local circumstance lacked any long history of warfare with them, and so it was easy for Euro-Americans to understand the original inhabitants of the area as unfortunate victims of change they had not wanted. Convention of the time nevertheless relegated the indigenous Northern Paiutes to segregated dwelling sites, often at the base of mine dumps far from the rest of the community. Still, American Indians may have sought out such locations, perhaps preferring to live in places of privacy. Scant primary sources combine with the archaeological record to suggest that the Northern Paiutes were in some ways successful in adjusting to the new circumstance. Their centuries-old culture provided the means to adapt to change in a harsh environment. Euro-Americans and the other newcomers had destroyed acres of piñon groves that had once produced the pine nuts essential to American Indians for winter sustenance. By disrupting a delicate desert ecosystem, the mining communities ruined habitat for game, which overhunting further decimated. The Northern Paiutes also suffered from the introduction of new diseases and the loss of access to lands that they had once regarded as their own.

In spite of the obvious degradation and painful disruption caused by the arrival of thousands of immigrants, American Indians found ways to adapt to the new environment while preserving much of their culture and sense of sovereignty. Northern Paiute women, who once roamed the land gathering food and other resources, now scoured the mining district for handouts and the discarded goods of a wasteful society. Occasionally, when they needed money, they worked as servants or laundresses. Men sometimes chose to become part-time laborers for the same reason.

While the American Indians adopted some Euro-American clothing, they often retained precontact face painting. They also raised their families apart from the intruders. The Comstock Northern Paiutes lived in buildings made of tin and wood gathered from dumps, usually in the round shape of indigenous precontact dwellings. The children played with marbles and dolls, but their parents continued the ancient practice of gathering for gambling tournaments, albeit often using cards instead of the traditional painted bones and sticks. Larger meetings, which Euro-Americans dubbed "fandangos," assembled hundreds of Northern Paiutes and provided an additional means of reaffirming identity. Newspaper accounts of solitary, drunken American Indians begging on the streets of Virginia City testify to the suffering that Euro-American settlement caused. Certainly hundreds died of new diseases or starvation, passing unnoticed in the written record. Nonetheless, many survived the onslaught and retained their families, their dignity, and much of their culture while exploiting the new opportunities that the Comstock offered.[33]

Between the examples of the Irish and Cornish on the one hand and the Chinese, African Americans, Hispanics, and American Indians on the other were dozens of immigrant groups that expressed their ethnicities in varying ways. The Comstock Scots, for example, never numbered many more than three hundred, and it is likely that they encountered no prejudice from the rest of the community. They hosted a Burns Supper as early as 1865 and eventually, in October 1873, formed a Caledonia Club. Their calendar included at least two annual celebrations. The Burns Supper, held each January 25, commemorated Robert Burns, Scotland's poet laureate. These were lengthy affairs that served mainly Scots, although

Northern Paiutes gather for a card game near E and Union Streets, ca. 1880. (Courtesy of the Fourth Ward School Museum)

the Caledonians welcomed all. The events included tributes to the poet, readings of his poems, and a marathon of dancing, eating, and drinking. Alfred Doten, who was always ready for a party regardless of the occasion, attended at least two of these celebrations and took special note that there was "a highland piper in costume." He also mentioned that the celebration lasted past two-thirty in the morning.[34]

The second gathering of the Scots was their annual picnic, usually held at a park in Carson City. A broad cross section of people attended, requiring the Virginia and Truckee Railroad to add extra cars to a special excursion train. The Comstock military companies attended to participate in shooting contests, the Irish usually taking the prize. The games included competition in throwing both heavy and light hammers, putting both heavy and light stones, short and long races for both men and women, sack races, both standing and running jumps, running high leap, vaulting with pole, tossing the caber, hurdle, archery for both men and women, the Ghillie Callum or sword dance, and the Highland fling. At one of these picnics, the Caledonia Club valued the total prizes as worth about $400. Entertainment included Professor Cara's brass band, and there was ample food and drink, as well as a multitude of other diversions.[35]

The Scots were not unique, however, in their ethnic celebrations. On the same page of the *Territorial Enterprise* that discussed the formation of the Caledonia Club, there are articles about the Grand Anniversary Ball of the Italian Benevolent Society, the first annual ball of the Washington Guard, and a Chinese festival.[36] Few issues of the local newspapers did not include at least as many references to such events and groups.

As early as October 1863, German immigrants founded an athletic club in Gold Hill called the Turnverein. Virginia City's German immigrants founded a similar organization on March 27, 1870, which, as Myron Angel observed a decade later, was "for the mutual improvement, athletic culture and recreation." He further noted: "The Society has about forty members and holds social dances about twice a month."[37] Although the Turnverein constructed a hall, fire destroyed it, and the organization subsequently rented facilities for its events. Germans also enjoyed an opportunity to socialize by patronizing General Jacob L. Van Bokkelen's Beer Garden, a brewery-saloon outlet. Like so many of the Comstock ethnic institutions, the beer garden attracted a wide variety of people who wished to enjoy the best of what diversity had to offer. More specifically, the patrons, regardless of nationality, sought to relish the fruits of centuries spent developing the art of brewing. Besides owning his drinking establishment, Van Bokkelen was a longtime Comstock resident, community leader, and businessman. He became a fixture of Comstock society, and although he was born in New York of Dutch ancestry, he came to represent the Germanic contribution for the rest of the community to see. His accommodating beer garden with its social, family-style drinking venue made him a cornerstone of the German community. When his pet monkey apparently set off a charge of explosives, killing itself as well as its owner on June 29, 1873, the Comstock mourned the passing of a prominent citizen, an exemplary friend of immigrants, and the purveyor of a good glass of beer.[38]

The beer garden was not the only manifestation of German association with the liquor trade. Although the immigrants never represented much more than 5 percent of the adult men on the Comstock, Germans who were making, distributing, and selling alcohol amounted to roughly a quarter of those involved in the industry. Indeed, they dominated Comstock beer making: in 1870, when there were twenty-six brewers in Storey County, all but four were German or of German ancestry.[39]

In March 1864, Jacob F. Hahnlen founded the short-lived *Nevada Pioneer*. Published in German, it was the state's first foreign-language newspaper. Two years later, he founded the *Deutsche Union*, which lasted only a couple of months.[40] That such efforts were even feasible is far more remarkable than that they were fleeting. Hahnlen attempted to serve a growing German-speaking population that, numbering more than two hundred in 1860, had tripled in ten years. In spite of its size, however, the community was not necessarily unified. These people were born in the Germanies, which were independent states before unification in 1870, each its own place of diverse history, culture, and dialect. The various Germans may have attempted to find the common aspects of their heritage, but they remained separate in both where and how they lived. They were the German-trained engineers who first invented Comstock mining; they were the merchants who built stores that made shopping in Virginia City the rival of San Francisco; they were the brewers who slaked the thirst of thousands of miners. The evidence of tombstones in the Comstock cemeteries clearly demonstrates that the natives of the Germanies were both Catholic and Protestant. There were also many Jews among their number, and although some may have taken their Germanic heritage as seriously as their religion and distinct origin, they nonetheless stood apart.

Besides those of German nativity, the Jewish community included people from Poland, Russia, and North America. Comstock Jews established a chapter of B'nai B'rith as well as the Eureka Society, which maintained its own cemetery. They never numbered many more than a hundred, but they still gathered to celebrate holidays.[41] Of more interest than looking at Jews as part of a group is the achievement of some of the individuals. Philipp Deidesheimer, inventor of the square-set timbering system, Herman Schussler, who designed the Comstock water system, Adolph Sutro, father of the famed Sutro Tunnel, and Joe Goodman, editor of the *Territorial Enterprise*, were all Jews.[42] The Jewish community also included numerous merchants, doctors, milliners and tailors, a police chief, a state assemblyman, a saloonkeeper, a teamster, a gambler, a jeweler, and a stockbroker.

In addition to the groups mentioned here, there were many more immigrant and ethnic organizations. Practically every ethnicity with more than a few representatives formed a fraternal organization, a military company, an athletic club, or a drinking society. Thus there was the British Benevolent Association, the Helvetia Society, the Italian Benevolent Society, the Welsh Club, a group called the Friends of Poland, and the Scandinavian Society. Each of these clubs and societies, together with those of the Irish, Mexicans, Cornish, Jews, Chinese, Scots, and all the rest, filled the calendar with balls, picnics, socials, and benefits. For the most part, all were welcomed at the gatherings. These organizations and their celebrations made the Comstock an international smorgasbord of events that attracted the open-minded and adventurous.[43]

As if this richly diverse population were not enough, traveling shows and exhibits brought additional expressions of international possibilities to the mining district. Circuses frequently passed through the area with their menageries of animals, and with an entourage of people from exotic places. The 1870 census, for example, records a circus in Gold Hill, complete with many performers from Morocco. An 1871 display at the National Guard Hall further exemplifies the possibilities. The exhibit included artifacts and people from Fiji. The Polynesians demonstrated how to use the various implements on display, and they performed dances and songs for onlookers. Included among their number was a young princess, who according to the *Territorial Enterprise* was "rather gawky, but then she is young and bashful, and having nothing to do but sit and be stared at, she doubtless does not appear here to as good advantage as she would in her native jungle."[44] Misunderstanding the island culture, promoters billed the Fijians as cannibals. In spite of its flaws and its exploitation of the Polynesians, however, this exhibit, and other similar events, broadened Comstock horizons even beyond the local diversity.

Not all groups had a distinctive neighborhood or a calendar filled with special events, but most had a particular place to drink that they could call their own. Lord points out that "men of different nationalities, working side by side under-ground, here first exhibited a clannish disposition: the Italians had their favorite meeting place; the French their 'Café de Paris'; the Germans their beer-cellars."[45] Outsiders were welcomed to varying degrees, depending on the background of both host and visitor. The Comstock was certainly a place where people from throughout the world could enjoy the best of what other ethnic groups had to offer, borrowing holidays, foods, and customs. Still, drinking can be a private thing, and according to Lord, there were places where ethnic groups drew the line on assimilation.

Language, however, was another matter, and the speech of the Comstock adopted liberally from various groups. Along these lines, Mark Twain observed, "As each adventurer had brought the slang of his nation or his locality with him, the combination made the slang of Nevada the richest and the most infinitely varied and copious that had existed anywhere in the world." This was particularly true in the mines, where the technical skills of various groups from throughout the mining world combined to address the resources of the West. The result was a technical language with an English syntax but an international vocabulary. The potential for confusion, particularly for the uninitiated, was profound, inspiring famed eastern mining expert R. W. Raymond to publish a glossary of mining terms complete with the country of origin for each word. His book documents thousands of terms borrowed from Spanish-speaking countries, Wales, Germany, Cornwall, and other counties in England.[46]

Along with the Comstock residents of foreign nativity, those born in North America asserted their own brand of identity. Although not normally thought of as an ethnic group, they lived in a community where most people were born elsewhere. This contrast certainly inspired many to form associations of one type or another. In keeping with other communities, North Americans founded organizations that either passively asserted claims to premier status or aggressively attacked the rights of others to citizenship, employment, or, in the extreme, immigration. Echoing the complexity and contradictions of the Comstock, membership in these groups sometimes included immigrants. The Anti-Chinese League of Gold Hill, for example, had some leaders of Cornish nativity. Its Virginia City counterpart was heavily Irish.[47] Racism against the Chinese provided common ground for a wide variety of people who might otherwise have been antagonists. The moment of ethnic unity that Asians inspired, however, was not lasting enough to prevent non-Catholics from forming an American Protestant Association.[48] In the midst of the swirl of hatreds, antagonism, and newfound alliances, the anti-Chinese initiative had its detractors among the Euro-Americans. In 1876, the *Territorial Enterprise* wrote of a newly formed Order of the Caucasians: "Their ostensible aim is to drive the Chinaman away; but compared with the men who could subscribe to such a creed . . . a chicken-thieving Chinaman appears like a high-toned gentleman."[49] Nonetheless, anti-foreign feeling remained endemic among those born in the United States.

Still, the native-born joined many other organizations that played a more positive role in the community: North Americans dominated the Freemasons, Odd Fellows, and other fraternities. They held public office and played prominent roles, but their control of society and local politics was not a given. Unlike established cities where immigrants arrived to find an entrenched native-born hierarchy, the Comstock constituted an ethnic free-for-all almost from the start. North Americans helped themselves to more than an equitable portion of the economic and political pie, but they had to share with others who were forbidden at the table elsewhere.

Figures 7.1 and 7.2 illustrate how ethnic groups varied in their composition (for comparison, see figure 6.1). It was a peculiarity of the mining West that the region did not attract all people in the same ways or for

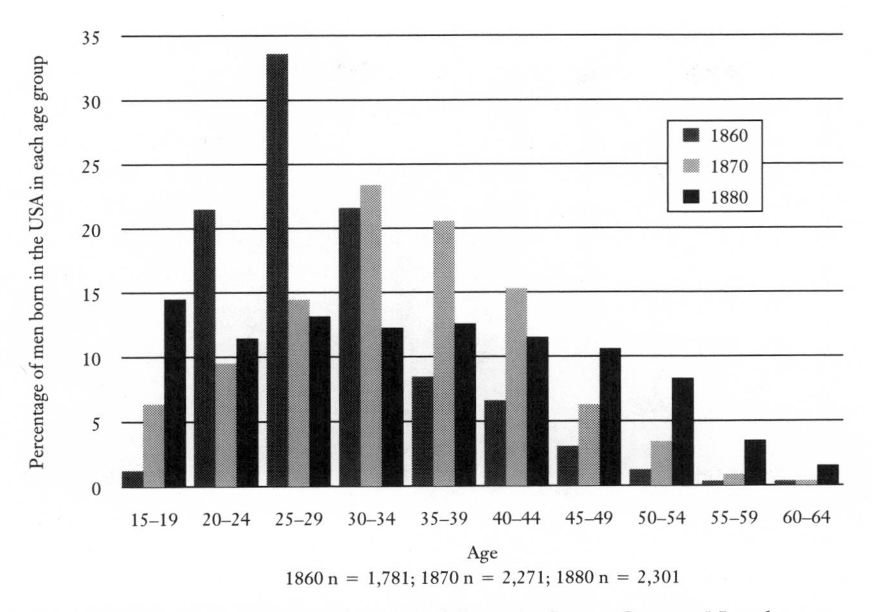

Fig. 7.1. Ages of Men Born in the United States in Storey County, Nevada, 1860, 1870, and 1880. (Sources: 8th, 9th, and 10th U.S. Manuscript Censuses of 1860, 1870, and 1880, respectively.)

the same reasons. The figures, which contrast the demographic distribution of Irishmen and their counterparts born in the United States, show how the groups changed. In 1860 young men dominated both the Irish and the native-born. By 1870 men born in the United States began to show a more even distribution of ages. There was still a preponderance of youth, but the majority was now slightly older and less exaggerated in comparison to other age groups. Ten years later, those born in the United States show a virtually even distribution of ages. The dominance of young workers is gone, and it appears that age was not a factor for those coming to the Comstock from other states.

The Irish show a different pattern. In 1870 there continues to be an uneven distribution of Irishmen, which had also occurred in 1860, but like those born in the United States, it has shifted to a slightly older group. This pattern is repeated in 1880, when Irishmen still demonstrate a population peak, though this most-represented age group is older and less predominant. Most groups are too few in number to examine in this way, but the example of Irish and native-born men suggests that ethnicity partly determined which age group was most likely to relocate to the remote mining frontier of the West.

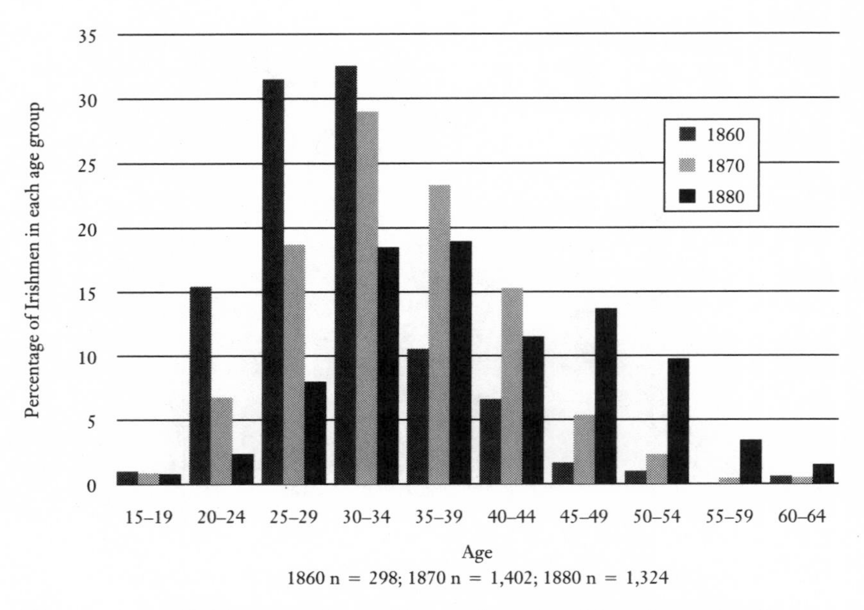

Fig. 7.2. Ages of Irishmen in Storey County, Nevada, 1860, 1870, and 1880. (Sources: 8th, 9th, and 10th U.S. Manuscript Censuses of 1860, 1870, and 1880, respectively.)

The identification of particular ethnic groups with single occupations was rare, and it diminished over time. The era of the Mexican packers, when almost everyone pursuing this line of work was Hispanic, quickly passed. The few anomalies, including the Irish laundresses and the German brewers, were rare, and the association was not monolithic: other ethnic groups also participated in the occupations, and most Irish women and German men pursued alternative employment. The Chinese, as always, proved the exception to the rule. Asians did not blend into the larger community as time went on. They did not become more diversified in employment and in the location of their homes. Instead, most clung to the stereotypical occupations that society dictated as appropriate: throughout the history of the Comstock, Chinese earned money as launderers, laborers, servants, cooks, and prostitutes. That is not to say that there were not exceptions, but most Asians working in other capacities served Chinatown itself, as doctors, priests, merchants, and housewives.

In spite of the assertion that it became increasingly difficult to identify specific ethnic groups with particular occupations, it is possible to arrive at some generalizations about employment pursued, residential choices, and lifestyles. Most of Comstock industry was based on mining, and so it should come as no surprise that practically all ethnic groups were largely dependent on that industry for a livelihood. There were differences, however, in how closely aligned each group was to working underground or to working with associated businesses. The Chinese, representing one extreme, provided no wage-earning miners for the Comstock. On the other hand, mining dominated the Cornish enclave, employing four out of five of that group's adult men. This contrasts with the English and Irish men, only about half of whom worked in the mines; the rest found opportunities in a variety of occupations, serving as business owners and laborers. Yet another distinction between these groups was that more of the Cornish men arrived without wives. The Cornish population consequently had thirteen men for every child, while for the English and the Irish, this ratio was three to one.[50]

Census enumerators record Comstock women overwhelmingly as keeping house, regardless of ethnic group. Exceptions are found among Hispanics and Chinese, almost half of whom appear as prostitutes, but prejudice and misunderstanding may often have been at the root of this designation. Of those recorded with employment, many Irish and African American women appear as servants. The Irish worked less frequently as seamstresses, a profession that U.S.-born Euro-Americans more commonly pursued.[51]

Additional generalizations about men and women of other ethnic groups are problematic because of the limited numbers. One of the more subtle and yet important differences was the tenacity exhibited by different groups. Some arrived at different rates or lingered on the Comstock longer than others, causing the ethnic weight of the district to shift with the passage of time. While U.S.-born Euro-Americans initially dominated Virginia City, the Irish, Cornish, and Chinese later arrived in such numbers that they became substantial block in the area. When the mines began to fail, however, the Cornish and the Chinese were the first to leave, abandoning the Comstock for more promising opportunities. The Irish, on the other hand, remained longer than most, apparently hoping more to build a community than to find the best-paying job.[52]

Any summary of ethnic groups on the Comstock is difficult because of the countless exceptions to every generalization. Thousands of people from throughout the world made the Comstock their home at one time or another, but it is not possible to arrive at concise ways to describe each group. In the final analysis, perhaps the most significant point to make is simply that each group existed, living side by side with other groups. Ironically, coexistence has been a rarely developed theme in explorations of these groups, probably because the ethnic clashes between them are what has provided the best fodder for both journalist and historian.[53]

There can be little question that while some ethnic groups lived side by side without comment in the historical record, others won a well-deserved reputation for ongoing prejudice and antagonism. Still, even those people regarded as being at the center of the greatest ethnic clashes managed to coexist most of the time. Violence among ethnic groups will always capture headlines and create images of deep-seated hatreds, rivalries, and resentments. Clearly, bigotry on the Comstock made life difficult for many by creating an ever-present threat of violence and discrimination, filling the atmosphere with tension. Nonetheless, one should not lose sight of the day-to-day existence of the people who failed to inspire newspaper articles, those whose lives represented mundane peace. In the rich ethnic mix that was the Comstock, most of the immigrants and the native-born simply got along.

8: THE MORAL OPTIONS

Sinners

There have been attracted to the Comstock range hundreds of gamblers of all grades, and men of all kinds who live by their wits. There is always a small army of men who haunt the saloons and gambling-rooms and by begging a good deal and stealing a little, and playing all manner of tricks and dodges, manage to pick up a precarious subsistence. There are in Virginia City about one hundred saloons, all of which have their customers.
—*William Wright,* The Big Bonanza

To call a place dreary, desolate, homeless, uncomfortable, and wicked is a good deal, but to call it God-forsaken is a good deal more, and in a tolerably large experience of this world's wonders, we never found a place better deserving the title than Virginia City.
—*Miriam Florence Squire Leslie,* California: A Pleasure Trip from Gotham to the Golden Gate *(1877)*

On the morning of January 20, 1867, the Chinese servant who occasionally tended to the needs of Julia Bulette, a prostitute of Virginia City, entered her crib and found her dead. An assailant had beaten her, slashed her throat, and taken some of her clothing and jewelry. Ironically, the incident set Bulette on the road to veritable sainthood, even though people might have regarded her as part of the darker side of the community because of her profession. It was her killer, however, who eventually assumed a role in local folklore as a clear manifestation of evil on the Comstock. John Millian became one of the few people hanged in Virginia City. An immigrant from France, Millian had worked on the Comstock as a laborer, most recently employed by Hall's Pioneer Laundry, a large steam operation that cleaned linens and pieces of clothing by the thousands per week.

Millian maintained his innocence, but the judge and jury convicted him and sentenced him to death. He continued to insist that two other men had committed the murder and that his role was minimal. Nonetheless, on April 23, 1868, jailers escorted Millian to the gallows at the north end of Virginia City, and the Frenchman ended his international odyssey in front of 4,000 or 5,000 spectators enjoying picnic lunches.[1]

As with so many things, nothing about this case appeared in simple terms of black and white. Although Bulette was an average prostitute and hardly the most famous one killed on the Comstock, a long procession turned out for her funeral. Starved for diversions, the community rarely missed the chance for a parade. On the other hand, some Virginia City ladies took pity on Millian, apparently regarding his deed as not that bad. They brought him food during his stay in jail, and their kindness moved the convicted murderer to thank them from the gallows. Sin, then, can be a matter of perception, and it remained for the Comstock to define the greater transgression: the popular court placed prostitution and murder on the scales and passed its own judgment on the two participants. In keeping with human nature, the thousands of sightseers who came to witness the execution of a fellow human being certainly did not judge themselves sinners for enjoying the hanging. Instead, history turned against the women who saw Millian as something of a hero. Over the course of the following century, he turned into a villain and Bulette became a saint.[2]

At the outset, Julia Bulette and John Millian appear to provide less than vivid contrasts of saints and sinners. Indeed, there were more-obvious examples of the darker aspects of Comstock life. Most people in Virginia City were terrified of the criminals in their midst. Particularly in the earliest years, violence was common. As noted earlier, Mark Twain maintained that the first men who died in Virginia City were murdered. Even if he was exaggerating or accepting local oral tradition without question, there can be little doubt that the first months during the rush to Washoe after 1859 were chaotic, with little inhibition against all sorts of antisocial behavior.[3]

Like any community of size, Virginia City and the other Comstock towns had their share of violence and crime. Murders, robberies, and assaults were all too common. Commenting on his Comstock of the early 1860s, Twain pointed out, "Vice flourished luxuriantly during the heyday of our 'flush times.' The saloons were over-burdened with custom; so were the police courts, the gambling dens, the brothels, and the jails—unfailing signs of high prosperity in a mining region." In the early days of the Comstock, before law enforcement and the more respectable part of the community had a chance to become securely established, highwaymen worked the roads throughout the region, and cutthroats and ruffians were commonplace in town. Colonel Henry E. Dosch recalled of his arrival in early Virginia City that he encountered a place where "the saloonmen, barkeepers and gamblers were the aristocrats of the community, and their vassals were blacklegs, road agents and thugs." After acquiring a job riding for Wells, Fargo and Company, Dosch witnessed an incident that demonstrated how suddenly violent the Comstock could be: "One of the saloon men in Virginia City came to his door to see me start out on my run, a man stepped out and shot the saloonkeeper through the heart. Instantly the barkeeper shot the murderer, who also fell fatally wounded, but before he cashed in his checks, he shot the barkeeper and also killed him, so inside of sixty seconds there were three dead men."[4]

The wealth of the mines caused many people to walk the district with inspiration for criminal activity jingling in their pockets. The Divide, the once-empty ground on a rise separating Gold Hill and Virginia City, was particularly dangerous. The same can be said about the toll road between Gold Hill and Silver City and on to Dayton. When complete, Geiger Grade down to Steamboat and the Truckee Meadows also became a favorite of robbers. One spot in particular was known as Robber's Bend because stagecoaches had to slow down there to take a sharp turn and thus they became vulnerable to thieves. For those who hoped to build a civilized community, there were far too many of these criminals. At times it must have seemed that hard men controlled the Comstock as much as, or more than, the mine owners did. Mark Twain described his Virginia City as a place where "the desperado stalked the streets with a swagger graded according to the number of his homicides."[5]

As the district matured, the notorious figures who had colored the first years of Comstock history found their brand of success more difficult to achieve. In 1865, for example, officials apprehended and convicted members of a notorious group known as the Harris, Waterman, and Haynes gang.[6] Still, there were times when the law never caught the criminals.

To handle this problem, people often took matters into their own hands. As Lord points out: "The conviction of a murder by an ordinary jury was an anomaly, and capital punishment by process of law had never been inflicted in Carson City or the new Territory. Still murders were commonly avenged in a rudely equitable fashion. Mining-camp justice demanded an eye for an eye, a tooth for a tooth, and the victor in a Washoe duel might expect to be killed himself by some comrade of his victim, who in turn would fall before the pistol of another." Lord maintains that such "consecutive deaths" could claim many lives and in one case more than a dozen men died before participants put the vendetta to rest.[7]

There were also those rare times when some local citizens formed a "601," a western term of disputed origin referring to vigilante groups that undertook to administer justice themselves. Western writers have given the subject emphasis more because it is exciting and fits the myth of the Wild West than because it was an integral part of society and law enforcement. After the organization of legitimate governments and courts, the need for lynch mobs lessened dramatically. All this having been said, protective associations did occasionally form and take command, enforcing their own standard of justice. Oddly, vigilante organizations thrived in the period after official law enforcement became available.[8] Dosch recalled another incident from a night in the mid-1860s: "My horse shied out to one side and would not pass a clump of several juniper trees along the road. I finally forced him back into the road and then I discovered why the horse had shied. From each of these small trees was suspended a man with a rope around his neck, and on each of their chests was pinned a notice from the Vigilance Committee. You should have seen the clearing out of Virginia City after that lynching bee."[9]

The most famous action of a Comstock vigilante organization occurred long after any period of frontier justice. In 1871 a series of fires panicked the community, and people began to fear that there might be a conspiracy of arsonists. Even though such a ring of criminals was not likely to exist, the vulnerability of Virginia City to fire was sufficient to inspire the "better" citizens of town to form a 601 committee. Authorities had recently arrested Arthur Perkins Heffernan for a shooting in a saloon. There was little reason to believe that the wheels of justice would move slowly or improperly in his case, but the vigilantes needed to make a name for themselves. When the rumor spread that Heffernan might also be responsible for some of the fires, he became the perfect means for the committee to acquire fame and credibility. On the night of March 24, 1871, members of the 601 gained access to an armory of a local Guard unit, borrowed muskets and bayonets, and seized control of the central part of Virginia City. They broke into the county jail, apprehended Heffernan, and escorted him to the old Ophir works.

Although officials never identified who participated in these events, word soon passed through the community about the victim's final moments. Heffernan, with a rope around his neck that was secured to some timbers overhead, stood where the 601 had placed him, on a plank over a shaft. According to diarist and newspaper reporter Alfred Doten, the vigilantes "kindly advised him to give a good jump straight up, when they would quickly remove or turn away the plank, thus allowing him a clear, effective drop." Not wanting to strangle slowly, Heffernan obliged his assailants and leapt into the air so that he might fall with sufficient force to break his neck and make his suffering short. Doten recorded his last words as "Turn her loose, boys!"[10] With the deed done, the vigilantes set off a cannon to signal their comrades to withdraw from their siege, and after returning their weapons, they all disappeared within the community.[11] The next morning found the victim with a small placard pinned to his lapel reading, "Arthur Perkins No. 601."[12]

A grand jury was unable to convict anyone for the lynching, but Judge Rising issued a resolute, if unenforceable, warning, when he said, "Such flagrant violation of the law cannot be suffered; the majesty of the law must be asserted and maintained, that law and order may prevail." For a community striving for respectability, organized ruffians were hardly better than individual criminals. According to local lore, the unavenged ghost of Arthur Perkins Heffernan subsequently haunted the Storey County jail.[13] Perhaps his appearance reflected society's misgivings for the act.

Nonetheless, the following summer, the 601 struck again. This time the group's leaders felt that a man named George B. Kirk was undesirable, having killed a man in California and being only recently released from the Nevada State Prison. The vigilantes gave him a "ticket of leave," a note suggesting that Kirk should move elsewhere or suffer the consequences. Recognizing the urgency of the situation, Kirk left town, but then he returned, presumably hoping either that the 601 would not notice or that they had less fortitude than needed to meet the challenge. Vigilantes who do not answer such defiance, however, can expect to have little influence over other bad men, and so on the night of July 18, 1871, the cannon sounded again.[14] It would be the last vigilante hanging for Virginia City.

The mining district was not as lawless as one might think. The sheriff, police, and courts were usually able to hold crime in check, removing most of the inspiration for private law enforcement as early as 1860. There were also times, however, when the officers of justice were worse than ineffectual. Mark Twain cited an instance of robbery in which the perpetrator was none other than the deputy marshal. He fled when exposed by one of his victims, but his case underscores the complexity of the early mining district as it struggled to find a sense of balance.[15]

The constant threat of violence could sometimes have the opposite of the expected result. Newspaper editor Wells Drury noted, "These were probably the most polite lot of men ever herded together in such a region of sagebrush and cactus. When every man is the judge of his own conduct and is swiftly resentful of even the slightest insult, the standard of behavior is necessarily marked by a high degree of punctilious courtesy." Nonetheless, exceptions with deadly outcomes sometimes seemed to occur on a daily basis. One particularly severe incident, in 1863, left six people dead or wounded. Beginning as a dispute over a nearby prizefight, the disturbance escalated until spectators drew their weapons and in a matter of seconds, bullets flew and victims lay bleeding.[16]

Eliot Lord cites crime statistics for the month of June in 1863 and 1880, providing an opportunity to compare Virginia City at two times separated by nearly twenty years. He astutely observes that there is a dramatic difference in the number of arrests per capita between the two periods. While Virginia City of 1863 had, according to Lord, no more than 7,000 people, the community had grown to roughly 11,000 by 1880.[17] Nevertheless, the number of arrests actually declined from 167 to 100. This means that while there was about one arrest for every 42 people in 1863, there was only one arrest for every 110 people in 1880. As Lord suggests, Comstock society had changed. Besides shedding its boomtown persona, Virginia City had also grown into a family town. With nearly one third of its population under the age of sixteen in 1880, it is not surprising to see that arrests declined.

The contrast between the two decades is even more striking when the statistics for individual crimes appear side by side. Table 8.1 illustrates which offenses increased and which declined over the years. Direct comparisons are problematic because the terms used vary in some cases. The difference between fighting in 1863 and assault, which is more common in 1880, is questionable. "Sleeping on the sidewalk" in 1863 is likely, but not certainly the same as vagrancy in 1880. Many of the terms apparently acquired a technical veneer over the course of the two decades. That problem aside, alcohol-related incidents, including drunk and disorderly, delirium tremens, and loitering in a saloon, declined dramatically. The additional consideration of disturbing the peace, a crime that might have had drinking as its cause, further underscores the improvement. With these combined categories, alcohol-related arrests declined from 103 in 1863 (1:68 per capita) to 30 in 1880 (1:365).

8.1 Crime Statistics for Virginia City, June 1863 and June 1880

Crime	Number in June 1863	Number in June 1880
Assault with battery	3	24
Assault with intent to kill*	1	2
Cruelty to animals	—	1
Delirium tremens	1	—
Disturbing the peace	36	11
Drawing deadly weapons	10	4
Drunk and disorderly	66	18
Fighting	13	—
Forgery	1	1
Gambling without a license	—	2
Grand larceny	6	3
Held to bail to keep the peace	1	—
Illegal voting	—	1
Loitering in a saloon	—	1
Malicious mischief	2	10
Petit larceny	7	9
Resisting officer	4	—
Robbery	1	—
Selling goods without license	1	—
Sleeping on the sidewalk	11	—
Threatening life	1	1
Vagrancy	1	7
Violating city ordinance	1	5
Total	167	100

Source: Lord, *Comstock Mining and Miners*, 210, 378.
*The 1880 statistics list "Assault with intent to kill"; 1863 lists only "Assault" (which generally appears here as "assault with intent to kill") as distinct from "Assault and battery."

Increases in crime between 1863 and 1880 are rare but are particularly noticeable in the categories of malicious mischief, petty larceny, assault and battery, and violating city ordinance.[18] The last term is a euphemism for prostitution. The increase probably suggests that officials were clamping down on sexual commerce outside the red-light district, and there is no reason to believe that there were more incidences of prostitution. The other categories may have shown higher numbers for several reasons. Since the number of children and adolescents had grown by 1880 with the emergence of a native-born generation, it is not surprising to see

Virginia City, ca. 1880. (Courtesy of the McCarthy Collection, Nevada State Historic Preservation Office)

small-scale crimes on the rise. It is also possible, however, that law enforcement officials had more time to pursue criminal activity that was less pressing than the murders and armed robberies that were all too prevalent in the early days of the Comstock. The more serious crimes would have diverted the attention of officials who could not be bothered by nonlethal assault, minor theft, prostitution, or any number of other smaller matters. Virginia City and Gold Hill combined to form a big city for their day, and as such, they possessed problems common to all cosmopolitan centers. When Miriam Leslie toured the Comstock in 1877, she was grateful for the company of "two policemen, who followed close at our heels [and were] by no means a guard of ceremony but a most necessary protection."[19]

Not all crimes were as flamboyant by nineteenth-century standards as murder, assault, and armed robbery, however, and many acts of wrongdoing tended not to capture newspaper headlines or find their way into early historical treatments. An "Amazonian 601," an all-women vigilante group, targeted wife beaters. The organization did not last long, but it gives testimony to an existing but infrequently discussed problem. It also underscores the resoluteness of women who insisted on a solution.[20] Predictably, wives were not the only victims of the misdeeds of evil men. Mary McNair Mathews recounted an incident when she fought off the advances of a lawyer—a "contemptible villain," as she put it—whom she had commissioned for assistance. In addition, domestic violence occasionally extended to children, who tragically endured such abuse.[21]

Crimes could be much smaller in scope but no less dangerous. An 1873 article in the *Territorial Enterprise* put pet owners on notice that Chief of Police Kelly would no longer tolerate unlicensed dogs. The newspaper pointed out that some people allowed dogs to roam free, and particularly those in the habit of attacking horses represented a genuine and sometimes deadly nuisance. It further reported: "One such dog is generally followed by a dozen yelping curs of all kinds, and the team, if at all skittish, is rendered almost frantic."[22] Officials had in the past rounded up and killed dogs that were unlicensed. The *Territorial Enterprise* advocated just as stern a measure against all public nuisances, tagged or not.

Dogs were not the only animals that roamed free. Mathews tells of a sheep that went out "every morning to the nearest saloon, put his forepaws upon the bar, and [bleated] for his morning dram. After he is waited upon he will go out of the door, look up and down the sidewalk until he sees some old friend coming. He will then go and meet him, and follow him about until he gives him a chew of tobacco." Usually fast to condemn the vices of alcohol and tobacco, Mathews had to concede that the sheep was "a fine fat fellow, and quite intelligent-looking."[23] Not to be left out, wandering hogs and goats found themselves the objects of law enforcement efforts. In April 1879, a local ordinance went into effect prohibiting "hogs and goats to run at large." The *Territorial Enterprise* noted that owners would do well to keep their livestock in pens, but it further observed that while officials would strictly enforce the law, trouble loomed: after all, "hogs are not as hard to catch as goats." Along the lines of minor crimes and sins, in the autumn the local newspapers regularly carried articles pointing out that with the rains there was a need to replace boards across streets so that mud would not consume the citizens, women in particular.[24] Invariably the newspapers complained that city government had taken too long to respond to a need that had become pressing. Like the situations with the licensed but nonetheless troublesome dogs and the noisome free-roaming hogs and goats, sometimes the smaller issues, even when no law is violated, cause the greatest consternation.

People despised the all-too-frequent practice of manipulating stocks to glean profits from unsuspecting investors, viewing it as unethical and reprehensible. Convention may not have defined such white-collar crime as breaking the law, but clearly there was an element of sin to such activity. Local oral tradition, for example, includes stories of mine owners keeping workers underground to create the impression of a rich strike.

With the resulting rise in stock prices, owners could reap huge profits, but they also won popular condemnation for the deception.[25]

The Comstock, like any other place, offered a wide variety of diversions that, while not illegal, were condemned by society. Prostitution, saloon life, and drug use each played a role, finding both adherents and detractors. Contemporaries generally scorned prostitution, but they often looked at the individual women as "soiled doves" or "fallen angels," seeing them as victims. Eventually, over the course of decades, this dichotomy caused popular perception to view many of these women as "whores with golden hearts," some of the most important heroines of the Wild West.[26] Regardless of the roles prostitutes played in reality or perception, as individuals or as a group, clearly they had their own significance in the society and culture of the nineteenth-century Comstock.

Mining in the nineteenth century generally attracted a large number of single men. Predictably, prostitutes recognized the opportunity this represented. As described in chapter 2, there is evidence that these women did not flock immediately to the Comstock, apparently fearing that it would fail, like so many other booms had. Nonetheless, prostitutes arrived soon enough and found a ready market for their services. By the early 1860s, they established a red-light district on D Street, just below the main commercial corridor of Virginia City.

A number of women were prostitutes, representing the full spectrum of ethnic possibilities and offering a range of prices and perceived quality. While some prostitutes worked alone in small shacklike cribs, others found employment in a brothel, run by a madam who served as business manager and protectress. With at least 161 Storey County prostitutes in 1870, there were far more than the 77 of a decade later, but these women never represented a significant part of the population.[27]

While prostitutes in Virginia City generally lived along the D Street red-light district, convention restricted Chinese women to Chinatown, several blocks downhill. For the convenience of Euro-American patrons who might have been reluctant to delve deep into the Asian neighborhood, most Chinese prostitutes probably situated their businesses on Chinatown's western edge, closest to the D Street district, where they were not welcome.[28] Arriving at a firm understanding of the extent of Chinese prostitution is extremely difficult, because there was apparently a tendency for census enumerators to identify Asian women as being involved in sexual commerce whether or not that was the case. As stated earlier, sometimes the census enumerators may have misunderstood the role of second wives and assumed that they were prostitutes.[29]

A few prostitutes were able to claim some notoriety in the community.

Certainly Julia Bulette did so, but not for the reasons frequently assumed in later years: she was far from the grandest prostitute of Virginia City, as twentieth-century folklore claimed. Instead, she gained lasting fame by being the victim of a sensational murder. What is even more important, officials hanged her killer in the community's first public execution.[30]

Jessie Lester and Cad Thompson were two of the better-known scandalous women during the Comstock's heyday. Lester ran a brothel and accumulated considerable property. She had a dispute on Christmas of 1864, presumably with her lover, which ended in her being shot and having to have her arm amputated. Reporter Alfred Doten visited the victim as she recovered from surgery and the effects of chloroform, and he recalled that she "was shrieking with pain—and in her delirium, calling on her mother."[31] She died a month later, in the care of the Daughters of Charity. Lester, former prostitute and madam, used the interim period to get herself in good graces with the Church. She donated much of her estate to the sisters, and upon her death she was buried in the sacred ground of the Catholic cemetery.[32] Perhaps because Lester refused to name her assailant and authorities could not bring him to justice, she failed to win posthumous notoriety, in spite of the significance she once had.

Cad Thompson was a long-standing madam in Virginia City, running the Brick, one of the better-known brothels. Born in Ireland, she came to Virginia City in the mid-1860s and operated her house there for more than a decade. There were, of course, other women who won fame through their association with the red-light district. Jenny Tyler's Bow Windows, for example, was also a well-known institution.[33]

The 1890 Sanborn-Perris Fire Insurance Map shows a shrinking red-light district during a time when the Comstock was seeing its overall population decline. Nonetheless, twenty-one separate addresses along D Street appear as "female boarding," clearly indicating the presence of prostitution. The map shows other small criblike structures in the area as vacant, suggesting that these had once housed prostitutes who had left for better markets. In keeping with other primary sources, the document includes both cribs and larger brothels, documenting the two most common options available to women involved in sexual commerce.

This was not the only place where one could find prostitutes, however. As mentioned earlier, in the 1860s and 1870s, a block on the western uphill side of South C Street became known as the Barbary Coast because, like its San Francisco namesake, it was "the roughest and worst place in town."[34] The area had served as home to numerous saloons that respectable people avoided. After the Great Fire of 1875 burned most of the red-light district, the Coast became even more of a haven for Virginia

City's prostitutes and toughs. Although the area had a reputation for attracting the darker elements of society, the fire exacerbated the problem with refugees, making the intrusion on C Street all the more noticeable.

Even before the Great Fire, the Barbary Coast hosted one of the more notorious murders in Comstock history. The problem began when Irish immigrants Peter Larkin and Nellie Sayers, who owned neighboring saloons on the Coast, ended a close relationship and replaced it with deeply felt animosity. Sayers took up with another Irishman by the name of Daniel Corcoran. Larkin subsequently hired Susie Brown, a young prostitute from San Francisco. He then quarreled with her, and she fled to Sayers, now Larkin's archrival. In the early-morning hours of August 4, 1875, Corcoran chased Larkin away as he was apparently trespassing in Sayers's house. Within seconds after Larkin fled, the report of pistol fire broke the quiet, and Corcoran fell to the ground with a mortal wound to his abdomen. The police arrested Larkin, tried him, and convicted him of murder. His hanging on January 19, 1877, was only the second public execution in the history of Virginia City, and so it caused considerable excitement, which was echoed in the local newspapers.[35]

Larkin's trial and execution occurred while the red-light district on D Street was rebuilding itself after the Great Fire, but many of the hardened inhabitants of the Barbary Coast lingered, enjoying their enlarged beachhead on the commercial corridor. The sordid case of Peter Larkin became a symbol for the community of the Coast's vile nature.

After 1877, the Coast was a particular concern for Virginia City's respectable citizens with the opening of the Fourth Ward School to the south. Children living to the north of the school would now need to walk past the tough quarter every day. The resulting community outcry was frequently given voice in the newspapers, and indeed, those who worked the area appeared in print regularly for minor as well as major infractions.[36] With the 1877 arrest and conviction of Nellie Sayers, for running "a disorderly house"—a euphemism for a brothel—the newspaper felt it possible in April to declare: "This writes 'finis' across the door of another deadfall on the Barbary Coast, and closes another den of iniquity on our public thoroughfare."[37] Sayers, however, refused to give up, continuing to run her place without a license. In June 1877 local police officers discovered that she had compounded her sins and reached a new low by offering her customers a child prostitute. The officials retrieved a thirteen-year-old girl, who told of druggings and forced sex. She asked officials to send her away from the place.[38] The Barbary Coast had gone too far, and the community forced the darker elements of society to retreat from the main street to the red-light district below. Within a month the *Territorial*

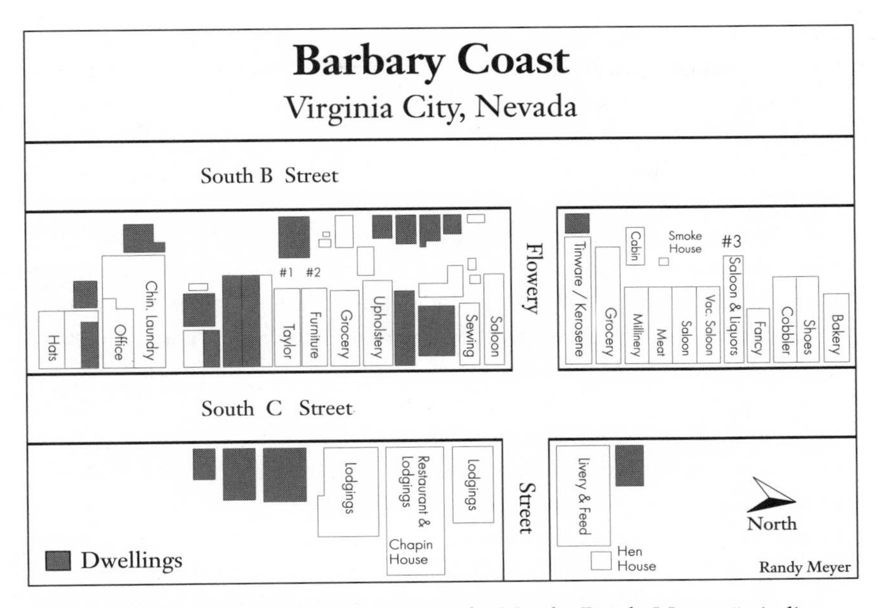

Map 4. Barbary Coast, Virginia City, Nevada. Map by Randy Meyer. #1 indicates Peter Larkin's saloon; #2 was the location of Nellie Sayer's saloon; #3 was Rose Wilson's saloon. Locations in 1875, before the community cleaned up the coast two years later. Map not to scale. (Source: Adapted from the 1890 Sanborn–Perris Fire Insurance Maps.)

Enterprise was able to report that carpenters were remodeling buildings in the area "once known as the 'Barbara [*sic*] Coast.'" With new fronts and overhauled stores and shops, there was space for a stockbroker's office as well as millinery and fancy-goods stores.[39]

There may have been a rehabilitation of the Barbary Coast, but the 1890 Sanborn-Perris Fire Insurance Map shows an echo of the district's former persona: within less than a block and a half stood six functioning saloons and two more that were vacant. The old core of the Barbary Coast, which consisted of little more than a dozen buildings, still included three functioning drinking establishments and an abandoned one. The map also shows a bakery, a cobbler, a meat store, a millinery store, a tinware and kerosene shop, a sewing machine distributor, and an upholsterer, as well as dwellings and lodging houses. Still, the cluster of saloons, unprecedented in number elsewhere except at the core commercial area of Virginia City, suggests that something of the Barbary Coast lingered after officials declared changes officially implemented (map 4).[40]

In its heyday, the Barbary Coast served as an example of how the line between brothel and saloon often blurred.[41] Even without the companionship of women, saloons provided a marvelous escape from a tedious, mundane, or lonely life. Although the early Comstock, as part of Utah Territory, faced a government that discouraged liquor sales, the mining district demanded its comforts. Historian Myron Angel, writing in 1881, suggested that "'whisky or death' would have been a rallying cry" for this early population had the territorial government pressed its case.[42] During the first winter on the Comstock, in 1859, people came to the saloons as much for warmth as anything else. A saloon offered a respite in many ways, but the issue of heat ceased to be such a great concern as the district matured. What grew from these first seeds were businesses that became an important part of the local economy and that featured a wide variety of institutions appealing to every class and ethnic taste.

J. Ross Browne, in his second visit to Virginia City in the 1860s, observed that saloons were so numerous that "the competition in this line of business gives rise to a very persuasive and attractive style of advertising. The bills are usually printed in florid and elaborately gilt letters, and frequently abound in pictures of an imaginative character. 'Cosy Home,' 'Miner's Retreat,' 'Social-Hall,' 'Empire,' 'Indication,' 'Fancy-Free,' 'Snug,' 'Shades,' etc., are a few of the seductive names given these places of popular resort; and the announcements are generally followed by a list of 'choice liquors' and the gorgeous attractions of the billiard department."[43]

Wright claimed that Virginia City had "about one hundred saloons" at the time of his publication in the mid-1870s. Lord echoes this number for both Virginia City and Gold Hill, but the figure, cited by all who followed in describing the Comstock, is conveniently rounded off and nearly impossible to verify. Nonetheless, in general there is no reason to doubt Lord. Indeed, Comstock editor Wells Drury later used specific figures, declaring that in 1876 there had been 100 saloons in Virginia City, 37 in Gold Hill, and 7 in Silver City. All of these authors marveled at the amount of drinking in their communities, implying that the number of saloons was evidence of the wild West character of the mining district.[44]

The 1890 Sanborn-Perris Fire Insurance Map, documenting much smaller communities, records at least 45 saloons in Gold Hill and Virginia City. With a population then of 8,806, this means that there was one saloon for every 197 people—a significant decline from 1880 when, according to Lord's estimate, there would have been a drinking establishment for every 160 inhabitants. Drury's assertion that there were 137 saloons in Storey County at a time when there were about 20,000 inhabitants in the area would mean that the ratio in 1876 would have been one saloon for every 145 people. Still, these statistics do not suggest a particularly inebriated society. Instead, the numbers are within the range of contemporary communities: historian Ron Ruthbart finds the ratio of people to saloons in Virginia's Shenandoah Valley to be 1:192 in 1890, with the businesses becoming even more plentiful over time; Perry Duis, also studying the history of the saloon, identifies Chicago's ratio as 1:203 in 1885. Duis, however, suggests that this contrasted with sober, regulated Boston, which had only one saloon for every 500 people.[45] It appears, therefore, that while the Comstock was no puritanical bastion of morality, it was not outside the bounds of other places at the time.

Most saloons, according to Wright, were "bit houses," where all drinks and cigars were twelve and a half cents each. Wright's book, *The Big Bonanza*, includes an elaborate discussion of the logistics of making change in a society that charged a bit but preferred to work with nothing smaller than a dime. This circumstance forced saloonkeepers to accept ten cents, or a "short bit," as full payment. They won back the difference, however, from customers who paid for a drink with a quarter; change would return as only a dime.[46]

The better saloons, typically located on two or three blocks of the "uphill" side of C Street or above, were classier establishments with prices to match. Known as "two-bit" saloons, they charged a quarter for each of their products, including the same drinks that cost half as much in the bit houses. Nonetheless, Virginia City was a place of contrasts. There was ample clientele to support the more dignified businesses that catered to those who preferred not to drink in the company of the Comstock's average saloon customer.[47]

As Lord points out, "Saloons of all descriptions, from the spacious rooms furnished with walnut counters, massive mirrors, and glittering rows of decanters to the cheap pine bar with its few black bottles, were to be found on every street and lane corner." An account in an 1867 issue of the *Territorial Enterprise* tells of the opening of one of "the *creme de la creme* of Virginia saloons" at 36 North C Street. This "most magnificently appointed establishment" included a solid mahogany bar, behind which hung a 7-by-9-foot mirror that the newspaper asserted was the largest in the state. It went on to report, however, that this was merely the "baby mirror," destined eventually to be located elsewhere with the arrival of the actual backbar mirror, a behemoth 10-by-13-foot piece of glass. Although this saloon stood on the downhill side of C Street, it was apparently a two-bit saloon.[48]

Still, the difference between a bit house and a two-bit saloon could be subjective, and reviews were not necessarily consistent. Wright describes an incident that has the appearance of being lifted directly from contemporary oral tradition of the 1870s. The story features a man who, after drinking a shot of whiskey at a two-bit saloon, paid only a dime when finished. The saloonkeeper objected, proudly pointing out that his was a two-bit house. To this the customer answered, "Two-bit house, eh? Well I thought so when I first came in, but after I had tasted your whisky I concluded that it was a bit house."[49]

The front-page advertisements of the *Territorial Enterprise* provide evidence of the diversity of saloons and the quality they boasted even if it was not delivered. Of course, only the better and more profitable businesses could afford the extravagance of buying space in a newspaper, but the half dozen or more that consistently appear there give a hint of the choices available. The newspaper of January 1, 1875, for example, lists seven Virginia City establishments. Most were on the uphill side of C Street, located near Union Street, in the commercial core of the community. The Magnolia offered the "best of wines and liquors" together with the "best brands of cigars" with "everything first-class." Gobey and Keely's Saloon, on the main floor of the Molinelli Hotel, boasted the "finest billiard room in the state" and a "Barber Shop . . . attached to the saloon." It featured brandies, wines, liquors, and cigars. The advertisement for the International Hotel claimed with good cause that its saloon was one of the "most elegant places in Virginia City." This prestigious saloon kept current stock reports and newspapers for its customers. The Washington offered a comfortable club room in the rear of the saloon.[50]

Of those advertising on that day, only the Sacramento Saloon was off C Street, being located on the northwest corner of D and Union Streets. Although D Street had racy connotations because of the red-light district that spread along the thoroughfare for several blocks, the Sacramento was situated in the respectable Frederick House, which fronted on C Street above. Its saloon, which was unique in advertising lager beer, was merely one of the many entertainment options available along D Street. Most of the saloons that could afford advertisements were two-bit establishments. The Yosemite, however, claimed to be "a first class bit saloon," situated on the uphill side of C Street. It pointed out that the "saloon having been newly built and fitted up in elegant style, is now one of the most comfortable places in this city."[51] O. C. Steel, the proprietor, apparently wished to cater to customers who sought the dignity of the west side of C Street but the prices of a cheaper establishment.

As pointed out earlier, some ethnic groups established themselves most clearly by their place of drink. Lines could just as easily become blurred, however. While there were certainly Irish businesses that did

Carleton Watkins took this photo of Virginia City in 1878. The large building in the left foreground is the Nevada Brewery. Across the street are the furrows of a vegetable garden. (Courtesy of the Nevada State Historic Preservation Office)

not welcome outsiders, for example, some Irish saloonkeepers sought as large a base of customers as possible. The so-called German beer garden of Van Bokkelen, while designed with a Teutonic motif, found appreciative consumers among a broad cross section of the community.[52]

Whether in a bit house or better, in an ethnic saloon or a more open establishment, the Comstock drank its share of beverages. Lord, for example, cites statistics indicating that Virginia City and Gold Hill consumed "75,000 gallons of liquor, chiefly whisky, exclusive of beer and wine" in 1880. He goes on to point out that the local breweries produced 147,996 gallons of beer that year. Together with imports including 67,800 gallons from California and 10,000 gallons from the eastern states, the Comstock drank roughly 225,000 gallons of beer in 1880. Lord further notes: "This is an average of 15 gallons per head for every resident of the county, in addition to the average consumption of 5 gallons of liquor. At 'a bit a drink'. . . , the usual price, the cost of the liquor per head was at least $40, and its total cost $600,000. . . . The price of the beer and wine at retail was probably half this sum, so that $900,000 was expended in quenching the thirst of 20,000 people."[53] The question these statistics pose, however, is whether they are extraordinary. Although the total alcoholic consumption of a community can be astounding, in fact it may not be much beyond that of other places. The Comstock, with its high proportion of single men, may have drunk more than residents of a city where families were more common, but there is no reason to conclude that its inhabitants remarkably exceeded the standards of the time.

One should not assume, of course, that a saloon represented nothing more than drinking and an occasional card game. These businesses catered to different needs and markets in highly specialized, often diverse ways. The key to the industry, and indeed the most important common denominator for a bar's success or failure, was its saloonkeeper. These businessmen had to be wise, shrewd, likable, and ready with sage advice if their establishments were to survive. Mark Twain noted that "the cheapest and easiest way to become an influential man and be looked up to by the community at large was to stand behind a bar, wear a cluster-diamond pin, and sell whisky. I am not sure but that the saloon-keeper had a shade higher rank than any other member of society. His opinion had weight." The complex, often subtle relationship of host and guest affected the role of saloonkeeper. While many Comstock saloons offered free lunches to their patrons, there was an etiquette that convention dictated customers should obey. To "eat and run" was to exhibit bad manners, and certainly one should not take too much or too quickly. Conversation with the bar owner was a polite necessity, and of course only liberal purchases of alcohol would keep the institution viable.[54]

The 1860 census records 49 people, including two women, working in saloons. That number climbed to 162 in 1870 and then dropped ten years later to 142. Three women appear in both censuses as employed in drinking houses. The number of American-born saloonkeepers declined from nearly two thirds of the total in 1860 to less than a third in 1880. An increasing number of business owners were from Ireland or the Germanies. Especially in the case of the latter, there can be little doubt that they did not run exclusively ethnic saloons since there were not enough Germans on the Comstock to support more than a few businesses that might cater only to their ethnic group. Instead, most German saloonkeepers sought to attract a broad clientele, hoping for profit from anyone they could lure through their doors. Census records indicate that these immigrants lived away from their businesses, supporting the conclusion that they were not running ethnic institutions. This was also true for those born in the United States. Others, principally from Ireland and the rest of Europe, frequently lived at or near their saloons, thus making the business more integral to their lives.[55]

Evidence from the newspapers suggests that while some saloonkeepers were quite successful, others failed quickly at their business endeavors. A few saloons became venerable institutions on the Comstock—the one in the International Hotel, as well as the Washoe Club, the El Dorado, Piper's Old Corner, the Delta, the Palace, the Silver Palace, the Capitol, and the Sazerac.[56] Most were less famous, frequently surviving only a matter of months before new ownership took over the location and attempted to succeed where others had not. Regardless of the reputation of the saloon, the business could represent a significant investment. When an 1865 fire destroyed Tom Carson's operation in Gold Hill, his loss in bar fixtures, liquor, and other items amounted to $2,500.[57] Although a saloon was a way for someone to enter the business world, there was considerable risk. Clearly the success of a business depended greatly on whether its owner could manage his affairs and attract customers.[58]

The survival of a saloon was also dependent in part on the ability of the owner to create an establishment that was unique, answering the needs of a particular part of the market. Some drinking establishments provided separate entrances for ladies who wished to attend a private club or restaurant but who preferred not to walk past the bar.[59] At the same time, other facilities offered female companionship of a different sort, including a full spectrum of possibilities from the flirtatious to the more intimate. Hurdy-gurdy girls, some of the region's more notorious characters, figured prominently in the myth of the wild West as the famous dance hall girls. These women either served as a form of stage (or floor) entertainment or danced with customers for a fee. Some of the earliest sources on the Comstock describe them, and author-artist J. Ross Browne provides a vivid illustration of the women at work. In spite of this, hurdy-gurdy girls do not appear in the census record. Perhaps the enumerator classified them as waitresses or bartenders, or he may have perceived them as prostitutes. If or how often they crossed the line into sexual commerce was clearly a matter of individual circumstance, however, and historical documents are not likely to provide much information. Intimations like those of Miriam Leslie, writing of her 1877 visit to Virginia City, for example, are more tantalizing than concrete. She pointed out, "Every other house was a drinking or gambling saloon, and we passed a great many brilliantly lighted windows, where sat audacious looking women who freely chatted with passers-by or entertained guests within." While the example of the Barbary Coast lends some insight into the role of women in the saloon, that area clearly represented an extreme expression of the possibilities. Some women may have found their way into the liquor trade more innocently by assisting the family or merely as a means of employment, with no prostitution expected or delivered. Nonetheless, the community often attached a stigma to women who worked in a saloon, granting only limited exoneration to a wife who helped her husband.[60]

Saloons contrasted with one another in myriad ways. As noted, many offered food, either in the form of free lunches and bar food or in the back at a small restaurant. Women played a variety of roles, both as customers and as part of the business. Some places specialized in whiskeys and other distilled liquor, while upscale businesses boasted fine wines and champagne. The working man's saloons, which only occasionally find mention in the local papers, featured imported and domestic ales and beers.[61] Some institutions catered to ethnic customers, while others sought a broader market within a specific class. At least one saloon, O'Brien and Costello's Barbary Coast business, combined drinking with a shooting gallery.[62] More than a hundred choices of drinking establishments allowed for considerable variation.[63]

Besides these differences, gambling provided yet another way to offer variety in a saloon. Many establishments featured billiards, and any place with a table and chairs became home to poker or other games. Some saloons, however, promoted these diversions more than others. At the other end of the spectrum, all gambling houses sold drinks, and the line between these businesses and saloons became blurred.

The history of gambling in the area actually predates the presence of the Euro-Americans, since the institution was an integral part of Northern Paiute culture. Newspaper accounts as well as historic photographs document gatherings of American Indians as they gambled among Virginia City buildings.[64] The other residents of the Comstock continued this practice in an elaborate and frequently pursued tradition.

Eliot Lord recalls the nature of gambling in early Virginia City, providing a vivid portrait: "Little stacks of gold and silver fringed the monte tables and glittered beneath the swinging lamps. A ceaseless din of boisterous talk, oaths, and laughter spread from the open doors into the streets. The rattle of dice, coin, balls, and spinning-markers, flapping of greasy cards and chorus of calls and interjections went on day and night, while clouds of tobacco smoke filled the air and blackened the roof-timbers, modifying the stench rising from the stained and greasy floors, soiled clothes, and hot flesh of the unwashed company."[65]

Excavations that archaeologist Donald Hardesty began in 1993 shed light on the breadth of life in saloons. One site included the foundation of a brick building that had served as home to the Shannahan and O'Connor Saloon as well as the Hibernia Brewery.[66] A second dig was undertaken at O'Brien and Costello's Saloon and Shooting Gallery. A few artifacts speak clearly for themselves. Among the thousands of pieces of bottle glass at the Hibernia Brewery are several broken bone poker chips, six ivory dice, cribbage pegs, several clay pipes, and a spent bullet. These, however, are isolated bits of the past that are in some ways less informative than the site as a whole.

Archaeological investigation of the Hibernia reveals a rectangular building, approximately 60 by 20 to 30 feet, with its long axis perpendicular to South C Street. Customers would have stepped down into a sunken room with a wooden floor. There, the single long room served all, heated in winter by one or more cast-iron stoves. Kerosene lamps and chandeliers provided light, and chamber pots, ceramic washbasins, clocks, coat hooks, and ornamental brass fish furnished other amenities and decorations. Patrons drank ale, beer, wine, champagne, and whiskey. There was also evidence that some enjoyed mineral or soda water, as well as bitters and other medicinal substances, including gingers and Ayers Cherry Pectoral. Like many saloons, the Hibernia served food. Lamb, mutton, and beef dominated the remains of meat found, but customers also ate pork and chicken, and occasionally turkey, duck, and fish. Along with the gambling artifacts, ceramic doll fragments, marbles, corset stays, glass beads, and perfume bottles indicate diverse activities and the likely presence of women and children. Whether taken together or individually, these remnants of the past vividly echo an aspect of the Comstock that often finds little elaboration in written records.[67]

One of the better sources of information on Comstock gambling comes from the U.S. Census records. Although no gamblers appeared in the 1860 census for the Virginia City area, ten years later there were 111 men who declared gambling as their livelihood. Representing almost 2 percent of the men, most were either Chinese or born in the United States. Significantly, more than 10 percent of the Chinese men declared gambling as their occupation. Ten years later, the number of professional gamblers had dropped to 60, or less than one percent of the men in Storey County. The total number of Chinese in the county had declined only slightly in the intervening ten years, and yet the number of gamblers among Asians dropped to 18, or about 3 percent of that group. In 1870, there were several Spanish-speaking gamblers, coming from Chile, Costa Rica, and Mexico. By 1880 they were all gone, so that only northern Europeans augmented the majority of Asians and North Americans.

Many of the gamblers did quite well, while others clearly made only a passable living. William Tamkin was a twenty-seven-year-old gambler from New York who in 1870 claimed to have $8,000 in personal estate.

His brother, I. K. Tamkin, and another companion, Richard Weldon, did less well in the trade but appear to have been involved in the business. Assets of gamblers in 1870 generally rank lower than those of their peers: the manuscript census reported a gambler's average real estate at $68 and personal funds at $306; these figures for all adult men averaged $344 and $488, respectively.[68]

There were diverse types of gambling, both in the saloons and in the more specialized gambling halls. Mary McNair Mathews, writing of the 1870s, noted: "The gambling dens are perfectly magnificent in style. They are gotten up to attract people to them. Some are kept very orderly, while others are not."[69] Chinese-operated businesses offered games of Asian derivation both for their neighbors in Chinatown and for intrepid visitors. Euro-American games included faro, roulette, keno, and poker.[70] There was also gambling outside the saloons and halls along C Street. Private games filled Comstock evenings with entertainment, as they did in other communities. In 1869 Louise Palmer protested that her husband regularly left home to visit a business associate, allegedly to discuss investments. She pointed out, "I have heard of games of 'poker' and such enormities in connection with these nightly business meetings, and I have a shrewd suspicion of billiards. But John tells me that though the rest of 'the fellows' do indulge in those recreations, his hours, torn from his wife and children, are religiously devoted to business."[71]

Similarly, Mathews complained bitterly of a man who lured young boys into gambling using sticks of candy and telling them that although they cost the exorbitant price of $1, some were wrapped with money. Faced with her threat of turning him in to the authorities, he fled the community. The incident serves as yet another example of gambling in the community at large. When Mathews sold small pieces of a cake at the high price of "four bits" each, enticing customers because one slice contained a prize, she too was promoting gambling. This would-be gambler intended the scheme to benefit the local temperance league, however, and she would certainly have objected to any assertion that it was a sin.[72] People also gambled on prizefights, wrestling matches, shooting contests, and horse races.[73] In spite of private gambling and other examples of such games outside drinking establishments, the image of the wild West will always link the legendary poker game with the saloon. Although gambling was not the principal function of these businesses, the sport provided an important distraction in the midst of drinking and other socializing.

Keeping dozens of saloons, stores, and restaurants supplied with beverages required an elaborate infrastructure of merchants and breweries.

The 1859 rush to Washoe caused an incipient brewing industry to blossom into a rich array of choices. While some breweries established themselves on the Comstock, others took root elsewhere, closer to water sources that were better and more reliable.[74] The 1860 census lists three German brewers on the Comstock, but their number grew to twenty-six over the following decade. The Germans dominated the industry, importing their expertise from Europe. Throughout most of the Comstock heyday, the district had six breweries, representing a $25,000 investment, according to an 1872 article in the *Territorial Enterprise*. Of these, the Union, the Nevada, the St. Louis, the Washoe, and the Philadelphia gave the Comstock popular local products.[75]

A network of merchants and teamsters handled the imports needed to augment the Nevada brew so that a thirsty mining district could be satisfied. Many entrepreneurs were certainly involved in the liquor trade in some form, but only a few listed that as their principal business in the census. That the eight liquor dealers in 1860 increased to only ten in 1870 suggests that as soon as the Comstock boomed, merchants realized that the thirst of the miners represented a significant opportunity for profit. Although Germans could not claim a hereditary right to this aspect of the industry, as they did with brewing, they nonetheless also dominated the importation of liquor, wines, and beers.[76]

Appropriately, much has been made of saloons and alcohol consumption in the mining West, but drug use on the Comstock was not restricted to alcohol. With the Chinese immigrants, the practice of opium smoking arrived in the region and became a habit that took hold among the general population. In addition, many people consumed opiates in the form of medicines that all too often made them addicts. This was particularly true of women, since it was often the prescribed drug of choice for the ailments of their gender.

In April 1877 law enforcement officers raided an opium den in Virginia City's Chinatown, arresting "four young men and a girl, all Americans." They were tried in court, to the shame of their families, at least one of which was highly respected in the community. As the newspaper suggested, "The sight of these young persons ought to prove a warning to all." This sort of flirtation with the imported Asian drug was hardly rare. Newspapers frequently reported on the arrest, death, or attempted recovery of Euro-Americans who delved into the opium dens. Some of these people were criminals or prostitutes, but many were respectable members of society who could not resist the allure of the exotic forbidden drug. Much of society regarded addiction to opiates in the form of patented medicines as far more respectable, failing to recognize the similarities between that use and smoking opium. Nonetheless, the use of the addictive medicines was widespread, indicating that the community, in keeping with national trends, was more involved with opium than people either recognized or cared to admit.[77]

The sinful distractions of the Comstock were certainly no more prevalent than in any other place in the mining West. Many of those who participated probably would have objected to the term "sin," since every vice has its adherents as well as its detractors. Although there might have been a consensus that most crimes were antisocial, there were people who found nothing wrong with lynchings, stock scams, prostitution, the consumption of alcohol, gambling, and drug use. "Sin" is a relative term, best understood in the context of its opposite.

9: THE MORAL OPTIONS

Saints

By the civic organization, the foundation of churches, and the establishment of schools the union of the people was more closely cemented.
—*Eliot Lord,* Comstock Mining and Miners

Perhaps at this point the reader will allow me a personal recollection. In 1986, while involved with the restoration of Virginia City's Fourth Ward School, I asked for the assistance of Don Dakins, my father-in-law and a woodworker of immense skill. Streaks of black scarred the solid wood railings of the school's staircases, and I asked Dakins how to remove the marks without damaging the finish on the wood. For some time he examined the blemishes, running his fingers along them, until finally he said in a tone of realization, "I know what that is." "I don't care what it is," I thought to myself. "Just tell me how to remove it." "That's where the kids slid down," he continued, "and braked themselves with the heels of their shoes. Look here, where they approached the knob at the end of the stair rail. The marks get bigger and darker."

It was one of those magical moments when the sounds and images of the past drift into the present. The Fourth Ward School, built for a thousand students, completed months after the Great Fire of 1875, graduated more than fifty classes before it was closed for fifty more years. In spite of decades-long abandonment, that day those silent halls rang with the screams of children hurling themselves down the steep balustrade. It was possible to hear their feet as they jumped off and ran away before a nearby teacher could catch them. For just an instant, part of nineteenth-

The Fourth Ward School, built for one thousand students, opened in 1877. (Courtesy of the McCarthy Collection, Nevada State Historic Preservation Office)

century Virginia City came alive. We resolved that the marks should remain.

One could argue that young would-be scholars sliding down a stair rail is an unlikely place to begin a chapter on Comstock saints. While it is true that students have acted better, the mere presence of children—indeed, of thousands of children—is more to the point. The fact is, Virginia City was a family town, where parents, motivated as most have been for all time, attempted to raise their young to be upstanding, successful members of society. The family and its children, even when the young were engaged in mischief, represented the first pillar of the moral side of Virginia City.[1]

There are limits as to how far the role of families should be taken, however. The common assumption holds that women played a role as civilizers of the West, softening the hard life that the rugged men who preceded them carved from the wilderness. It is part of the myth of the Wild West, and like so many, it warrants reassessment. There is no question that many women who settled in places like the Comstock were important in founding communities. Nonetheless, they were not alone in these efforts, nor did women pursue them uniformly or in a one-dimensional manner. Like all people, the women of Virginia City were as diverse as their numbers. While women played critical roles in shaping the Comstock, stereotypes do little to illuminate the real contribution or the variety of options available and pursued.[2] Like some men, there were women who drank too much, smoked cigars, and occasionally committed crimes. Some did not even go to church. More than one woman became addicted to opiates. The immediate assumption would have prostitutes best fitting this profile, but other women also slid here and there into the netherworld of sin.[3]

In the same way, the stereotype of the unkempt, uncivilized miner of the West is often unfair. Some men worked hard to establish schools and churches, to help raise families, and to make the Comstock a stable, respectable community. An 1878 lithograph in *Frank Leslie's Illustrated News* shows a miners' changing room. Among the furnishings is a large mirror, before which one of the men is combing his hair. Again, this image does not fit the stereotype that would have miners as ruffians with no sense of civilization's virtues. With comb in hand, one miner asserted his concern for grooming and appearance.

With or without combs, there can be little question, however, that families went a long way toward supplying the cement to bind society into a community. The birth of a baby to the Tiltons in April 1860 marked the first Euro-American born in the mining camp, and her parents named her Virginia in honor of the place. Indeed, she became a sign of the permanence that many perceived—or hoped to perceive—on the Comstock. The Tiltons remained in the district for years as Virginia grew into a young lady, forever symbolizing the transition that Virginia City made in 1860 from camp to community.[4]

The Tiltons were not unique in their longevity in the district. Many arrived and left in short order so that there was a constant current of humanity working against permanence. Others lingered, apparently hoping to put down roots and build a place and a life that would last. These people sought continuity in an industry that created change. Still, the Comstock was unique, and it inspired reason to believe that its mines could be productive for decades.

In 1880 there were 3,599 women twenty years or older in Storey County, of whom 73 percent appear in the census as married. Of more importance than the simple existence of families was their longevity in the area. An analysis of the ages of the oldest child born in Nevada for each household shows that some stayed for years. Ethnicity apparently influenced the inclination to seek a permanent place on the Comstock:

The miners' dressing room. *Frank Leslie's Illustrated Newspaper*, 1878. (Courtesy of the Nevada Historical Society)

the Nevada-born children of Irish mothers were much older than their English and Cornish counterparts, for example, indicating that Irish immigrants lingered longer. This detail is crucial in understanding the question of community stability, because the Irish contributed nearly half the women of the Comstock.[5]

In perception or in fact, families may have been bastions of permanence, standing against sin, but there were also times when marriages ended in court. These situations left the judge or the public to redefine who was a sinner and who was not. From 1861 to 1880, Storey County courts granted more than 350 divorces, the plaintiffs of which were usually women. Those suing to end marriages often claimed cruelty as the cause, although desertion and neglect were also frequently cited reasons. Cruelty could be mental or physical, and court records recount details of both. For some Comstock residents, divorce may have seemed a sinful epidemic. Reporting on the 1870s, Mary McNair Mathews observed that there was "hardly a week but there is an application for divorce. Virginia City is truly the city of divorces."[6] Louise Palmer, in her scathing 1869 review of the mining district, wrote along a similar vein that divorce was all too common. She noted that it was as if "there were some hidden law compelling ladies there to obtain divorces from their first husbands and to choose others."[7] Although these two social critics saw divorce as all too common, it was actually an anomaly. The vast majority of marriages remained intact.

Where there were marriages, children soon followed. To attend to the educational needs of the Comstock, there were many schools; the Fourth Ward School is merely the sole survivor of what was once a fleet of facilities. As William Wright pointed out in 1876, "In the several wards of the city are handsome, commodious, and comfortable schoolhouses."[8] The 1865 annual report of the State Superintendent of Public Instruction shows that Storey County already boasted twelve public schools—seven in Virginia City, three in Gold Hill, one in American Flat, and one down Six Mile Canyon in the Flowery District. Enrollment in the county increased from 420 in 1863 to 895 in 1866. By that time about 85 percent of the district's children attended school.[9] The number of pupils on the Comstock increased dramatically over the subsequent years. Teachers correspondingly became more common on the Comstock. In 1870, there were fifteen in Storey County. Ten years later, that number had increased to more than seventy.[10]

Along with the public schools, there were many private facilities, run by both churches and individual entrepreneurs. Mary McNair Mathews, for example, describes how she opened a school to accommodate her son and others. That enterprise, one of the many she pursued, proved to be profitable.[11]

As will always be the case, there were undoubtedly both good and bad teachers. For the children who enjoyed the nice ones and endured the mean-spirited ones, there could have been no better demonstration of the difference between saint and sinner. John Waldorf, for example, kindly remembered his first teacher as "a great hearted little woman, who refused to believe that boys were savages." He further tells of how she attempted to kiss her little scholars each Friday at the end of the school week. While the girls consented, the boys rebelled, seeking every opportunity to escape unscathed. "They sneaked through and trickled through and rushed through, and when such schemes didn't work they formed a flying wedge and upset her. She never succeeded in kissing more than ten percent of the boys, despite the fact that our fathers advised us to submit; but still, she never would agree with anyone who called us savages."[12]

Besides the teachers, many others provided needed services, creating the fabric of society and making the community a better place. These included the sheriff and his deputies, the police, constables, and the U.S. marshal, all of whom held crime in check. Oddly, the number of those involved in law enforcement did not increase with the population: in 1870, there were 25 officers, or one for every 453 Storey County residents; five years later, during the population boom associated with the big bonanza, there were only 2 more sheriffs, constables, and police officers, for a ratio of one to 723; and in 1880 there were 24, for a ratio of one to 667. The change appears to provide yet another expression of how the Comstock matured in those ten years. Dozens of others contributed to the community infrastructure as well—postmasters, water company employees, toll road agents, judges, librarians, telegraph operators, and other government employees.[13]

Those elected to public office provided another type of cohesion for their community. Throughout history, people have vilified politicians as self-serving, unscrupulous, evil predators on society. Still, some tried to do a good job. It is clear, for example, that after the mid-1860s the miners' union ensured that elected officials either came from their ranks or supported their cause.[14] A successful Comstock politician, then, served at least part of his constituency. In a setting where a single industry dominated the economy, local leaders also tried to create an environment conducive to the success of business. Except in those instances when the interests of labor and capital conflicted, elected officials usually accommodated two masters, both miners and the mines. As always, because of the demographic and economic importance of the Comstock, the representatives from the mining district virtually controlled what happened in state politics. This remained true even long after the mines failed, since until reapportionment in 1966, Storey County had legislative representation far exceeding its population, a vestige from its long-past glory days.[15]

Because Virginia City was an incorporated community and a county seat, ample offices existed for aspiring politicians. City government dated to January 1861 under Utah Territory. It had five trustees elected by wards, and this group in turn elected a president of the board. The Nevada Territorial Legislature created Storey County in November 1861 and a year later authorized the incorporation of Gold Hill. Three commissioners directed the course of the county. Each of the local governments had a variety of other elected officials, including clerks, recorders, police chiefs, the county sheriff, treasurers, secretaries, a marshal and street commissioner, the justice of the peace, the county surveyor, a public administrator, an assessor, a district attorney, tax collectors, and the county superintendent of schools. Once these positions were established in the early 1860s, there were few additional opportunities for would-be local politicians, but the number of offices provided many chances for election.[16]

An elected position on the Comstock provided an excellent springboard from which local politicians could seek higher office, both in the legislature and in the executive branch. Because the legislature developed a habit of returning the same senators to Washington, D.C., these men assumed power surpassing that warranted by the numbers of people they represented. Especially when dealing with bills that affected mining, Nevada's two senators were inordinately effective. For example, Senator William M. Stewart's involvement with the 1866 passage of the National Mining Law, together with its 1870 and 1872 amendments, guaranteed a generous environment for his state's predominant industry. The act, having now lasted for 125 years, has perpetuated the Comstock's influence on the nation long after the district passed its prime.[17]

If bringing order to society is to work on the side of good, then local, state, and federal elected officials were saints of sorts. They helped to create a sense of community where there had been nothing, and they worked to show that civilization had arrived. At times, however, they clearly fell from grace, compromising decisions to benefit special interests and fighting among themselves, employing less-than-honorable political tactics.

Doctors, dentists, midwives, and nurses also made their community better, by addressing human suffering. Unlike those involved in governing and policing the Comstock, these public servants increased in number with the population. In 1870, for example, there were 31 doctors, nurses, and dentists. Five years later their numbers had increased to more than 50, but because of the increased population, there were roughly 360 people living in Storey County for every health care provider in both 1870 and 1875. In 1880, the number of health care professionals grew still more while the population declined, leaving about 239 people for every health care provider. Dramatic change occurred with dentists in the county, who numbered two in 1870 and eight in 1880. Similarly, the number of nurses increased from 2 to 33 during that ten-year period. In addition, there were several "doctresses" in Virginia City who made an important contribution to the community's well-being. Clearly services for the sick and injured were becoming more common.[18]

Some who were not doctors nonetheless played an important role in health care. Gorham in his Comstock recollection tells of Jim Francis, whom miners called "Doc." The mines employed him because of his knack for dealing with illness and injury, and as Gorham remembered, "'Doc' Francis' face was a welcome sight many times. . . . In urgent cases the first thought was of that kindly man." Gorham writes of an accident in a shaft that killed several miners, mangling their bodies terribly. As he puts it, "'Doc' and kindly Walter Cobb, the undertaker . . . , those two gentle souls had worked all night to repair and soften as much as possible the destruction of that terrible fall."[19] In this way, a miner who knew first aid and an undertaker became saints.

Mary McNair Mathews regarded herself as a nurse, but not because she had training. Instead, she too had an instinct for caring and healing. Intuition and a background in folk medicine guided her. When one of her son's fingers was accidentally cut off, she placed it in a brandy jar, calling on a background in sympathetic magic, so that it would not hurt him in the future. Indeed, Mathews was not the only one who delved into folk remedies. Western historian Duane Smith, in his treatment of Comstock medicine, notes, "Wood ashes or cob webs would stop bleeding; a bag of asafetida worn around the neck could remedy a cold; a salve of kerosene and beef tallow softened chapped hands." And so on.[20]

Scores of men belonged to the various volunteer fire departments, standing ever ready to risk injury or death in the midst of flames. Along with the volunteers, there were paid professionals. Fourteen of these appear in the 1870 census, a number that increased to twenty-four in 1875 and more than thirty in 1880.[21] Together, they surely stopped hundreds of fires that could have spawned the sort of epic inferno that occurred in 1875. Their struggle underground during the Yellow Jacket disaster, fighting against formidable odds and chancing death at every moment, won them a permanent place in Comstock legend. Many of the volunteer and professional firefighters were no doubt also sinners of various sorts. Nonetheless, they won the mantle of sainthood for their willingness to gamble everything for their neighbors.[22]

If any group of people on the Comstock deserved the title of saint with little or no qualification, it was the Daughters of Charity of Saint Vincent de Paul. In 1864, Sister Frederica led three other sisters to a still-young Virginia City and thereby added a new component to the increasingly complex society. Over the course of the next thirty years, the Daughters of Charity founded and administered an orphanage and school, as well as the best hospital in the district. For three decades, they attended to hundreds of children and sick and injured people. Soon after their arrival, they obtained their own house, with a single room. It was a meager abode, but it served as a beginning for their orphanage and boarding school, accommodating 12 children. Within the following year, attendance grew to 25 resident children and a total of 112 students. Because the Daughters were fulfilling a need, in 1867 the State of Nevada provided them with funds to build a larger orphanage. They eventually constructed a complex of buildings that over the years gave a home to hundreds of children.[23] Their orphanage no longer survives, but it is possible to see large mauve stones, locally quarried, that once supported the institution, itself a cornerstone of Comstock society.

Early in 1876, the Daughters of Charity opened their St. Mary Louise Hospital, a four-story brick building that could accommodate seventy patients. Marie Louise Mackay, wife of bonanza king John Mackay, played a pivotal role in establishing the facility; she gave the sisters the pastoral setting of Van Bokkelen's Beer Garden (the proprietor had died a few years earlier). The imposing hospital that they built dominated the base of Six Mile Canyon, providing something of a boundary for the eastern extremity of Virginia City.

The facility remains a substantial architectural anchor of the historic district. Rooms are spacious and well lit, each originally outfitted with marble-topped washstands and hot and cold running water. There was a chapel on the third floor of the hospital, complete with what the *Territorial Enterprise* described as "a handsomely decorated altar." Accommodations included a reading room that was "large, well lighted and ventilated . . . [with] a good stock of books, newspapers and magazines."[24] The sisters retained quarters on the top floor, in the attic where the roof sloped in. Their rooms were nevertheless comfortable, and the height of the building afforded them an excellent view of the community. In the distance, St. Mary in the Mountains Catholic Church still stands on the hill in full view of their windows. The fourth floor also includes a small room with bars on the windows for the mentally unstable who needed to be confined. By today's standards it was nothing short of horrifying, but in spite of the darker aspects of nineteenth-century health care, St. Mary Louise Hospital stands as a testimony to the humility and hard work of the Daughters of Charity. Affording themselves few comforts, those who worked in the facility were promised long hours filled with unending tasks through which to express their devotion to God.

Besides the Daughters of Charity, there were members of the clergy. The first to appear in the record as working on the Comstock was the Reverend Jesse L. Bennett, a Methodist. He preached on the streets in Virginia City and after passing the hat, found a heap of gold and silver in it. Bennett stayed long enough to establish a small congregation, but the next year he passed the pulpit to S. B. Rooney, a defrocked minister from California who continued his work. Rooney built Virginia City's first

The Methodist church was once an important part of the community, but as Virginia City lost its Cornish immigrants to more prosperous mining districts, their church declined. Today it no longer stands. The steeple of St. Mary in the Mountains Catholic Church rises in the background. (Courtesy of the Nevada Historical Society)

church in 1861.[25] In the same way, the San Francisco Catholic Diocese was not slow to respond to the presence of a new boomtown. In 1860 Archbishop Sadoc Alemany sent Father Hugh Gallagher to the area. He built a church in Virginia City, but a strong wind blew it down.[26]

Although each of the early members of the clergy was important in his own right, none assumed the mythic proportions of Father Patrick Manogue, priest to the Virginia City Catholic community from 1862 to 1885. Manogue, an Irish immigrant of enormous stature, had been a miner in the California goldfields. In the 1850s, he heard the call to minister to his compatriots and returned to Europe for training. After ordination, he found himself assigned to the fledgling parish of Virginia City. He quickly built a more substantial church, naming it St. Mary in the Mountains. With much of the Comstock population coming from Ireland and many of the rest being Catholics from other places, Manogue became the most important religious figure in the district. It was not merely the numerical significance of his congregation that elevated the priest within the community. He was such a powerful presence that he often mediated in affairs outside his church. The imposing priest governed his flock, affecting the community as a whole, with a forceful but compassionate hand. Stories of his leadership and kindness circulated in Virginia City for decades after he left to become a bishop and after his death in 1895.[27]

Of course there were many others who followed in the footsteps of Bennett and Manogue. An Episcopalian congregation formed in September 1861, eventually building St. Paul's in 1863. The Presbyterians established their community in 1862 and constructed a church five years later. Baptists and African Methodist Episcopalians formed congregations of varied durations and success. The Jewish community in Virginia City struggled unsuccessfully for years to build a synagogue, but it lacked the size to support such a facility. Nonetheless, the Jewish community held religious observances.[28]

The 9th U.S. Manuscript Census of 1870 records eleven clergy in Storey County. They included priests, ministers, preachers, a sexton, and two attendants to the joss house, the temple serving Chinatown. Their ranks increased to thirteen in 1875 and eighteen by 1880. The priests, ministers, and sisters all had their own stories to tell, and it would be unfair to assume that they were all saints. Still, most, if not all, played important roles in making the Comstock a better place, providing the cohesiveness that religion affords for the people as they strove to build a community.[29]

An example of how these church leaders affected life illustrates their success. The sons of an Irish immigrant named Timothy McCarthy wrote to a relative in the early 1880s, proud to declare that they had joined a newly formed sodality, a lay society within the Catholic church. The charitable group boasted more than one hundred boys, each wearing a badge "with the picture of the Blessed Virgin Mary . . . and bearing

The Presbyterian church was built in 1867 and still serves a small congregation. (Courtesy of the McCarthy Collection, Nevada State Historic Preservation Office)

the following inscription: 'St. Mary's Sodality, organized October 9th, 1881.'"[30] It may have been a small success, but for about one hundred young men, it was a declaration that on occasion the side of good won the day.

While many attended church, Virginia City, like all places, knew the disparity between the sacred and the profane. Mathews recalls how she sat in church listening to a sermon, all the while hearing "the snap of the whip, or the oaths of the driver."[31] In its frantic attempts to harvest gold and silver, the Comstock seldom stopped its labor, further underscoring the contrast between those who worshiped and those who worked on the Sabbath.

The cliché of saints and sinners is integral to the myth of the West. For members of the Temperance League, those involved in the trade of alcohol—whether as producers, entrepreneurs, or consumers—were on the side of sin. Among those who drank, however, the temperance movement, with its reputation for meddling and imposing an unwanted standard of morality, was hardly on the side of good. Many saloon patrons must have seen the self-righteous judgment of an innocent gathering of friends over beer as less than saintly. Hemmann Hoffmann, a Swiss immigrant who worked in a brewery, wrote that many temperance advocates drank alcohol-laced medicines until they were intoxicated but saw no contradiction between their message and their actions. Hoffmann went on to assert: "These Temperance followers who cling to their hypocrisy are boring, pedantic fellows, and they lack the vigorous, fresh edge that otherwise typifies Americans. It seems to me more practical if everyone sets his own rules and stays true to his own principles, enjoying God's gifts happily with or without alcohol, and at the same time avoids excess since too little and too much ruins all fun."[32]

Temperance advocates appear to have met as early as 1864.[33] In March of that year a Virginia City lodge of the Sons of Temperance boasted "some 250 members." The *Gold Hill Daily News* was able to comment that the "saloon keepers in Virginia say that Sunday is now their poorest day for business," whereas before, the Sabbath had been one of their best. The newspaper further commented that with temperance in the wind, "This is a good indication, and shows that we are advancing in morality."[34]

One year later, yet another temperance organization, the Independent Order of Good Templars, or IOGT, organized for "ladies and gentlemen" to gather for "great sociability and pleasure, as well as for the benefit of its members in a temperance point of view."[35] The group subsequently elected officers and scheduled regular meetings. By the end of two weeks, the movement seemed to be gathering such momentum that the *Gold Hill Daily News* observed, "Several of our prominent bar keepers are sighing for the sight of their late prominent patrons . . . and a few on B Street seriously contemplate closing their establishments in view of the benign influence of the I.O.G.T. seriously affecting their business."[36] Whether this assertion was serious or in jest, the future of the saloon industry turned out to be less dire than predicted. Still, there can be little question that the Comstock proved fertile soil for an ongoing temperance movement that helped keep at least some from excessive drink.

Mary McNair Mathews, who wrote of her life in Virginia City during the 1870s, joined the Sons and Daughters of Temperance, an organization that by then held weekly meetings on the Comstock. Mathews participated in fund-raisers and collected names on petitions for the state legislature. These were appeals to change the laws related to drinking and gambling. Although not seeking to prohibit such activity, proponents hoped to discourage these vices by moving gambling houses to second floors and by preventing "a dealer from selling over $5 worth of spirits to one man."[37] Temperance organizations also sponsored lectures. In 1875, Miss Sallie Hart, a nationally known proponent of abstinence, spoke in Virginia City.[38] The temperance movement may have had success on the level of the individual, but evidence of community-wide change is elusive. Opposition to saloons and alcohol on the Comstock ultimately had little effect on the institution of drinking and the businesses that supported it.

Again, archaeology can provide insight. While saloons continued to play an important part of Comstock society, life went on, juxtaposed against the world of alcohol consumption. Donald Hardesty excavated two other building sites next to the Hibernia Brewery in Virginia City. One of the buildings had a number of commercial uses, and the archaeological record preserved remnants of a Singer Sewing Machine sales office. Next door to that structure was a lodging house, and indeed all three structures had rooms for rent at one time or another. The remnants of decades of use included hundreds of household items. Some of the most telling of the artifacts were marbles and other toys, including a metal locomotive and a metal wagon complete with driver. Now dirty, crumpled, and rusted, they had once brightened the day of children who filled the streets with play. By their presence, these families challenged those who would assert that because the Comstock had saloons, it was a sinful place.[39]

From its early days, the Comstock had a growing number of unions, military companies, fraternal organizations, and social clubs. These groups played a crucial role in developing the mining district, providing entertainment and distractions, and giving new immigrants a means to fit in. Members of the Sazerac Lying Club, for example, wore little hatchets and attempted to earn the title of Monumental Liar of America. Although this could hardly be regarded as a saintly goal, the group was harmless enough, providing a relatively innocent distraction. They were kin to those of the community who were always willing to play practical jokes, but like the fabrications of the Sazeracs, such antics were usually harmless. As journalist Wells Drury pointed out, they usually involved nothing "more serious than breaking a man's leg."[40] Other organizations, including the Freemasons, the Knights of Pythias, and the Odd Fellows, were social, philanthropic groups intended to improve the community and to provide a forum where like-minded men could gather.

Along the same line, many of the major occupations on the Comstock had unions that promoted a sense of community. As described earlier, a number of ethnic societies functioned in much the same way. The Fenian Brotherhood, for example, was an Irish nationalist organization that raised money for a free homeland. Although the Brotherhood was secretive in the United Kingdom, on the Comstock members could be more casual about their associations. Consequently, the Fenians operated more or less in public. Their connection with several military units was less well defined but nonetheless clear: the Emmet, Sarsfield, and Sweeney Guards gave Irish exiles a chance to prepare militarily for their struggle.[41] For all their drilling, however, they never fired their weapons in battle, failing to attack England or even Canada, for that matter, as their eastern counterparts did. Ultimately, the guards existed more for camaraderie and show than anything else. Newspaperman Wells Drury, who was a member of the Sarsfield Guard, recounted meeting President Grant during an 1879 visit to the Comstock. The former Civil War general responded to Drury's salute "with a tolerant smile [and said], 'Young fellow, I never had as fine a uniform as that all the time I was General of the Armies!'"[42]

Fraternal organizations provided a means for people to gather and develop a sense of community, but nationalist groups that promoted hatred for others served also to create barriers. Thus the Irish revolutionaries of the Comstock did little to foster good relations with the Cornish and English immigrants to the mining district. Indeed, as discussed earlier, there are abundant newspaper accounts of Cornish-Irish antagonism.

In the eyes of many, the Comstock Chinese and their Chinatown represented the pinnacle of sin and evil. Many perceived their neighborhoods as places of opium dens, prostitution, and deceptive thieves and murders. Mary McNair Mathews, who reserved a particularly venomous racist hatred for the Chinese, spewed forth diatribes against them in her published account of Virginia City in the 1870s. She complained about what she saw as their custom of spitting milk on pastries and starch on clothes. Mathews also perceived them as dirty and disease-ridden and as capable of undercutting the wages of Euro-Americans. These were common complaints about Asian immigrants to the mining West, and yet there were those who rose up in defense of the Chinese. None other than Mark Twain pointed out: "They are harmless when white men either let them alone or treat them no worse than dogs; in fact, they are almost entirely harmless anyhow, for they seldom think of resenting the vilest insults or the cruelest injuries. They are quiet, peaceable, tractable, free from drunkenness, and they are as industrious as the day is long. A disorderly Chinaman is rare, and a lazy one does not exist." Certainly the Chinese were no worse or better than the rest of the Comstock. Like most others, they came to profit from the mining district. There were criminals and prostitutes among them, but they also established a community where peaceful coexistence was the norm, not the exception. Chinese families raised children, priests tended the joss house, and Chinese servants and launderers worked tirelessly to address the needs of others.[43]

A lithograph of Virginia City's Chinatown published in *Harper's Weekly* in 1877 depicts various facets of the immigrants' lives, not one of which is clearly antisocial behavior. Instead, the images show everyday street scenes and a great deal of hard work. Everybody from laundry workers, sausage makers, and cooks to a street cobbler, a barber, a fortune-teller, and a pharmacist is included in the spectrum of possibilities in this illustration (see first illustration in chapter 7).[44]

The manuscript censuses reinforce the image of a vibrant Asian community of hard workers. Clearly many Chinese immigrants spent time in local jails, but there is no reason to believe that they did so more often than others. Indeed, the 1880 census documents twenty-nine people residing in the county and city jails. Of these, only two were Chinese. The rest were Euro-Americans, of whom five were women and the others were boys or men. This places the incarceration rate at one for every 318 males, with the Chinese not significantly deviating from this trend.[45]

Limited surveys of Chinatown archaeology depict a community not unlike the rest of Virginia City. Pottery and other items bear the distinctive stamp of Asian imports, but they have counterparts in Euro-American refuse. Rather than giving evidence of a sinful place, they conjure up images of everyday lives in an immigrant neighborhood.[46]

At the other end of the social spectrum, the Comstock created one of its most vivid contrasts between a saint and sinner. John Mackay, richest of the bonanza kings, was beloved in Virginia City as an excellent example that wealth could not corrupt a truly good man. People saw Mackay, even after he accrued millions, as remaining humble, charitable, fair, and honest. The story of Mackay in the midst of the Great Fire, promising Manogue that he would rebuild St. Mary's Church remains in local oral tradition, as an example of his generosity. Along this line, John Waldorf, who wrote of his childhood recollections on the Comstock, tells of how he and other children waited outside Piper's Opera House in the evening, hoping that Mackay would take pity on them and pay the fifty cents for each to see the show. The bonanza king was not known to disappoint them.[47]

In contrast to Mackay, James Fair assumed the mantle of dishonest, niggardly, cruel sinner. In an ironic play on his name, he was popularly regarded as unfair. Fair became the ultimate sinner in local legend, counterbalancing the reputation of Mackay and illustrating how wealth could corrupt a man with flaws. One popular story told how Fair tracked down some miners who were inclined to smoke underground during breaks. Company policy strictly prohibited the practice because of the threat of fire. One day, Fair, dressed in common miner's garb, came upon several miners relaxing in an adit. He sat down with them and asked for a smoke. They brought out their tobacco and everybody lit up, but when they ascended the cage at the end of their shift, they found the company had dismissed them.[48] Local folklore regarded Fair to be within his rights, but it nonetheless judged his trickery as underhanded and his unwillingness to confront the miners directly as cowardly.

Fair's divorce in 1883 added scandal to his dark reputation, confirming in the minds of many that he was less than honorable.[49] Popular opinion sided with Theresa Fair, seeing her as the opposite rather than the complement of her husband. She was known to participate in charity fundraisers, and people regarded her as generous, whereas her husband was stingy. A comparison of Theresa Fair with the popular assessment of Marie Louise Mackay heightens the contrast. The latter bonanza queen preferred to use her financial ascendancy as a means to leave the Comstock. She spent much of her marriage in New York or Paris, where she held court, becoming part of high society. Her husband joined her on occasion, and by all accounts theirs was a mutually satisfactory arrangement, but people regarded Mrs. Mackay's choice as an act of snobbery. Even though she also gave to local causes, she won a reputation that stood in opposition to that of Theresa Fair. The two couples found themselves cast in a Comstock morality play that illustrated, as far as the local residents were concerned, what happened to people when confronted with sudden riches. Clearly, wealth could not taint a truly good man or woman, but it made a sinner's dark soul even more apparent.

While people watched the Fairs and the Mackays act out their parts, they also enjoyed theater as it appeared onstage. Indeed, going to the theater was often a more wholesome activity than peering into neighbors' lives and passing judgment on them. The stages of the Comstock provided theatrical diversions for a community starved for entertainment, making the mining district a better place. Theater was one of several institutions that offered alternatives to the saloons, gambling halls, brothels, and opium dens where workers might otherwise spend their wages.

Virginia City was home to several theaters of regional and even national fame. As early as the winter of 1860–61, a business owner had established a thousand-seat hall. In February 1861 the Melodeon, associated with the Grunwalds, a German touring group, opened its doors. Tom Maguire built a theatrical empire in the West with his San Francisco business and with his ability to book the top acts of the day for regional tours. Not surprisingly, Maguire's Opera House became a popular source of entertainment in Virginia City during the early 1860s. And there were others—Topliffe's Theater, E. W. Carey's La Plata Hall, the Temple of Comedy, the Niagara Concert Hall, Henry Sutliff's Music Hall, the Gold Hill Theater. All added to the spectrum of possibilities as residents of the Comstock pondered the choices for an evening's distraction. Finally, Piper's Opera House rose above them all and became a required venue for any touring musician, lecturer, actor, or performer of any sort.[50]

The 1870 census documents a wide range of performers who made theater a going concern on the Comstock. It includes ten actors, a "tragedian," and three others who were involved in the theatrical business. In addition, the Comstock employed many people who supplied the music for the singers and musicals that made the district's stages the envy of the West. In 1870 there were fourteen professional musicians in Storey County, together with two music teachers. They came from throughout the world, supplying the census enumerator with nativities that ranged from Spain, Germany, Ireland, and Austria to several of the states.[51] By 1880 the music industry had grown much more complex. There was almost the same number of professional musicians, but teachers had increased nearly tenfold, reflecting the intent of a more stable community to see its swelling ranks of young people properly schooled in the arts. In answer to the growing appetite for the arts, the Comstock of 1880 included at least two employees at a music store.

Unfortunately, there were also times when the stage slipped into the risqué or became a place of low entertainment. The community calendar included many prizefights, wrestling matches, and other forms of combat both in and out of the formal theaters. The contests pitted man against man and animal against animal. A fight in 1866 attended by about one hundred spectators involved a 75-pound cinnamon bear cub against "three good bull dogs." According to journalist Alf Doten, who watched the contest, the bear "came off best" and was "not much hurt." In 1871 Doten described yet another fight at an opera house, a "big wild cat and bull dog fight." In this case, the dogs won, leaving two cats dead.[52]

Besides theater, the Comstock offered many other harmless forms of entertainment. Circuses frequently came to town, and John Waldorf recounts how important they were to a young boy. One Virginia City schoolchild wrote to his cousin in 1882 about a circus and the wonder it brought into his life: "There was a good many animals. There was a rhinoceros got out of the cage and made the crowd scatter. There was a man on the seat of the cage when it broak [*sic*] and he jumped of [*sic*] just in time to escape. They caught the rhinoceros and put ropes around him and tied him to two poles. . . . They had two rings in the circus and there was three eliphants [*sic*] and they walsed [*sic*] and turned round on barrels and laid down until the ring master could stand on them."[53]

Doten records the arrivals of circuses with regularity in his diary, and the enumerators from the 9th U.S. Census of 1870 even managed to capture one in their document.[54] Twenty-two circus performers appear in the manuscript census for Gold Hill. Of these, only one was a woman. Ages ranged from eight to forty. Most came from places as far away as Morocco and Australia, but there were also three Canadians and five people born in the United States. The enumerator recorded a smaller group of men related to the circus in Virginia City. The performers had camped in Gold Hill, where they set up their tents, unable or unwilling to haul their larger wagons up Greiner's Bend, the steep hairpin separating most of Gold Hill from the Divide and Virginia City. Instead, they sent a few of their number as an expedition into the larger town to sell tickets and foment interest. Thus the Virginia City contingent of the troupe consisted of three "hustlers," two drivers, and a "circus man." These people probably worked at a number of tasks, ranging from selling tickets to hawking food and other things at the circus itself. In all, this circus was presumably fairly small, but like performances in an opera house, these "horse operas," as they were called, provided a wholesome diversion for the Comstock, helping build it into a better place.[55]

Complementing performers onstage or in the ring were others who entertained the Comstock. Again, the census records a variety of people

A view of Virginia City looking down its busy C Street, which was decked out for the Fourth of July. The International Hotel stands in the background, ca. 1880. (Courtesy of the Fourth Ward School Museum)

who provided diversions. In 1870 two gymnasts from Baja California were making a living at their trade. Whether they performed onstage or in the street is not known. The same was true of the numerous musicians, who were certainly not restricted to opera houses. Indeed, they found ample employment with the community dances and occasions that would call for their services. Others found the street an acceptable venue: Frank Lobo and F. Nagel, from Portugal and Italy, appear in the 1880 census as itinerant musicians, and it is likely that they were street performers. Similarly, D. Blackburn, a woman of twenty-nine from New York State, operated a melodeon on the streets of Virginia City in 1870.[56]

Numerous artists and photographers completed the spectrum. Shortly after the first strikes in 1859, J. Ross Browne arrived as an artist-writer for a national magazine. African American Grafton T. Brown also earned money as a Comstock artist in 1861. Cyrenius B. McClellan, from Pennsylvania, became Virginia City's most prominent artist, finding enough employment to stay for more than ten years.[57] There were others, of course, and there were also several photographers—those who made the Comstock their home as well as dozens who came through town in search of more temporary markets.[58] These people rounded out the artistic scene of an increasingly sophisticated mining district.

Art could take many forms, however. The Comstock also employed several gardeners, who helped the more affluent citizens tailor their yards into the nineteenth-century ideal of a country garden, despite the parsimonious soil and meager rainfall. Together with the many people who worked without professional assistance, they created places of beauty that softened the harshness of the mountain. Mary McNair Mathews, for example, described a house in 1869 that had a "neat little yard in front . . . surrounded by a white fence. The yard had several fruit trees, each loaded with fruit nearly ripe; climbing rose bushes and several other plants adorned the yard, and a portion of which was covered with young wheat." She went on to note that hanging from the veranda were two birdcages with canaries.[59] The descendants of locust and fruit trees, stands of wheat, and scatterings of flowering bulbs, none of which are native to the region, continue to cling to the upper slopes of Mount Davidson, remnants of attempts to transform the district. The canaries, however, have long since vanished from the Virginia Range.

While most tombstones in the area's cemeteries came from California, the Comstock employed a few stonecutters who provided the mining district with a most personal form of art. Hugh Muckle, for example, was a local favorite for years, and his work survives, marking numerous graves. He and his compatriots were also responsible for mantels, tables, and washstands for the nicer homes, adding yet another dimension to the production of art on the Comstock.[60]

Wholesome entertainment could also manifest itself in a variety of ways. The rugged character of the mine occasionally contrasted with the activities in its depths. Cornish choral groups sometimes put on underground concerts. On April 19, 1879, the New York Mine of Gold Hill hosted a dance 1,040 feet below the surface. About one hundred people enjoyed music, dancing, and food in a subterranean chamber 36 feet long, 14 feet wide, and 36 feet high.[61] In the same way, activities on the surface could sometimes defy the popular image of a mining town. By 1880 Virginia City employed three professional baseball players, the core of a local team that provided yet another diversion for the increasingly complex metropolis. Indeed, as early as 1869, Alfred Doten recorded going to a game "on the [baseball] . . . ground down near the Gas works." Sadly, the Virginia City Base Ball Club lost to the Star Base Ball Club of Carson City, 54–17.[62] Virginia City redeemed itself nearly a decade later when its Bonanzas defeated the Gold Hill Actives, 9–7.[63]

These groups, clubs, and establishments added to the diversity and complexity that gave the Comstock its flavor. Despite the wide range of possibilities, many continued to classify their society in terms of sinners and saints. Even that could be problematic, however. Prostitution, for example, was a cornerstone of Comstock society, and yet the community had mixed feelings about the institution. The response, discussed above, of Virginia City's proper ladies to John Millian, the convicted murderer of Julia Bulette, underscores the problem of balancing murder against prostitution. Along this same line, Alfred Doten records an incident in November 1866 when, in the early-morning hours, a group of men took a fire company's engine down to Cad Thompson's Brick House and pumped the hoses inside.[64] This symbolic attempt to wash out the brothel was little more than the nineteenth-century equivalent of a fraternity prank, making the already strenuous lives of a few prostitutes even more miserable. Again, it is difficult to discern the difference between saints and sinners. Still, the event reveals something of the undertone within a community that preferred to have no red-light district.

It is all too easy to discuss those people who appear to have been saints, or who professed to adhere to sanctified causes. What is more difficult is to identify those responsible for the quietly performed good deeds that help constitute a community. Certainly there were hundreds, maybe thousands, of people with saintly inclinations among the multitude who called the Comstock home at one time or another. Recollections published in the twentieth century of Comstock childhoods in the 1870s tell of the good people who filled the place. John Waldorf, for example, remembered "Old Doc," who spent his retirement in a shanty in the midst of sagebrush that he called his garden. He entertained neighborhood children with candy and nuts and told them stories, as they surrounded him on his porch.[65]

Harry M. Gorham recalled an incident, repeated hundreds of times in various forms throughout the Comstock's history, involving heroism in the mines. Selfless acts in the treacherous underworld brought with them a mantle of sanctity. According to Gorham, there was a disaster in the Alta Mine in 1882 that typified bravery. Miners were working in a shaft 2,000 feet deep. One of the drifts tapped a steady flow of water, so carpenters constructed a wooden bulkhead to hold the flood back while miners explored other areas. Several men working a drift that inclined slightly were trapped when the bulkhead broke, cutting off all avenue of escape and leaving the men in a dry pocket. The mine superintendent started the pumps to drain the shaft. A pump rod broke but was repaired, and as Gorham recounted, "Hours had given place to days. I think four days had elapsed as men and machinery struggled against that flow of water." Two young miners decided to attempt to reach the stranded miners by rowing a boat up the tunnel, but bad air and hot water overtook them, and they died, having gone only a short distance. At that point a pipe fitter in town constructed a helmet that would hold air and into which air could be pumped. John "Yank" Van Duzen volunteered to try out the invention. He walked up the submerged drift and found the famished miners waiting at the end of an air pipe that had kept them from suffocating. Van Duzen managed to make it back to tell of his discovery, thus renewing hope and inspiring the means to rescue the men. Gorham concisely summarized the heroism that day when he pointed out that Van Duzen "was not ordered in there; he volunteered. And it was a grand action."[66] Most of the "saints," having struggled against all odds to save others and working tirelessly to make the Comstock a better place, failed to have their tales recounted.

Edward Lovejoy serves as yet another powerful example of someone who subtly took the side of right but largely eluded the historical record. Lovejoy arrived in Virginia City in 1877, having been a newspaper editor and a judge in Trinity County, California. Before coming to Nevada, he had been known for his editorials defending the rights of Chinese immigrants and African Americans. As a member of the Freemasons and the International Order of Odd Fellows, Lovejoy sought to improve his community.

After his mother died in 1870, Lovejoy came to the Comstock to settle anonymously among the thousands of others. The degree of his anonymity, however, was unknown to all but his closest relations. He was, in fact, the orphaned son of Elijah Lovejoy, who had died in 1837 defending his abolitionist newspaper in Illinois. President John Quincy Adams called the elder Lovejoy "the first martyr to the freedom of the press and to the freedom of the slave."[67]

Elijah's widow and infant son toured the North after the murder, held up by abolitionists as an example of how slavery was victimizing not only the slave. Eventually tiring of the role, mother and son began a western migration to claim obscurity, one of the many commodities that the region had to offer. Anonymity became a family tradition. Elijah Lovejoy was a household name throughout the nineteenth century. Mother and son could have exploited his memory in the pro-Union West, but instead they settled for unremarkable lives. Edward Lovejoy, however, had his father's heart, and he too eventually spoke out on injustice. Still, his would remain a more subtle contribution. Whereas his father soared into fame as a martyr to a great cause, Edward Lovejoy faded away from history with his 1891 burial in Dayton. Nonetheless, his editorials remain as testimonials to a man who, though the victim of a struggle for justice, insisted on standing on the side of decency. He could have exploited his name, but he chose the more honorable path of anonymity, standing for what was right in a quieter fashion that gained him no recognition. On the Comstock, he continued his involvement with the lodges, joining in their efforts to create a better place.[68]

The Edward Lovejoys of this world create a problem when writing history, because of the role of the written record. Those who perform humble acts of good, who succeed in seeking no fame, are the noblest of all, and yet history will fail to mention them. One can only guess at how many other Edward Lovejoys lived on the Comstock, but there can be little doubt that thousands occasionally stood for justice, fairness, and dignity and yet will never appear in the story of the mining district. These saints, or at least their saintly acts, built a respectable place where hardworking citizens could raise families and live contented lives.

10: PRINCES AND PAUPERS

Contrasts in Class

SUBUNIT: *Virginia City;* STREET NAME: *County Hospital Road;* NAME: *Mary McEntire;* AGE: *56;* MARITAL STATUS: *Divorced;* PROFESSION: *Pauper;* SICKNESS: *Debility;* PLACE OF BIRTH: *Ireland*
—*From the 10th U.S. Manuscript Census of 1880*

Labor is positively powerless without capital, and the extraction of the precious metals from the bowels of the jealous earth cannot be accomplished unless the gloved hand of the millionaire joins the horny fist of the laborer.
—Frank Leslie's Illustrated News, *April 13, 1878*

Ethnicity and morality made the Comstock diverse and complex, providing the district with the means to stratify itself. Economics represented yet another way in which to evaluate and amplify the differences among people. Most treatments of the economics of the Comstock focus on the rich, analyzing the phenomenon of the common man made into a millionaire.[1] Scholars have not given as much attention, however, to the overall class structure of the Comstock, and in particular to the lowest end of the economic ladder.[2]

Mining can create fabulous wealth. That potential invariably attracts far more people than can actually become rich through the endeavor. Nonetheless, the Comstock manufactured millionaires almost as if they were the product rather than one of the by-products of the industrial machine. Even the first prospectors, who sold out for thousands, felt they possessed wonderful profit in a time when workers regarded $4 a day as a good wage. From the start, the district erupted with capital, affecting a wide variety of people. Indeed, many of the earliest investors, miners, and entrepreneurs became richer than they probably had thought possible. Most residents of the district, however, joined a wage-earning middle group that enjoyed the benefits of flush times. Excessive wealth eluded them, but home ownership and a good life were within the grasp of typical workers, making their existence better than that of their industrial counterparts elsewhere. At the same time, a few people failed on the Comstock, and they came to occupy the bottom rung of the economic ladder.

The mining district's children keenly recognized, but certainly also often misunderstood, the differences between the rich and the poor. John Waldorf remembers: "We children made the class line clear and distinct. If a boy gathered wood on the dumps, he belonged to the common people, which was my crowd. If he didn't, he belonged to the tony bunch, or wanted to. In any event, we looked upon him as altogether too stuck-up to make good company."[3] Collecting wood dumped from ore cars carrying rock and other debris from the mines could be both exciting and dangerous for the boys of Virginia City. Injury was always possible as contenders slid along the ever-growing heaps of mine refuse, dodging rocks dislodged as they struggled to reach the prizes. The boys often fought for the right of retrieval. Their efforts were more than sport, however. The wood they collected was an important means of heating many homes. Because underground water frequently saturated support timbers, adding to their weight, they often would not burn without first being stacked and dried. The whole process was an arduous task, but it was one that many Comstock youths pursued as they contributed to the upkeep of their families.

Waldorf correctly recognized the gathering of wood as a reasonable line of demarcation between children of upper-class and working-class families. The author also described a few adults who scavenged for wood. Some competed with the boys during the day, while others came to the dumps in the dark early-morning hours to glean the refuse that the night crews had left. Much like the children, adults recognized wood scavenging as an important class marker: working-class adults would not normally engage in the activity that only the most destitute pursued. Mary McNair Mathews, on the other hand, might have regarded child care as a better means of distinguishing the upper class. She frequently accepted $5 and a meal as compensation from society's ladies for baby-sitting their children while they attended parties.[4]

Unfortunately, the U.S. Census lists no people as professional babysitters. It does, however, record several people on the Comstock who claimed scavenging, in one form or another, as their occupation. In fact, the profession was likely quite common among the poorest adults as both a part-time and a full-time endeavor. These were the recyclers of nineteenth-century society, making certain that the community did not discard anything that still had a use. The 1880 census lists scavengers and ragpickers, who, while employed, were on the lower end of the distribution of wealth. The scavengers were all single Italian men between thirty-five and fifty years old. The ragpickers were all Chinese men whose ages ranged from eighteen to sixty years. Mary McNair Mathews, observing them in the 1870s, noted that the Chinese were helpful in the way they collected rags, bottles, and old cans. "The bottles they use," she wrote, "and the cans they melt in order to get the solder."[5] Presumably the Italians were collecting material for Henry Robinson, a fifty-five-year-old immigrant from Austria who operated a junkyard on South I Street, where he lived with his wife. Although the Chinese apparently operated their own businesses, they may also have sold to Robinson, since his facility was the most successful of its kind in the community.[6] There are also abundant sources that tell of the Chinese scavenging the hillsides around Virginia City for firewood. They were particularly noted for digging the stumps of trees, so they could then sell even that material to be used as fuel in Comstock stoves.[7]

John Waldorf describes children scouring the neighborhood for glass, metal, and cloth to sell to Robinson's junkyard. The Italians and Chinese had a great deal to contend with, competing as they were with an army of young people constantly on the prowl for the same opportunities. Robinson paid half a cent for a pound of rags, a rate that required considerable gathering if one were to make a living of the task. Although Robinson and the adult scavengers dealt in junk, there is no reason to assume that they were at the poorest end of the economic spectrum. The Euro-American community may have looked down on their occupation as undesirable, but profits were certainly available. From a child's point of view, Robinson "had a financial rating second only to Mackay and Fair," as Waldorf observes, but he based this evaluation more on youthful naïveté than reality.[8] Still, gathering the refuse of society was certainly a difficult, time-consuming activity.

In the same way, many may have looked down on the Chinese and the Northern Paiutes, but it would be wrong to assume that they failed economically. Both ethnic groups had their own, largely insulated means of gauging success, and economic and social standing in the Euro-American community was not usually an important factor. The American Indians, in particular, placed little value on having excessive wealth, regarding money as necessary on occasion and working only here and there to obtain it as needed.[9]

The Chinese, on the other hand, had a reputation for hard work and for saving. For them, financial success meant independence from harsh employers. It could also mean the ability to send money home to family in China or eventually to return there with wealth. Primary sources as well as the census manuscripts document a highly stratified Chinatown, with merchants, doctors, and other professionals, along with their wives, at the top and servants, laundry workers, and prostitutes at the bottom. An analysis of ages for various occupations of the Chinese confirms the economic hierarchy within the Asian community. Servants tended to be the youngest, many of them still in their teens. Laundry workers and cooks were typically in their twenties and were apparently better situated. Those who appear as laborers were usually between thirty and forty-five years old, and so presumably they had access to even more lucrative employment. Chinese professionals and gamblers were evenly distributed among age groups, suggesting that they relied on special skills rather than on seniority, and presumably their abilities afforded them some prestige.[10]

Almost without exception, however, Euro-Americans regarded a Northern Paiute, an Asian immigrant, or an adult scavenger of any ethnicity as occupying the lower rungs of the economic ladder. Nonetheless, these people were not alone in that status. Mining claimed hundreds of lives, leaving widows, many of whom had the responsibilities of raising children with little means of support. The same was true of those who ended their marriage in court or whose husbands abandoned them for other opportunities. Women without dependents could always become servants in the households of others, but those with children did not have this option, since employers were usually willing to house and feed only the hired help. A few lucky single women had enough property that they could take in lodgers.[11]

The poorest of the women who were left with children had few choices. Circumstance often forced them to work at home in order to care for their very young. If they lacked skills, they might, for example, resort to washing clothes for a living. Mary McNair Mathews tried that occupation for a short time and found it to be backbreaking labor with insufficient rewards. She quickly decided that sewing, nursing, and opening a private school were more lucrative.[12]

Besides the fact that cleaning clothes before the invention of the washing machine was physically taxing, there was a limit to the amount that women could charge for the service. The Comstock washerwomen competed with corporate laundries that employed dozens of workers and used steam-driven machinery that could process clothes quickly and efficiently. Hundreds of Asians also operated Chinese laundries known for their efficiency and low cost. These factors placed a cap on what women could charge, usually driving them out of the business at the earliest opportunity. The exception was the Irish widows and single mothers, who apparently could rely on their ethnic group as a guaranteed market. The Irish showed enough preference for their laundresses over the Chinese to encourage the women to remain in business, even allowing them to charge a little more for their service. In this way, the ethnic neighborhoods could subsidize the widows of men killed in the mines, protecting their fellow immigrants and creating a safety net that their own wives and families might someday need.[13] Comstock newspapers all too frequently reported on fund-raisers to help the widows and children of miners killed on the job. Fraternal groups offered some aid, but at least a few families found themselves in a desperate situation as they faced the need for survival with few skills and little property.

Even some families that continued to have a man in the house faced hard times, particularly during the periods of economic decline. During the bonanza periods, free-flowing money meant that earning a livelihood was usually not a problem for most people. During times of depression, however, an increasing number of people might find it difficult to survive. Mining means being a transient. The trick is always to leave at the right time. Leave too soon and you might miss an opportunity to exploit the next boom; linger too long and you might be caught in the worst of the bad times and miss the next boom elsewhere. The 1880 census, the first to record both months of unemployment and location of residence, reveals a pattern of increasing gloom as the community slipped into depression.

On the mountainsides above Virginia City, there remained cabins, many dating from the early 1860s, built at a time when it appeared that the community would drift uphill. As it turned out, the Comstock Lode plunged down the mountain slope, shifting the town with it and leaving abandoned property at the higher elevation. As more people became unemployed, many retreated to these humble homes where they could live without paying rent, waiting for the next boom. The promise of employment and the fact that the first on the scene usually won the best jobs inspired the gamble. Still, the 1880 census of the hillside cabins reveals dozens of impoverished laborers squatting rent-free with their families, often enduring months of unemployment.[14]

The downward slope of Virginia City, with the Combination building in the foreground, 1878. Photograph by Carleton Watkins. (Courtesy of the Nevada State Historic Preservation Office)

The poorest prostitutes rivaled the widows and the unemployed in their confrontation of poverty and desperation. These were the crib dwellers who lived on the line, in contrast to the better prostitutes, who worked in brothels and could discriminate among customers while also receiving higher wages. The poorest faced an emotional and economic low point. Marion Goldman in her monograph on Comstock prostitution estimates that roughly half of the women engaged in sexual commerce were what she called "working women." These were the common prostitutes who usually sold only sexual contact, unlike the higher-class brothel workers who spent more time entertaining men in parlors than in the boudoir and who earned more for their efforts. The working women were usually alone. For the most part, they had a poorer, less-refined clientele and served far more customers each night to make a meager living. Violence, alcoholism, drug addiction, and suicide were common among the residents of the cribs.[15]

Even the women of the line, however, did not meet the desperation of those whom the 1875 state census labeled "prostitutes of the lowest order."[16] These were vagrants whom age, disease, drug use, or a combination of afflictions, had driven into squalid living conditions. They no doubt offered sex for money when they could find a customer, but those occasions did not offer consistent employment. There are abundant accounts of how these women ended their hopeless lives by their own hand.[17] Some Asian prostitutes faced an even worse situation. Many had been sold to Chinese export companies for sexual commerce in North American Chinatowns. These slaves served masters who owned all proceeds of their labor. Their only hope was for someone compassionate enough to purchase them for marriage and end their slavery. Indeed, this did occur on occasion, but for most, it was a dismal existence.[18]

Besides the scavengers, the widowed laundresses, and the lowest of prostitutes, there were also those whom illness, injury, age, or other factors left with almost no means of support. The record of Mary McEntire from the 10th U.S. Census quoted above documents at least one of the impoverished. She was a divorced Irish immigrant, fifty-six years old, who could not work because of a disability. In 1880 McEntire apparently lived at the county hospital, where the local government cared for her and another immigrant, an older man named John Bagley, also from Ireland. Yet another pauper, this one from France, lived in Gold Hill, but he was only willing to give the enumerator the name of Joe Doe.[19]

Waldorf recalls "Uncle Jimmy" Anderson, who was too old to work. A minister had once stolen his life savings, leaving Uncle Jimmy to the kindness of others. Storey County placed him on its list of indigents so that he could receive $20 a month to keep him from starving to death.[20] In contrast to Uncle Jimmy, "Old Mother" Dildine may not have received assistance. As Nancy Nelson she moved to Virginia City in 1860 and ran boardinghouses until she married Abram S. Dildine in 1868. Their relationship was troubled, however, and she left him, retreating to the hills, where she lived in a cabin to tend a herd of goats. Dildine shunned visitors and became increasingly reclusive. In 1878 the *Territorial Enterprise* described her place "as a hut, where she keeps a few chickens and where she has a small stock of provisions." The newspaper also reported that "she makes temporary camps among the rocks. Her dress is simply a long sack, with holes cut in it through which to thrust her head and arms."[21] Whether Dildine was poor or merely deeply disturbed is unclear, but there can be little question that she, like many others, lived in an impoverished setting in one of the world's richest mining districts.

The poorest end of the Comstock economic spectrum included others as well. Illness and injury claimed the lives of many parents, leaving their children as orphans without means of support. Circumstances forced these youngest of the poor to rely on the Daughters of Charity or on the state for food and shelter. The census manuscripts document some of the hundreds of girls who depended on the Daughters for care from the 1860s to the 1890s. Eighty children appeared in 1870, 118 in 1880. The few young boys listed in the orphanage without exception appeared with at least one older sister; orphanage rules usually required that boys go to Carson City as wards of the state.[22]

There were, of course, examples of people rising from the depths of poverty. Richard Jose, for example, was a widow's son from Cornwall who fended for himself in Virginia City, trying to augment his mother's income by singing on the street. He eventually emerged as one of the nation's best-known tenors, famous for such favorites as "Silver Threads Among the Gold." He became a wealthy man who could clearly maintain that he had created his own success.[23] The story recounted earlier of John Mackay and his partner, Jack O'Brien, arriving in Virginia City with no money became the basis of one of the more famous local rags-to-riches tales. Still, such triumphs were the exception, and most people found it difficult to break out of a downward spiral. Mary McNair Mathews told of establishing a soup kitchen for the unemployed during times of depression, maintaining that she was able to feed hundreds.[24] The county also stepped in to help the poorest, and the Daughters of Charity and other groups were able to assist the needy. It was certainly a loosely woven safety net, but its mere existence shows some sense of compassion just as it serves to document the nature and scale of the poverty that existed in the midst of immense wealth.

While there was clearly destitution on the Comstock even during the best of times, most people belonged to the working class, and many others were affluent merchants and professionals who found life in the mining district quite lucrative. An examination of the distribution of wealth in the 1870 census demonstrates that while most people did not declare either personal or real estate, those who did were doing reasonably well. Of the 8,312 adults twenty years or older in Storey County, 2,109 gave the enumerator an assessment of wealth.[25] Figure 10.1 shows the distribution of property for both men and women. It documents that men of progressively older age did well, but those who were fifty or more appear not to have had either the ability or the inclination to be as prosperous on the Comstock. The distribution of assets among the women does not fall as neatly into patterns, perhaps in part because of the limited population. Women did not tend to declare accumulated wealth in the same way that men did: personal estate shows little variation, indicating that women were not able to accumulate moveable property as easily as men. It may also have meant that they were less inclined to claim ownership, presumably often in deference to their husbands. Real estate, on the other hand, shows a dramatic increase for women as they became middle-aged. A review of women with the most assets in this age group shows that they

All sorts of occupations made the Comstock a complex, diverse place. This organ grinder earned his living on Virginia City's C Street. (Courtesy of the Nevada Historical Society)

were almost all lodging house operators, hotelkeepers, or brothel owners. These women faced circumstances that encouraged them to acquire property for businesses. While there is a tremendous difference between engaging in prostitution and operating a boardinghouse, there is also a continuity in the two uses of real estate. The owners of both, after all, acquired profit by renting rooms, the one to lodgers and the other to women who operated their business out of that establishment. On a practical level, both kinds of property owners pursued similar strategies. The handful of laundresses and dressmakers who also declared larger amounts of real estate complete the statistical picture, providing diversity and demonstrating that there were several options. Still, there can be little doubt that most women who acquired wealth on the Comstock did so by

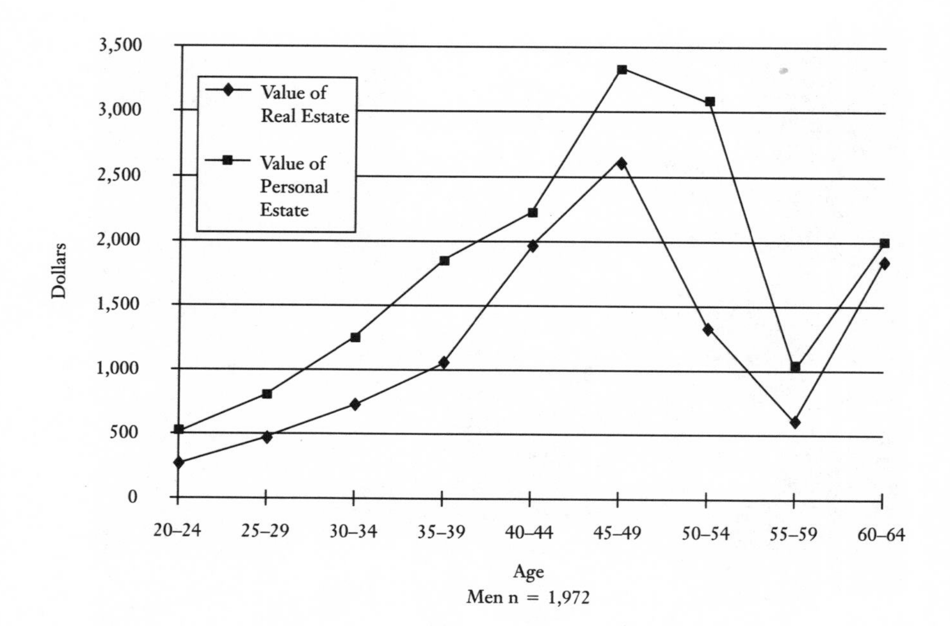

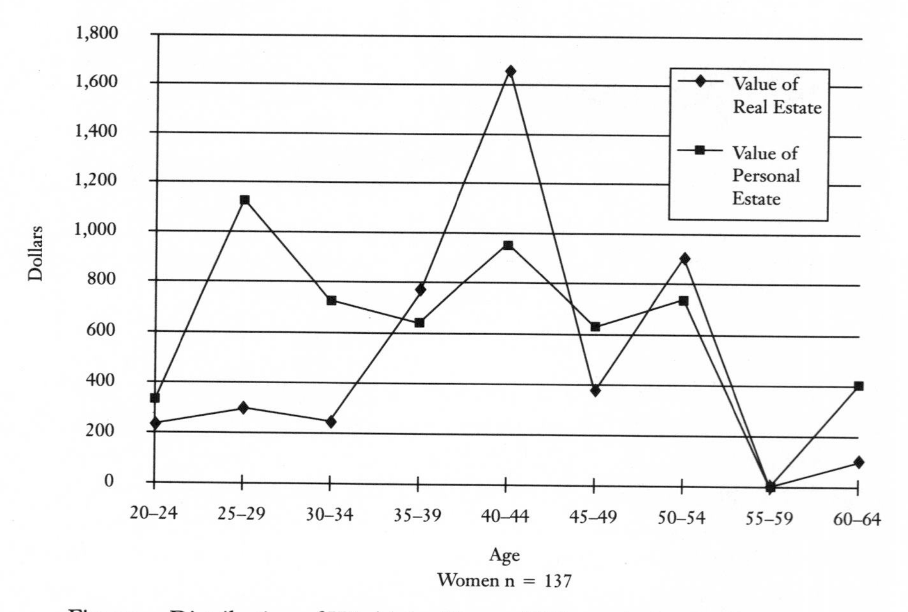

Fig. 10.1. Distribution of Wealth by Men and Women in Storey County, Nevada, 1870. Note difference in scale. (Source: 9th U.S. Manuscript Census of 1870 for Storey County, Nevada. Averages are only of those who declared wealth in either category.)

managing the place they lived in and owned so that its rooms produced profit.[26]

Society was capable of stratifying women by assessing the value of their real estate and the nature of their occupation, but more often it passed judgment based on the economic status of their families. A husband's profession, for example, usually determined the place of his wife. Still, the mining district had a quality about it that blurred economic station. Louise Palmer, who published a review of Comstock society in the *Overland Monthly* in 1869, was the wife of an important member of the community, and she apparently would have preferred to see people pay her proper deference. Western culture, however, with its strong undercurrent of egalitarianism, did not bow to her wishes. Palmer wrote that "ladies may envy me for living in [a] stone mansion . . . but nevertheless they are not ashamed to receive me in their cloth and paper dwellings." Interaction between the classes apparently was freer on the Comstock than in the East or in England.[27]

One of the best expressions of a woman's status has always been dress, but again the Comstock made this diagnostic tool less useful. Mary McNair Mathews commented on how she had "never lived in a place where people dressed more richly or extravagantly than in Virginia City. It is not only a few millionaires who indulge in it, but every woman on the Comstock who has a husband earning $4 or $6 a day, up to the superintendent, who gets his $500 or $1,000 a month, as some of them do who have two or three mines to look after. And many families live up to every cent of their wages or salary." This meant that women of less means often dressed as though they were affluent. At the same time, the mining West bred a freedom that further blurred the lines of economic status. Again, Mathews observed that when she first visited the house of a wealthy merchant, the woman of the house opened the door "wearing a black and white calico wrapper [and] . . . being dressed so much more plainly than the rich and gaudily-dressed ladies I had met with at other houses, I took her for the hired help instead of the lady herself." On the other hand, the Comstock remained small enough, with only a few stores, that it was possible to evaluate the cost of one's peers' clothing. Palmer wrote: "It is curious that we all protest against lunch parties, yet continue to give and attend them. It is stupid to dress in one's newest silk, and handsomest corals to partake of chicken, creams, ices, and Champagne, with a dozen of one's own sex. Who can help being painfully conscious that each and every one of them have priced the silk at Rosener's, and the corals at Nye's, before they came to one's wearing. It is a trying thing for one's dress to be subjected to the test of its value, not of its adaptability, or becomingness."[28]

The Comstock sewing trades exhibited a stratification that was as well defined as the apparel they produced. Milliners were specialists in producing women's hats. By necessity, they maintained a sizable inventory and commanded premier status in the industry. Dressmakers followed, having the training and skill to cut patterns and produce dresses from cloth. Next in line were the seamstresses and sewing women who only stitched seams and did not cut patterns. A recent study of these occupations demonstrates that while women were able to support themselves with their needles, only the most skilled could expect to gain some affluence for their efforts.[29]

The occupations of men who declared wealth in 1870 are much more diverse than those of women. Men with the least property represent a broad cross section of the population, with an ample representation of miners and other workers. Those who told the enumerator that they had more property tended to be merchants, professionals, and other business owners. Still, a number of carpenters and teamsters, a "Daguerrean artist," a coffee vendor, and the city tax collector were able to declare between $5,000 and $9,999 in personal estate. The professionals and the few workers who used tenacity, luck, and shrewd choices accumulated enough property to represent a fairly substantial middle class, and some of these became affluent. Abram M. Edgington, for example, began his Comstock life in the early 1860s. He was a native of Ohio and quickly found opportunity, serving as a deputy sheriff for Storey County. In 1865 he became a deputy internal revenue assessor. He also served as an accountant for the Morgan Mill in nearby Ormsby County. By 1867 his experience and expertise led to his assuming the prestigious position of superintendent of the Union Mill and Mining Company. Edgington had ascended the social and economic ladder. To prove this, in 1872 he built a house for his new wife, Mary Bailey of Gold Hill. It was an elegant three-story French Second Empire structure perched on South B Street in an area that would later be called Millionaire's Row. Whether Abram Edgington ever did become a millionaire is unclear, but there is no question that he managed to maneuver himself into progressively better positions and, with wise choices, became affluent. His social status was such that his stepdaughter, Lillie Edgington, was able to marry one of the wealthiest merchants in Nevada. Mark Strouse, a local Jewish butcher, had exploited the opportunity of the Comstock himself. With financial acumen, he combined terms as city treasurer and chief of police with business interests and made a small fortune. He and Lillie were married in St. Paul's Episcopal Church in front of a crowd so big that it spilled into the street. By participating in a ceremony in the Episcopal church, often regarded as the pinnacle of high society and economic standing, the young couple asserted their position in the community.

When Abram Edgington died in 1875, his family had a sizable marble headstone erected on his plot in the Masonic cemetery. Lillie supplied the footstone. Her mother went on to Butte, Montana, where she built a hotel. Lillie and Mark divorced in 1878 over her alleged indiscretions, and she went to San Francisco to pursue interests in the theater. Mark also relocated to the Bay Area and opened a new business there. The Edgington family was typical of the Comstock's wealthy merchants. Abram, as a Protestant Freemason born in the United States, was able to take advantage of opportunities and contacts. With tenacity and hard work, he established his place of prominence in the community, affording his daughter the chance of an economically advantageous marriage as well as the luxury of divorce and the arts.[30]

Not all the affluent were merchants and non-mining professionals. There was also a group of mining professionals who, although they never became bonanza kings, nonetheless ascended the economic ladder to local prominence. Frank F. Osbiston, for example, turned his Cornish ethnicity and his ability in the mines into a succession of mine superintendent jobs. Virginia City recognized his contribution by naming one of its prominent shafts after him. Through shrewd investments, Osbiston amassed interests in mining properties from Colorado to Australia, and while he remained on the Comstock he was one of the more important members of its upper middle class.[31] The case of Osbiston contrasts with that of Philipp Deidesheimer, whose brilliance in engineering gave the region square-set timbering and made exploitation of the Comstock Lode possible. In spite of the opportunity his newfound fame afforded him and regardless of the fact that mining companies repeatedly offered him employment in prestigious positions, the German-born engineer made poor choices in his investments and was never able to ascend beyond the rank of wage earner.

As in any community, the Comstock had its economic winners and losers. Those who were able to succeed represented a fairly broad cross section of society, a consequence of a new, less-rigid hierarchy where there was considerable opportunity. Predictably, there were many Euro-American males among the professionals and merchants, but there were also others who did well. Dr. Song Wing and his wife, Choney Wing, appear living near Dr. Son Haong and his wife, Pooty Chin, in the 1880 census. Both couples were affluent by Chinese standards and likely lived in the most prestigious part of Chinatown.[32] The Yee family, through its ownership of Quong Hi Loy and Company at 100 Union Street, were able to parlay the Chinese yearning for imports into a family fortune. Other Chinese immigrants did equally well, as did many others who were not Euro-American males. In spite of all her complaints about economic setbacks, Mary McNair Mathews found her Comstock experience profitable. Hard work and intelligent choices allowed this widow and single mother to amass property and run a lucrative business. Amanda Payne, an African American who was mentioned earlier, was probably the daughter of slaves but eventually acquired Comstock saloons, a boardinghouse, and a popular restaurant.

Map 4 of the former Barbary Coast illustrates about 100 yards of South C Street in Virginia City as it appeared in 1890. The community had changed considerably since the heyday of the Comstock, but much remained the same, and within this block, still active at the time, the echo of the 1870s is clear. The area included both commercial establishments and dwellings. Shops sold tinware and kerosene, meats, groceries, baked goods, shoes, liquor, and furniture. Separate establishments offered hats, one catering to men and the other to women. A Chinese laundry, an upholsterer, a cobbler, a seamstress, and a tailor provided services. There were four saloons in the area, a remnant of the neighborhood's once torrid reputation. There was also a stable, as well as three lodging houses, one of which was large enough to boast a restaurant. The stable sold feed and kept a henhouse out back, providing the neighborhood with eggs and chickens. The meat shop operated its own smokehouse. There were several other sheds and outhouses scattered throughout the area.

If this block ever served as home for a bonanza millionaire, nothing in the humble nature of the buildings suggests it. Instead, map 4, like so many of the Sanborn-Perris Fire Insurance Maps upon which it is based, documents a community where thousands of people lived the mundane, meeting the challenges of earning wages and shopping for food, clothing, and other necessities. The neighborhood, like the rest of the Comstock, offered its own entertainment, and life could be pleasant, but the glaring brilliance of the silver kings, which often blinds those who look back on the mining district's history, rarely cast a glimmer on South C Street.

In the same way, visitors to the mines often remarked on the egalitarian nature and leveling effect of work underground. Miriam Leslie noted in 1877: "The mines embrace every class of men, socially speaking, from the lowest grade of laborer to the ex–United States Senator, or man of title." One of the reporters for Leslie's husband's newspaper pointed out: "Titles and professions also find a place in the Comstock. There is the scion of a noble English house, in whose veins runs the blue blood of Norman earls, while French and Spanish counts are not singular." Still, not everyone on the Comstock came seeking the anonymity of the working class. There were also people on the Comstock who retained family wealth and sought to reproduce the rigid class structure of their eastern homes. These would-be aristocrats provided the capstone of society's economic pyramid.[33]

Only two women appear in the 1870 census with $10,000 or more in either real or personal estate. Alice Baldwin, a twenty-six-year-old housekeeper from Massachusetts, had $15,000 in personal property but claimed ownership of no land. No husband was recorded in her household, which included her seven-year-old son. Anna Wiley, a widow with two children, was a forty-year-old boardinghouse operator who claimed $10,000 in real estate and $5,000 in personal estate. Women in general either had or claimed less property than men. A wide assortment of people were able to do well in the Comstock economy, but all things were relative and barriers existed that ordinarily prevented women from becoming members of the bonanza rich. Marriage remained the most likely means for a woman to gain access to the millions of dollars that the mining district offered.

An exception to this norm was Eilley Orrum, a boardinghouse keeper shrewd and lucky enough to obtain several feet of Gold Hill's richest ore during the first Comstock boom in 1859. Marriage to Sandy Bowers, owner of the neighboring claim, helped her finances, of course, but much of what she had, she won by herself. Similarly, she could take much of the credit for squandering the fortune through mismanagement and extravagant expenses. After Sandy died in 1868, Eilley Orrum Bowers wandered the Comstock and the West, marketing herself as a fortune-teller. She died in poverty in 1903.[34] Her early success owed much to the freewheeling nature of a society without the established conventions that would later restrict the ability of a woman to strike it truly rich. By the early 1860s, however, enough of a traditional society was in place to limit women's access to wealth.

None of the fabulously rich people in the 1870 census for Storey County were women. Only eleven men who appear in that document listed $50,000 or more in personal property or real estate. Of these, three were miners and another was a mill proprietor. The remaining seven, including three bankers, a merchant, a toll road operator, a real estate dealer, and a newspaperman, were part of the community infrastructure. Some were more directly linked to mining than others: William Sharon, for example, had amassed hundreds of thousands of dollars while on the

Comstock to manage the Bank of California's interests in mining and milling.

All but two of these eleven men were born in the United States. Their birthplaces ranged from New York and Pennsylvania to Ohio, Kentucky, and Virginia. There were none from the Deep South or the northern New England states. Only two of the richest miners were from Ireland. The distribution of wealth does not correspond to the community's overall statistics for nativities. Even though most of the rich men were born in the States, roughly a quarter of all adults in Storey County in 1870 were born in Ireland, and fully two thirds came from someplace other than the United States.

This, of course, was a decade after the initial strike, long after society's conventions had crystallized. In the earliest days, the range of possibilities was wider. The same wide-open community that afforded opportunity to a woman such as Eilley Orrum Bowers placed little restriction on most other groups. While the door was shut from the beginning to Asian involvement in the mines, Mexicans could join North Americans and Europeans in claiming a part of the wealth of the Lode. It was in this fresh environment that the Maldonados struck it rich, exploited their mine, then sold out and left before the property turned worthless. The early 1860s gave the Hispanic community a unique window of opportunity that was not to be duplicated later.[35]

Many of the area's successful early entrepreneurs were well known in the community and appear in the histories of the area. H. S. Beck was a prominent merchant whose first Virginia City store appeared in the 1861 bird's-eye view poster of the community.[36] Dr. D. M. Geiger amassed a fortune by operating the toll road leaving Virginia City to the north. Philip Lynch of Gold Hill was a well-known newspaper editor who owned interests in profitable properties. It is a story repeated many times over. Most of the successes were associated with people who invested in community infrastructure: of the wealthiest men appearing in the 1870 census, only two could claim to be bonanza mining kings. Significantly, however, no one else claimed $100,000 or more in both personal and real estate. William Sharon, representing the Bank of California and the first Comstock monopoly, was clearly one of the fabulously rich. James Fair was wealthy in 1870, but his star was still rising. It would be three years before he and his Irish business associates would strike the big bonanza, propelling their finances to unimagined heights.[37]

The wealthiest of those on the Comstock were an elite group who won the equivalent of a nineteenth-century lottery. Talent and previous status could affect how well one was able to benefit from the district, but good luck was the most important factor for most as they ascended the economic ladder. A willingness to manipulate the economy to the detriment of others could also be a valuable attribute as one scrambled for success.[38] Having made millions, the Comstock princes and their behavior spoke volumes about the nature of their society. With only rare exception, those who made their fortune left the Comstock. They became the carpetbagger rich, interested in stripping the area of its wealth and then leaving for California, the East, or Europe. Even when John Mackay, richest of the bonanza kings, insisted on remaining most of the time in Virginia City to tend to his mines and his property, his wife fled Nevada, preferring the high society of New York, London, or Paris. Others, including James Fair and John P. Jones, used their wealth to persuade the state legislature to reward them with Nevada's seats in the U.S. Senate. Between living in Washington, D.C., and visiting California, they rarely appeared in the state they represented.

One exception to this was William M. Stewart, the lawyer who became a wealthy man during the early Comstock lawsuits over the multiledge controversy. Not only did he accrue a financial empire, but he emerged with considerable political power. Like so many of those who won riches in the mining district, Stewart was able to influence the course of Nevada politics. Whether these people left for California or remained, many continued to exert control over what happened in the state. After winning the passage of the 1864 state constitution, Stewart won a seat in the U.S. Senate. While representing Nevada, he affected national policy regarding railroads and, of course, mining. His 1872 National Mining Law provides the legal structure for the industry to this day. He fit in well in Washington, even renewing old western friendships with people such as President Ulysses S. Grant. Nonetheless, Stewart was always drawn home to Nevada.[39]

Still, most of the Comstock's princes sought to leave the area and invariably built their mansions elsewhere. The Virginia City houses that residents commonly referred to as mansions are humble by comparison to what the rich built in other places. These were functional abodes, fashioned slightly better than others to communicate success and to provide comfort while staying in the mining town. California came to have most of the true mansions. James C. Flood, for example, built a gargantuan architectural beast known as the Wedding Cake, in Menlo Park, and he constructed yet another mansion in San Francisco.[40]

One of the more significant exceptions to the tendency to build outside the state was the Bowerses, who chose to build their mansion in Washoe Valley, not far from the mines. Although modest in the context of other mansions, their home was substantial. New to the world of wealth, they apparently missed the word that a proper member of the nouveau riche built in California or elsewhere but never had a primary residency in the Nevada outback.[41]

John Waldorf recalls a time when his father's investments in stocks were rising and the family was convinced that they had managed to secure a part of the big bonanza for themselves. "I was a real king," Waldorf remembers. The family's hopes to abandon Virginia City and buy a farm in the East with their new wealth reveals much about the community as a whole. A stock crash dashed all plans, and like most, the Waldorfs were not able to leave the Comstock in style.[42] It does not bode well for a society when people hope that once they have money, they can go somewhere else to fulfill their dreams. Regardless of whether they went or not, those who held this attitude set a tone that would dominate Comstock and Nevada society for decades.

The migration away from the mining district, both the real and the dreamed of, gave its towns an unfinished, impermanent aspect. In spite of the optimists who believed that Virginia City would prosper for decades and in spite of those who sought to build a community, the prevalent wish to leave with prosperity in hand tainted the place. It reinforced an aspect of the Comstock's self-image that could be characterized as colonial, making it a region to be exploited but not truly settled. The ramifications of these imagined riches were at least as important to the nature of the society as was the actual distribution of wealth.

The idea of prosperity also affected the way the community presented itself. Again, image shaped the place as profoundly as reality. Although most of the Comstock's wealth left the district, people still attempted to make their homes opulent, in keeping with its reputation. The International Hotel, particularly in its third incarnation after the 1875 Great Fire, was one of the grandest structures west of the Mississippi. Besides boasting a well-known restaurant and saloon, it featured the first hydraulic hotel elevator in Nevada.[43] Indeed, local folklore maintains to this day that it was the first such lift west of the Mississippi. Reporters for *Frank Leslie's Illustrated Newspaper* stayed at the hotel shortly after its 1877 reopening. They found it to be "a new and handsome brick building. . . . Its newness is apparent as soon as we cross the threshold, so fresh and spotless is its white paint, so immaculate all its appointments, from the flaming red velvet and snowy lace curtains of the parlor to the least accessory of the large dining-room."[44] Such displays of wealth were not restricted, however, to just one place. All along C Street, business owners attempted to manifest the riches of the famed Comstock Lode. The Molinelli Hotel, two buildings to the north of the International, continues to boast finely appointed rooms that have given many guests a pleasant stay. The hotel still has iron-railed, oval openings in the floors of its halls to allow natural light to filter down to the lower levels in a most distinguished way.

Similarly, the numerous saloons that continue to operate along C Street to this day preserve declarations of wealth in the form of crystal lights, mirrors, and elaborate walnut backbars. Of these, the Washoe Club was one of the more direct attempts to communicate the financial achievement of the area. Situated above an opulent two-bit saloon, it was a private retreat for Virginia City's wealthiest people. This exclusive institution catered only to the richest of the mining district, providing a lounge for drink, cards, or simply conversation with one's peers.[45]

Virginia City's domestic architecture also demarcated differences in class and communicated wealth even though this was not a community filled with palatial estates. The streets directly above the commercial corridor exhibited house after house that boasted grand columns, elaborate woodwork, towers, spacious parlors, and rich appointments and furnishings. Perched on the higher slopes of Virginia City, these dwellings fea-

Virginia City, ca. 1880. (Courtesy of the McCarthy Collection, Nevada State Historic Preservation Office)

tured eastern windows that looked out on a grand view of Six Mile Canyon and the Carson River beyond. On a clear day, it was possible to see the Toiyabe Range, about two hundred miles away.[46] Although the custom in working-class neighborhoods was to build the houses cheek by jowl, the affluent could afford greater space between structures, resulting in side yards that, though not enormous, did give occupants more privacy. As a further expression of wealth, the homes of Virginia City's economic elite were often more square than rectangular. Working-class houses were typically longer than they were wide, maybe around 30 feet long and 20 or so feet wide. With neighbors close by, each house had little street frontage.[47] The larger domiciles of the rich dominated more of the street, and builders made little effort to economize by extending the structure back into the block.

Even in death people found ways to declare their wealth and assert their place among the classes. The grandest grave markers are invariably on the highest ground, a clear indication of wealth and elevation. Rich people sought to guarantee themselves the best view even during their eternal repose.

Certainly the Comstock was not unique in its expression of differences in economic standing. What touched it more than many other places, however, was the realization that the mining district produced fabulous amounts of wealth. With luck and shrewd choices, one could ascend the economic ladder and join others who also had newly made fortunes. The nation's aristocracy might shun those who had only recently soiled their hands with physical labor, but on the Comstock, striking it rich could mean acceptance among the local upper class. The magic of a place where fabulous wealth was within one's grasp affected the outlook of its citizens, shaping the community's point of view and its folklore about itself.

11: OVER TIME

Bonanza and Borrasca *(1877–1942)*

After celebrating . . . the golden anniversary of the Comstock for three days, Virginia City last night closed the carnival.
—Reno Evening Gazette, *July 6, 1909*

It is heartening in the midst of obvious decline to know how glorious the past has been.
—Dorothy Nichols interview

On September 1, 1878, Adolph Sutro's workers completed the last of the three-and-a-half-mile tunnel, breaking into the Savage Works at the 1,640-foot level. The sudden source of ventilation and the difference in air pressure in the two chambers released a rush of hot, putrid air into the Savage, and the fierce wind persisted until the pressure equalized. Such are the peculiar dynamics of underground mining. Lord estimates that the cost of the project was more than $2 million.[1] Clearly, the Sutro Tunnel was yet another dramatic technological achievement of the Comstock. Still, it came too late, and substantial profits would remain elusive. Always the visionary, Sutro recognized the situation early and quietly dumped his stocks while prices remained high. The mining entrepreneur retired to San Francisco, where he lived in luxury, became mayor, and pursued such monumental projects as the building of the Sutro Baths.[2] Ironically, Sutro's most impressive Comstock achievement, his excavation, pierced the Lode after shafts had probed much deeper than the point of intersection. The mines still needed pumps to raise water to the point where it could flow down the tunnel.

Of more profound significance, however, was that the big bonanza was ending. A subtle economic chill began to descend on the Comstock in 1877. Even before that time, most of the mines had not been profitable for months or even years. They relied on stock assessments and other supplements from foolish investors. As long as Virginia City captured headlines and imaginations, however, euphoria dominated the district. Optimism drove the underground search elsewhere for further successes. The spectacular nature of the 1873 discovery masked the failure of other claims for a while, but eventually a different reality was destined to rise to the surface.

The big bonanza had also attracted more workers to the district than could find employment. Even in the most prosperous times of the mid-1870s, the economy was not growing. In early 1877 word began to spread that the end of the flush times was in sight. Mackay's firm had sufficient ore to remain profitable for a few more years, but the idea that miners were about to exhaust the richest concentration of gold and silver ore sent ripples through a community that thrived on hope. Some of the more marginal operations slowed or stopped work. Unemployment rose steadily. The mines, as well as other parts of society, felt the effect of economic decline: the construction trades, for example, depended on growth and new development, and so even the anticipation of depression was devastating for the many carpenters, masons, joiners, and others.[3] With every job lost, there were fewer dollars circulating, and so restaurants, saloons, stores, and other businesses were continually finding themselves with one less customer. It was a slow downward spiral, and in such a single-industry economy, nothing could step in to provide balance during the moribund decline.

Still, there was always the hope that the next bonanza was right around the corner. The Comstock had seen fluctuations before. The faint of heart had left, only to miss the opportunities of the next boom. Conventional wisdom led the experienced people on the mining frontier to view mining booms with skepticism, but the Comstock had proved itself different. In the past, *borrasca* had not led to complete failure. Repeatedly, one mine or another had come upon the next rich body of ore just in time to save the district. The Comstock's treasures, while occasionally elusive, seemed inexhaustible. History had shown it was appropriate to maintain faith in the district. Optimists felt justified in believing that the next big discovery would prove the folly of leaving. In 1882, well into the decline, a regional business directory could note: "When 'stocks are down' it will be observed that Virginians do not hang their heads like devotees of Dame Fortune in other towns less elevated. The atmosphere will not admit of it. A little growl, that hardly suggests complaint, is all that is to be heard, while good humor and pleasant sociability prevail."[4]

This was not a fickle boomtown population, ever ready to lose faith and switch attention to the next strike. It was a stable community supporting an industrial giant. Considerable investments, personal and corporate, emotional and economic, inspired most to cling to hope. Ultimately, however, that optimism proved ill-founded. From a historical perspective, it is now clear that ore production was destined to continue its decline. With every month, more and more people gave up hope and left the Comstock, thus eroding the district.

Events occurred, however, to cloud the contemporary perspective, providing evidence to bolster the faith of the optimist. In November 1877, miners in the Ophir discovered a vein at the 1,900-foot level. Initial signs indicated a situation reminiscent of the Consolidated Virginia bonanza. Further exploration demonstrated that the analogy was false. Nevertheless, miners found that it grew into a small but rich concentration of ore, between the 1,900- and 2,200-foot levels, which improved the economy. The discovery was profitable on a limited scale for the next two years, yielding 20,000 tons at $66 each. Unfortunately, it was not large enough to deter many of the unemployed from eventually abandoning the district. As had occurred before, the predictable cycle of escalating stock prices, driven higher than their real worth and then followed by a crash, disheartened even more people.

In late 1878 a rich discovery in Bodie, California, to the south, attracted as many as a thousand Comstockers. Others went elsewhere.[5] The famed Cousin Jacks, now boasting Comstock credentials, went everywhere from Idaho to Colorado to Australia, looking for new mining opportunities.[6] Many of the Irish went to Butte, Montana, where they were known as "hot water plugs." This title of distinction recalled the searing steam, erupting unpredictably from Comstock clays, that made the deepest mines places to be feared. Butte miners often shortened the name to "hot waters" or simply "hots." They became some of the most respected of the miners in the newly opened mines of Montana. The Comstock Irish traveled there by the hundreds, inspiring one Butte resident to write a friend in Virginia City: "The hotwater people stand in very good, and all are working. Joe Nevin is foreman of a mine here and all of the Virginia boys are working in that mine. I see in . . . [the Virginia City] papers that Sandy Sullivan is coming up here, and a few others, mostly all of [Virginia City] will be up here soon."[7]

In spite of those who left, Virginia City still had some life in it, and there were times when the Comstock silver could be polished to its old shine. The arrival on October 27, 1879, of former president U. S. Grant and his family created considerable excitement, distracting the mining district for a while. Grant and his wife, son, and daughter-in-law were conducting a two-year worldwide tour,[8] and Virginia City spared nothing to honor the visit. Its buildings draped with bunting and its streets filled with parades, the town took on the appearance of an Independence Day celebration. Grant and his party stayed for three days, enjoying banquets and all the entertainment that the Comstock could muster, including a dignitary-led tour of the Consolidated Virginia and California Mines. Adolph Sutro received them at his mansion in Sutro City, and then the party traveled back to Virginia City along his tunnel.[9]

A year later, President and Mrs. Hayes, together with several officials including the secretary of war and famed Major General William T. Sherman, visited the Comstock.[10] Again the district celebrated, receiving the dignitaries as they had the Grant party. Both groups, like so many other famous people and the not-so-noteworthy, came to see the mighty Comstock, home of the big bonanza. If they got any inkling of the shadow that hung over the mining district, it probably mattered little to them. Virginia City was a "must-see" stop on any excursion through the region, and its economic trouble was a private illness, of only minor interest to tourists. Comstock contemporaries hoped that the district would continue to warrant such visits, but without realizing it, the two presidents were actually commemorating accomplishments of the past rather than anticipating achievements of the future. In 1881 the Comstock produced only $1,075,600.[11] Historian Grant Smith points out that the market value of the mines dropped to less than $7 million, losing $293 million in six years.[12] Further adding to the calamity, the celebrated Consolidated Virginia and California caught fire underground on May 3, 1881. Miners sealed the area, hoping that they could starve the blaze of oxygen, but clearly no further work there would be possible for a long time.

A subtle but no less profound indication that the Comstock had changed appears in the 1880 federal census records. Figure 11.1 shows a population pyramid for Storey County. The dominance of working-age men is characteristic of the West, especially of the region's mining towns. The fact that it was not as dramatic in 1880 as it was earlier serves as evidence that the Comstock had matured in its twenty years, with women and families settling in significant numbers. An expression of economic decline, however, occurs in the statistics for the fifteen-to-twenty-four age group. These young people included more women than men, an anomaly that hints at changes within the economy and the community. A detailed analysis of unemployment trends and of where these young peo-

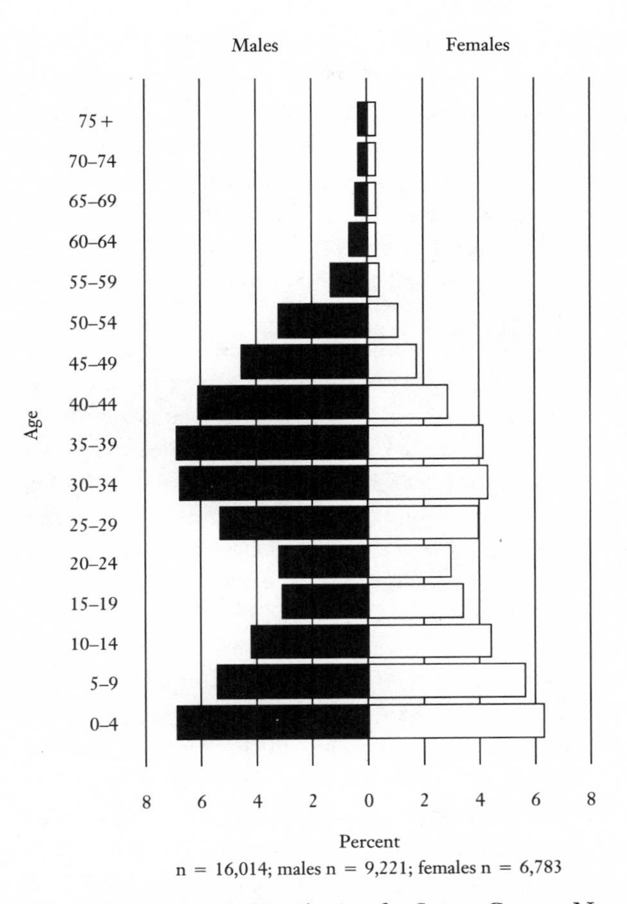

Fig. 11.1. Population Distribution for Storey County, Nevada, 1880. (Source: 10th U.S. Manuscript of 1880.)

ple were living suggests that many found it difficult to secure work in the depression economy. Older men were the last to be dismissed, and those beginning their careers were finding it increasingly difficult to find new jobs. Young women, on the other hand, were more inclined to remain with established families. Proper young ladies, after all, would stay with their parents rather than leaving for the next economic opportunity. In addition, the thousands of bachelors meant that single women could find husbands if they wished. Those who did marry probably favored those with jobs, and so they too remained on the Comstock longer.[13]

A concrete sign that the boom was over occurred in July 1880, when the owners of the Virginia and Truckee Railroad decided to tear up the tracks on the spur to Silver City, salvaging the material for the Carson and Colorado Narrow Gauge. The *Virginia Evening Chronicle* carried an article on the turning point, using the headline SWITCHING OFF A CITY, and indeed there was a finality to the event that seemed to declare that the Comstock's fortunes were fading quickly.[14] It was only one of many such benchmarks to come that dismantled the once vibrant and proud mining district. The people of Gold Hill felt it just as keenly when their only two-bit saloon lowered its standard and joined the ranks of what locals called one-bit "shebangs."[15]

In spite of declining prospects, mining continued on the district. Whereas many boomtowns disappeared as quickly as they had emerged, the Comstock lingered, as further testimony to the spectacular nature of its silver and gold deposits. Low-grade ore in the upper levels provided marginal subsistence for the industrial complex. The 1882 directory for Virginia City depicts a vivacious community with no hint of becoming moribund.[16] A debate raged regarding how best to extend the life of the mines, to provide a more lasting bridge to the discovery of the next bonanza. Some thought technology would be the answer, while others argued that the miners should have lower wages. Both options had the potential of reducing costs, making poorer ores profitable. Technology would require imagination. A retreat from the $4-a-day minimum would mean abandoning a standard that was higher than the pay for most industrial workers at the time. The Comstock miner's salary had long been a source of pride for the local unions, standing as it did for the success of the Comstock and the professionalism of its workforce. Eliot Lord took on the topic in his 1883 publication, suggesting that labor concessions could extend the life of the mines, ensure continued employment, and extend prosperity to the unemployed. All that was required was that the miners accept the same sort of economic reductions that circumstances forced mine owners and investors to endure. The search for new technologies, just like the exploration for undiscovered ore bodies, continued, but miners were slow to entertain the idea of a reduction in the minimum wage for their dangerous occupation. A history of repeated successes convinced the union that the Comstock would endure. Even the Consolidated Virginia and California saw a revival three years after the fire, when miners injected carbonic acid gas into the sealed areas, finally suffocating the flames.[17]

The larger mines continued deep exploration for as long as they could maintain financial support: to do otherwise would have been to concede the end and to abandon an investment built over years. By 1884, however, investors' enthusiasm for assessments had withered. Once, the prospect of wealth inspired stockholders to respond to the demand for assessments (fees for the right to hold shares). Now, in the 1880s, more and more investors surrendered their shares. Lacking the funds to continue, managers began to cease the expensive pumping, leaving the mines to flood up to the level of the Sutro Tunnel, which would continue to provide a passive means of drainage. With this action, a share of the Consolidated Virginia and California, once the most productive mine on the Comstock, dropped from fifteen cents to five cents. On October 16, 1886, the owners shut down the last of the deep-mine pumps.[18]

Still, mining of low-grade ores yielded modest profits throughout the rest of the 1880s. Excavation even uncovered narrow bands of high-quality ore in 1886, 1891, and 1894. In 1888 miners discovered relatively rich ore used to fill abandoned stopes. Speculation suggested that during the bonanza period, the haste to reach even better deposits might have inspired the manager to set aside what was comparatively poor material. Comstock productivity again climbed, but except for an 1886 boom that ended in a crash, there was little excitement. The age of unguarded optimism was all but gone. A few clung to the hope that each discovery of rich ore, no matter how small, might lead to the next bonanza, but investors remained slow to act even when indications looked good. There was, however, enough optimism that at least one capitalist, Captain J. B. Overton, felt justified in building the last of the Washoe Pan Process mills in 1887. As a sign of the times, the mill used electricity, generating its own power with a wheel that turned with water from the water company's flume.[19]

The Comstock mines performed well, if not remarkably, for the decade following 1884, extracting low-grade ores at a profit and consistently giving dividends to investors, but 1895 would see the last of those payments. Mackay, having worked the Comstock for more than thirty years, sold his stock and retired. His last visit, with D. O. Mills in August 1895, signaled an end to the era of nineteenth-century Comstock mining. Coincidentally, the U.S. Mint in Carson City had issued its last coin two years before, in 1893.[20] That same year, the revered *Territorial Enterprise* ceased publication. In recognition of the event, newspapers across the nation published its obituary, and the Society of Pacific Coast Pioneers in San Francisco flew its flag at half mast. The final words of the *Territorial Enterprise* as it closed its pages were "For sufficient reasons we stop."[21]

The last entry in the ledger book for St. Mary Louise Hospital in Virginia City simply notes the end of an era: "The Sisters of Charity left for good, Sep 7, 1897."[22] It was the culmination of a long, tortured process.

The sisters, who had made such a difference in the community throughout most of the Comstock heyday, found themselves in a difficult situation, dependent as they were on contributions from a steadily declining population. For months the sisters sank into destitution, with little to eat and less to do. They repeatedly asked the church hierarchy to release them from their Virginia City assignment, since the usefulness of their presence had ended. Finally, in 1897, they received permission to leave, and the Daughters of Charity completed a chapter of Comstock history.[23] That same year the Washoe Club closed its doors.[24]

Even as the district experienced this transition, mining experts were determined to see a new period dawn for the Comstock. In 1896, Robert D. Jackson, a professor at the University of Nevada School of Mines, successfully used a new technique on mill tailings around Washoe Lake. The method depended on cyanide to extract the minute particles of gold that the old milling process had left behind. It worked wonderfully. This new approach, dating to a British patent granted in 1887, employed a compound of cyanide that dissolved silver and gold from crushed ore. Zinc then drew the precious metals out of the mixture, producing a sludge for further refining. Cyanide found its first application with gold ores, but eventually miners demonstrated that it was also effective with silver. Ultimately this new process held the promise of finding profit even in lower-grade ores, as well as in mine dumps and mill tailings. In 1901 Charles Butter built a cyanide mill down Six Mile Canyon,[25] and many hoped that technology could come to the Comstock's rescue.

Other technological changes made mining cheaper and more efficient, drastically changing its character. Electrification and modern engines minimized noise. As John Waldorf recalled, "I used to go to sleep to the clang of machinery and the puffing of steam. . . . Now all is different. The mines do their work quietly."[26]

In spite of promising changes, modest discoveries, and the persistent attempts at revival, thanks to the faithful, every year seemed potentially the last for the district as the Comstock gasped for each breath. By 1897 one Virginia City resident wrote: "All you hear on the street is the old cry 'hard times.'" A year later, another resident wrote: "This place will soon be deserted if they go on as they have been doing. Every other house is boarded up as they could not sell them."[27] In 1898 Comstock mines produced only $205,000, an amount far below the cost of production.

Disregarding the obvious signs of decline, some stockbrokers and mine owners, organized as the Comstock Pumping Association, decided to support the dewatering of the mines on the north end of the Lode.[28] Ostensibly this strategy would allow exploration for new ore bodies, but

Charles Butter built this cyanide mill in 1901, down Six-Mile Canyon to the east of Virginia City. (Courtesy of the McCarthy Collection, Nevada State Historic Preservation Office)

clearly the group hoped to raise the value of stocks and to attract and manipulate investors. The cyanide mill, however, also presented an opportunity to exploit lower-grade ores from deeper levels. Perhaps the brokers were as shocked as anyone when the exploration yielded a small but rich pocket of ore. Further work from 1901 to 1913 resulted in the discovery of more deposits, each unassociated with the next, each yielding limited returns. This find helped keep the Comstock profitable and sustained the optimists. Visiting the town of his childhood in 1905, Waldorf was able to observe: "The truth is that the camp is better off now than it was twenty years ago. . . . Now, with up-to-date mining and new ores in sight, the future is bright with hope. Even at present there are five hundred men at work, which is more than any other camp in the state can boast."[29]

Adding to the success on the Comstock, local mills found a new means to keep their machinery employed. A boomtown in central Nevada spawned a surge of people and economic success after 1903. Tonopah and Goldfield burst into existence as Virginia City and Gold Hill had forty years before. Within months they were the largest communities in the state, and as young places with insatiably growing industries and society, they offered opportunity. The entire region swung into action, benefiting from the growth. Miners flocked to the new cities, but mills were slow to follow. Owners of these claims to the south recognized that the Comstock's infrastructure of rail links and dormant facilities presented an opportunity, and before long they were shipping much of their ore to the older mining district for processing. This provided an added boost to the Comstock economy at a time when the optimists believed they had reason once again to see success and permanence on the horizon.[30]

The booms in Goldfield and Tonopah did more than transform the Comstock by providing employment in the mills. Ore was not the only thing moved from place to place in Nevada. The statewide boom created a need for housing elsewhere, and local tradition retains memories of Comstock houses being hauled to various locations. Folklore places Senator Fair's mansion in Reno, where it later burned; Senator Stewart's house is said to be in the Yerington area; other houses also traveled to Carson City and throughout the Truckee Meadows of Reno and Sparks. Local tradition also has many structures going as far as Tonopah and Goldfield. There are even those who maintain that with the end of the central Nevada mining boom, some Comstock houses moved once again to Las Vegas as it grew in the 1930s.[31]

The U.S. census reports document what Waldorf intuitively recognized: the Comstock had declined, but it had also stabilized into a mature community. The 1880 Storey County population of 16,004 was only 8,806 ten years later. By 1900 it had dropped to 3,560, but in another decade the 13th U.S. Census of 1910 shows the county with almost the same number of people. By then the population numbered 2,976, only slightly less than before (table 11.1). Significantly, the relative number of women peaked in 1900, representing 47 percent of the total. At the turn of the century, the Comstock had lost the male dominance that was part of its mining town character and had transformed into a distribution that was closer to the national profile. Subsequent years would see a greater disparity between the sexes again, and indeed it was not until 1950 that there was again a nearly equal split.

Further analysis of the 1900 and 1910 censuses demonstrates that Storey County had changed in other profound ways. Figures 11.2 and 11.3 are population pyramids illustrating the distribution of people by sex and age group as recorded in the 1900 and 1910 censuses. Storey County at the turn of the century had relatively even representation from all age groups. The large number of children is evidence of a permanent

11.1 Population Size for Storey County, Males and Females*

Census Year	Number of Males	Percentage of Males	Number of Females	Percentage of Females	Total Number
1860	2,857	95	159	5	3,016
1862	3,843	85	655	15	4,498
1870	7,814	69	3,505	31	11,319
1875	13,415	69	6,113	31	19,528
1880	9,221	58	6,783	42	16,004
1890	5,144	58	3,662	42	8,806
1900	1,874	53	1,686	47	3,560
1910	1,748	59	1,228	41	2,976
1920	803	55	666	45	1,469
1930	378	57	289	43	667
1940	709	58	507	42	1,216
1950	354	53	317	47	671
1960	295	52	273	48	568
1970	343	49	352	51	695
1980	767	51	736	49	1,503
1990	1,250	49	1,276	51	2,526

Sources: U.S. census manuscripts and reports, 1860–1990; the Nevada territorial census of 1862 and the Nevada state census of 1875.
*The statistics for 1860 summarize the communities of Gold Hill and Virginia City, Utah Territory before the creation of Storey County in 1861. Figures for 1870, 1880, 1990, and 1910 are based on the census manuscripts because they are more accurate than the census reports.

society with good prospects for the future. That young people in their early twenties remained suggests that there was sufficient opportunity and optimism to keep them from looking elsewhere for employment. This profile clearly reflects the new stir of activity at the turn of the century. Indeed, both the 1907 and the 1923 Sanborn Fire Insurance Maps for Virginia City reveal a small but vivacious community with a full range of stores and services.[32] The Comstock had survived, but clearly it had also changed. The decline in population, particularly among men between twenty-five and forty-four years of age corresponds to the economic doldrums that had dominated the region since roughly 1878: men of this age had left in their early twenties, unable to secure employment in the shrinking economy. The large number of people older than that, especially the men, represents the workforce that managed to retain the few jobs available in leaner times. Some of these people may also have been unable or unwilling to leave. The numerous old men echo the Com-

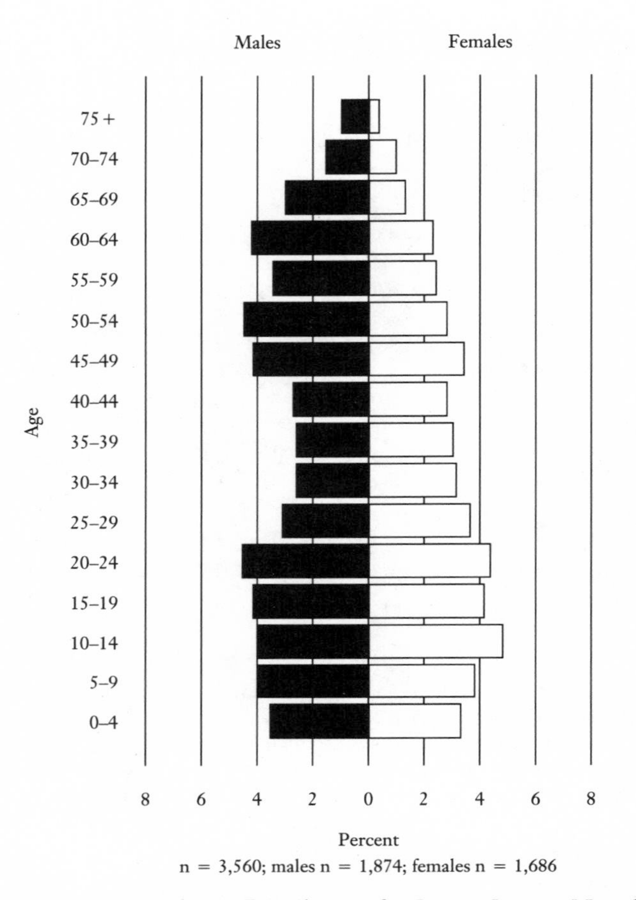

Fig. 11.2. Population Distribution for Storey County, Nevada, 1900. (Source: 12th U.S. Manuscript of 1900.)

stock boom of the 1860s and 1870s, when single miners and other workers constituted a large part of the community.

Figure 11.3 illustrates a population that had in some ways resumed its older, boomtown profile. The bachelor workforce had returned, responding to new employment in the mines and mills. Indeed, in the ten years since the turn of the century, adult men had increased from a third of the population to more than 40 percent. At the same time, the relative number of adult women twenty-five years or older remained almost unchanged, representing about a quarter of the population during both census years. The number of mine workers employed in Storey County is particularly telling. From 290 in 1900, their numbers increased to 552

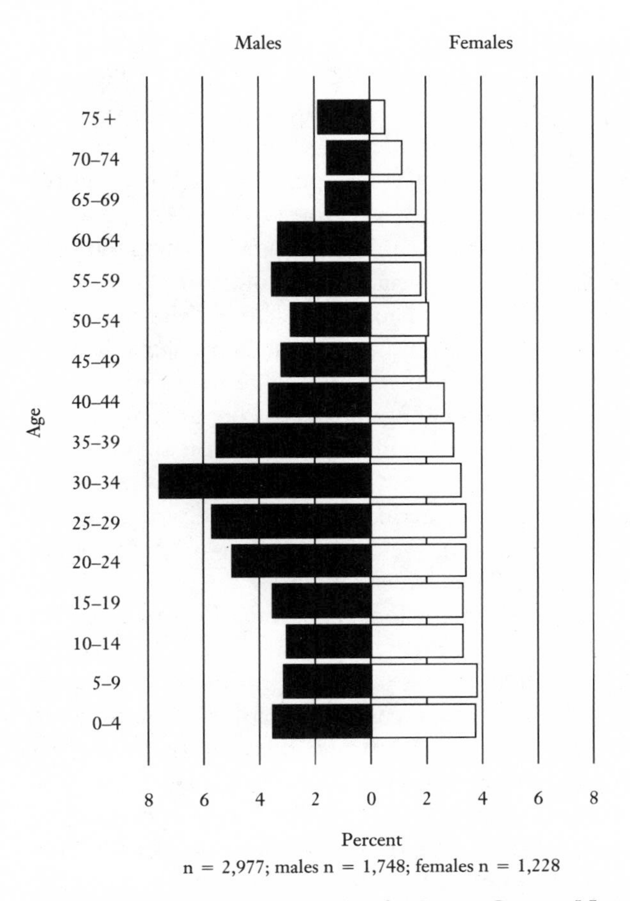

Fig. 11.3. Population Distribution for Storey County, Nevada, 1910. (Source: 13th U.S. Manuscript of 1910.)

ten years later, a shift from less than 20 percent to almost 40 percent of the adult men.[33]

Again, Waldorf offers an observation lacking statistical precision but less easily divined from the manuscript census. He points out that when he returned to Virginia City in 1905, most of the Paiutes were gone, having died or left the district. Only a few remained, and as he puts it, they had abandoned their "wigwams" in preference of "a battered shanty." In addition, some of the women had "sewing machines in the homes."[34] An older era had passed in many ways.

Some people felt that this turn-of-the-century mining would never be more than a shadow of the former bonanza days. Indeed, investors clearly did not receive remarkable returns, but the revival sustained the community and provided an important bridge that prevented Virginia City from becoming a ghost town. Shortly after exhausting these discoveries, economics again forced mine owners to abandon pumping, and by 1920 water had returned to the level of the passive drain that the Sutro Tunnel provided.[35] There was, however, ample reason for the optimists to endure, as fluctuations cyclically raised and then dashed hopes. Visiting Virginia City in 1924, Waldorf was able to declare: "The old town is far from gone yet. A world, pulling itself together after a demoralizing war . . . , grows hungrier for the gold and silver on which stability depends. Perhaps in the train of that growing demand will come new shafts, more glory holes and great mills, and the long-looked for boom come at last, not only come but stay and another wonder city rise on the steeps of old Mount Davidson."[36]

Throughout this period of alternating between steady decline and limited success, people of the mining West began to grow nostalgic about Virginia City. The "graduated" Comstock alumni had populated countless other towns and camps, and many looked back fondly on a time and place that rang with the vibrant excitement of a new, fresh frontier. Although the Comstock no longer sent as many people out into the region as it had during the 1870s and 1880s, there were still some who could boast of their Comstock years after the turn of the century. In keeping with the Virginia City flair, some acquired fame. Jimmy Doolittle, who became famous as the commander who led the first bombing raid on the Japanese mainland during World War II, was earlier known locally for a daring deed in the Comstock mines. He came to work in Virginia City during the summer of 1917, while attending the University of California. During his stay, there was an accident in one of the shafts and the cage was ruined. Doolittle volunteered to descend by rope to inspect the debris and look for survivors. Workers lowered the future aviator 300 feet, a dangerous undertaking. Both miners involved were dead, but through this episode Doolittle, like so many of the famous people who had passed through the mining district before, won his mention in the Comstock epic.[37]

After the turn of the century, Virginia City also became the site of field trips from the Mackay School of Mines at the University of Nevada in Reno. During their senior year, students would ascend Geiger Grade, cavort around the district, descend into its mines, and experience something of that classic institution of their profession. With each year, the mining community welcomed fresh recruits, who were then able to claim something of a Comstock connection for themselves. In short, the place

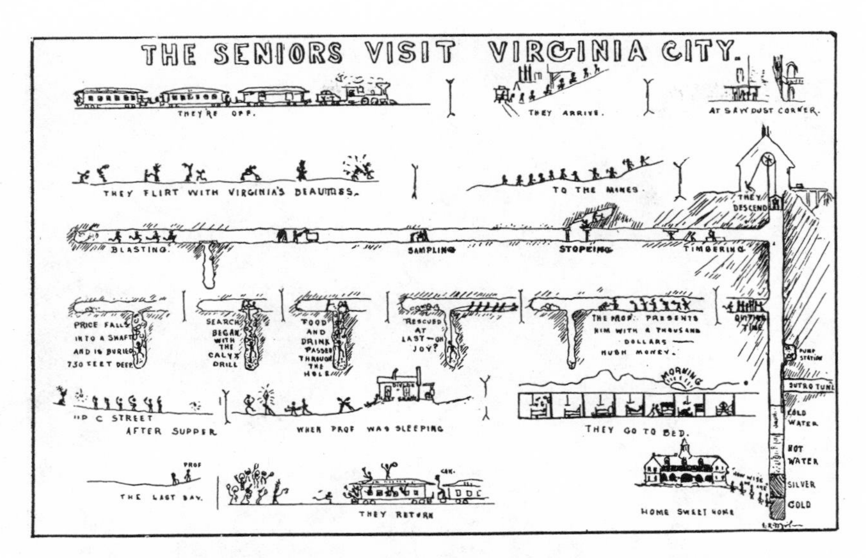

The Seniors Visit Virginia City, humorous illustration by graduating senior Frederick J. DeLongchamps from the 1904 *Artemesia*, the University of Nevada yearbook. (Courtesy of the Nevada Historical Society)

survived in many ways as a powerful force in the mining world, lasting far longer than other districts, surviving even the collapse of its mines.[38]

Even while Virginia City and Gold Hill were enduring the convulsions of a dying economy, Silver City was enjoying a quieter form of success, exploiting large low-grade ore bodies at shallow depths. Silver City did not capture headlines, failing as it did to serve as home to a remarkable bonanza. Still, just as it never managed to ride the crest of good fortune, it also avoided the depths of a severe *borrasca.*

The Fourth of July of 1909 saw the celebration of the Golden Jubilee, fifty years since the discovery of the Comstock Lode. The day attracted people from throughout the region, giving cause for a reunion of those who were there in the district's "palmiest days," as the *Reno Evening Gazette* described it. Town organizers contracted "with an aeronaut . . . [to] make two balloon ascensions and parachute jumps." Chinese lanterns decorated the electrical lights, and plans called for a twenty-one-gun salute at sunrise. The celebration lasted three days and included a parade and several dances. As the *Evening Gazette* noted: "It brought back to the minds of the old timers the good golden days when Virginia City's mines were producing thousands daily, and when the Queen City of Nevada was filled with thousands of money-mad men, who lived like princes and made Virginia one of the most exciting places in the United States." It

This float was part of a parade at the turn of the century honoring the fiftieth anniversary of either the original Comstock strike or the founding of the miners' union. (Courtesy of the McCarthy Collection, Nevada State Historic Preservation Office)

was one of those melancholy moments common to the mining world. Most communities can look back on their first half century as a period of discovery and growth, and they can simultaneously imagine the next fifty years optimistically with hope. Mining can breed glorious booms, but more often than not once the ore is spent, it leaves remote communities with too few resources to conceive of any future at all, let alone a prosperous one.[39]

Piper's Opera House, the Comstock's premier theatrical establishment, epitomizes the community's struggle to survive the decline of the mining district. John Piper, its German founder, died on January 3, 1897, but his son maintained the family tradition. The spectacle of performances continued even though the performers' footfalls sometimes echoed for lack of a full house. The prestige of the venue that had once served Edwin Booth, Henry Ward Beecher, and Buffalo Bill continued to attract the likes of Emma Nevada, John Philip Sousa, and Al Jolson.[40] James J. Corbett trained there in February 1897, in anticipation of his fight with Bob Fitzsimmons, a contest that promoters advertised as the "Fight of the Century." Top-billed plays continued to appear at the opera house, and Richard Jose, a Virginia City native and now a world-famous tenor, sang at Piper's in 1905. The following year the hall featured one of the first motion pictures. With a hint of irony, the subject was the Gans-Nelson fight that had taken place in Goldfield, now the largest city in Nevada and the home of the state's last frontier gold rush. Thirty years before, Goldfield had not existed, and a fight of such stature by rights would have belonged to the Comstock.[41]

One of the more obvious benchmarks of the closing of an era came in the form of the destruction of the famed International Hotel. The pride of Virginia City, built three times, with each incarnation grander than before, boasted an elevator, 160 rooms, a fine restaurant, and distinguished guests. In the early morning of December 12, 1914, a fire began in the kitchen. In the following hours, it consumed the building, leaving ash and rubble where there had been magnificence. Heat from the blaze burst windows at the Sawdust Corner across Union Street to the south, at the Frederick House across C Street, and at the Molinelli Hotel, two buildings to the north. Eyewitnesses reported that flames reached 100 feet into the air above the hotel and sparks flew everywhere. Snow on the roofs and the diligence of firefighters prevented a repeat of the Great Fire of 1875.[42] Nonetheless, one by one, fires and decay in the early twentieth century claimed many other buildings throughout the district.

The Chinese community of the Comstock serves as yet another example of changing times there. The 9th U.S. Manuscript Census records 749 Chinese in Storey County in 1870. Five years later, during the height of Comstock prosperity, a state census placed the number at 1,362. Another five years saw the number drop to 642. By 1890, there were only 245 Storey County Chinese reported by the 11th U.S. Census. At the turn of the century, the number had diminished to 76, and in 1910, the census showed only 44 Chinese in the county. One of the more striking transitions among the Chinese between 1870 and 1880 was the decline in the number of prostitutes. In 1870 census enumerators recorded 94 Chinese prostitutes in Storey County. Ten years later the number had fallen to about a dozen.[43] More than other groups in the society, these women sought someplace else to live, apparently being more quickly affected by a faltering economy. Chinese prostitutes were only part of the picture, however, and most of the Asian community eventually took leave of the Comstock as its fortunes declined.

In 1933 Virginia City saw the passing of the last of its Chinese residents from the mining heyday. Seventy-five-year-old Charlie Ching, born in 1858 in China, had arrived on the Comstock in 1876 to work as a young laborer. He eventually became a cook, and he invested in various stocks. By the turn of the century he had become a chef at the International Hotel, but when that monument burned, he opened the City Restaurant on C Street. Because of prejudice, customs, and laws that discouraged the immigration of Chinese women and also interracial marriages, Charlie Ching was a bachelor and died without children. The *Nevada State Journal* mourned his passing, recognizing that his death meant the end of an institution, not only because of his importance as an individual but also because of what he represented as the last of the original Chinese in the district. The newspaper pointed out that Ching had a "heart . . . as big as the mountain on which he lived. His generosity and kindness were a tradition, and his optimistic outlook on life unfaltering." The first Comstockers were destined to fade away, through either migration or death, but at least a few stayed long enough to create a new generation that would ensure some continuity. For the Chinese, however, this was not to be the case, and for an essential part of the Comstock community, the death of Charlie Ching represented extinction rather than the passing of the baton.[44]

One of the exceptions to the general demise of community and economy came in the form of chalk mining in the nearby hills. This new source of revenue emerged at an opportune time and helped prevent the Great Depression from silencing Virginia City. On a smaller scale, a thriving local bootlegging industry, tucked into the mountains of the Comstock, provided a steady income for at least a few intrepid souls who were willing to resist federal edicts.[45] Still, the 15th U.S. Census of 1930 recorded only 667 people in Storey County.

Those who remained fell upon a survival strategy that forever changed the appearance of the mining district. It became common for residents to buy the abandoned, dilapidated house next door for back taxes. During the winter they would cannibalize it for firewood, eventually leaving themselves with a larger yard. The original builders of Virginia City had placed many houses side by side, as they are yet today in the older parts of San Francisco. The Depression-era strategy of house demolition gave homeowners more space and was consistent with the changing ideas of domestic land use in the twentieth century.[46] Ironically, it was also early in the Great Depression that the federal government first acknowledged the importance of the Comstock, listing it on what was a predecessor of the National Register of Historic Places. Still, the shadow of the original residential arrangement remains visible in the property boundaries: invariably each house sits on the edge of its property, with the yard extending to the side until it meets the next house and its property line.

The pattern of demolition became so common that houses remaining

Virginia City's St. Mary in the Mountains stands in a community falling into ruins, ca. 1925. Photograph by Timothy J. McCarthy. (Courtesy of the McCarthy Collection, Nevada State Historic Preservation Office)

side by side, once the rule in the core of Virginia City, have become a rare curiosity. One of the best surviving examples is known on the Comstock as the "spite houses." Local folklore maintains that they were built close as an expression of a dispute between neighbors. Memory of the original arrangement of houses has faded so completely that an oral tradition has grown up to explain what has become an anomaly.

Twentieth-century cannibalization of the past was not restricted to houses. Veteran Virginia City resident Ty Cobb recounts an incident from his childhood in the 1920s. He and some other boys stripped a buggy they found in an abandoned shed and converted it into a precursor of a go-cart, a perfect vehicle for coasting down the Comstock's steep hills. After a few trial runs along short tracks, the boys decided to challenge Union Street, one of the steepest and longest thoroughfares. Their freewheeling car shot down the hillside, barely missing wagons, trains, people, and buildings, until it finally toppled, leaving the intrepid adventurers bloodied and bruised. In recounting the tale, Cobb describes a Virginia City that included the railroad, Northern Paiutes playing cards on a blanket on D Street, Chung Kee's Chinese store, and a joss house. In short, much of what made the Comstock a rich, complex place in the bonanza years remained for a new generation. Although fading, the Comstock was still a vibrant place with ample opportunity to reuse what the nineteenth century had left behind.[47]

Ironically, the early twentieth century was also a time of new construction on the Comstock. The majestic four-story Fourth Ward School, built in 1877 to house a thousand students, graduated its last class in 1936, yielding to a WPA-funded successor that bore little relationship to the nineteenth-century heritage of Virginia City. Similarly, numerous small houses, wooden bungalows now bearing the colorful pastels of asbestos and asphalt shingles, line a few of Virginia City's streets as testimony to changing needs and tastes. For the most part, however, the Comstock receded into a weather-worn shadow of its former self as it teetered into the Great Depression. When a limited resurgence of mining brought some new people into the district, they shunned the older, dilapidated houses and built in contemporary styles.

Prominent among the new ventures was a project that longtime mining entrepreneur William J. Loring initiated in 1933. Loring combined forces with a group of out-of-state investors to form the Arizona Comstock Company, which leased part of the Lode along the claims of the Chollar, Potosi, Savage, and Hale and Norcross. Initially Loring intended to work low-grade shallow ores with well-planned underground excavations. He built a mill downhill and began work, but the endeavor was unprofitable. Loring decided in 1934 to take a new approach: he initiated open-pit mining just above C Street on the Divide separating Gold Hill and Virginia City. The Loring Cut, as it was called, is a highly visible scar on the side of Mount Davidson, originally measuring 900 feet long, 300 feet wide, and 200 feet deep. The project's effect on the landscape was far more dramatic than its contribution to the economy, however. The Arizona Comstock Company shut down its operation and retreated from the district in debt.[48]

About the same time, other mining companies excavated open pits elsewhere on the Comstock, hoping that the inexpensive retrieval methods of such endeavors would compensate them for the poor quality of the ore. Even when that was not the case, the projects meant jobs and new money flowing into the communities, sustaining the district during the dark days of the 1930s. The tangible result of these undertakings is a long series of holes extending from Silver City, through Gold Hill, to the north end of Virginia City. Mining to this day continues to enlarge these pits, exaggerating an alignment that was well defined by the end of the

A miner harnesses a tractor to hoist a bucket at a Gold Hill mine, ca. 1925. Photograph by Timothy J. McCarthy. (Courtesy of the McCarthy Collection, Nevada State Historic Preservation Office)

Depression. Open-pit mining became the hallmark of twentieth-century technology, contrasting with the underground hard-rock excavations of the nineteenth century. The latter sank back into a romanticized past. Ironically, or perhaps appropriately, the northernmost of the surface excavations is also the oldest, dating to 1859 and the beginning of mining on the Comstock Lode. This is the Ophir Pit, expanded and changed over the years but still marking the location where miners scooped ore from the surface before deciding they would need to pursue their quest underground.

The *Reno Number 11* pulls out of Gold Hill, ca. 1925. Photograph by Timothy J. McCarthy. (Courtesy of the McCarthy Collection, Nevada State Historic Preservation Office)

In 1938, after sixty-nine years, the Virginia and Truckee Railroad phased out service to the Comstock. On May 23 a new timetable came out, listing no passenger runs from Carson to Virginia City. Special excursion trains to Gold Hill ran throughout the following summer, but the Comstock was clearly seeing the end of its railroad. The last freight train left Virginia City on June 4. Scrappers, who claimed much of the Comstock's heritage through the war years, eventually removed the tracks and rails. They even jerked the large timber supports from under tunnel roofs, judiciously retreating as ceiling boards and loose rock tumbled to the ground.[49] Although the Virginia and Truckee would continue service in other parts of the area, the company sold much of its older rolling stock to movie producers. Vestiges of the glory days of the Comstock became increasingly rare.

The cycle of limited new mining ventures followed by collapse inspired the 1940 population of Storey County to increase to 1,216, nearly doubling itself in ten years, but the role of mining in the district was about to change forever. In early 1942, the War Production Board, acting with the authority of the White House, rated gold and silver mining as a nonpreferred activity, excluding the endeavors from certain benefits. Then on October 8, 1942, the board issued Limitation Order L-208, giving gold mines seven days to close. The intent of these actions was to encourage the extraction of strategic metals for the war effort, but the effect on the Comstock was devastating. Now only the scrappers who came to scavenge iron for war industries gave the Comstock life, but it was that of vultures picking apart carrion.[50]

The year 1942 also saw fire return in a way that could remind its oldest residents of the scourge of 1875. Starting above Gold Hill, the Comstock's well-known Washoe zephyrs fanned the conflagration until it rolled up and down hills at a sprint. The Divide, which had remained a thriving community, fell victim. More than twenty families lost houses and all their possessions. Little survived.[51]

A final blow to Comstock society came when the War Department pressured Storey County to close its Virginia City brothels and cribs. The federal government maintained that prostitution represented a health risk and with so many soldiers traveling back and forth across the nation, Virginia City's businesses might spread disease and hamper the war effort. Lest Hitler's victory be hung around Virginia City's neck, its houses of ill repute closed. Stripped of its mines and of its prostitutes, the Comstock ended what seemed to be its final chapter.[52]

12: THE SEQUEL TO THE BIG BONANZA

Tourism and Television

The Comstock's Flame had been quenched.
Perhaps someone will find another match.
—Program for the April 22–24, 1993,
world premier of Mark Me Twain,
presented by the Nevada Opera

Beginning as early as the 1930s a strange transformation occurred on the Comstock. What had been a dormant, even dying mining district slowly blossomed into a magnet for artists, literati, and others who wished to experience something of the fast-disappearing Wild West. They were attracted to Virginia City because they believed that the place had not changed. Of course it had. Indeed, the irony of their attention was that promoting the Comstock as a tourist mecca did more to transform the district than the failure of the mines. For every person who came to experience the Wild West, the Comstock became even farther removed from its nineteenth-century roots. It mattered little, however, to these latter-day westerners; they apparently felt that to watch the last grains of sand slip through their fingers was better than not to have seen them at all.

One of the first to arrive was Duncan Emrich, a well-traveled scholar. Born to missionaries in Turkey in 1908, he was educated at Brown, Columbia, the University of Madrid, and Harvard. Before he was forty, he became the founding director of the Folklore Section of the Library of Congress, now known as the American Folklife Center. Emrich probably visited Virginia City for the first time in 1937, and apparently his love affair with the place took hold immediately. He returned whenever he could. Eventually, Emrich claimed to be a Comstocker and maintained a Virginia City address. Two of his marriages even began with a local ceremony.[1]

Emrich proved to be ahead of the curve. In 1940, Warner Brothers released *Virginia City*, starring Errol Flynn, Randolph Scott, and Humphrey Bogart. The film premiered in Reno and Virginia City. While the Reno affair went well, there was a problem on the Comstock. Warner Brothers executives judged the crowd as too intoxicated and potentially rowdy, and so they retreated from the hill, returning to Reno even as the film was ending. The locals, expecting to see the stars in person, were furious and left the theater to look for celebrities—or anyone—for revenge. The theater manager roughed up a representative from Warner Brothers, but in spite of numerous threats, there were no lynchings. The premier brought $50,000 to Comstock bars, and *Newsweek* covered the event. *Virginia City* did not reach artistic heights, but it did remind the nation of the importance that the Comstock had held little more than a century before.[2]

Others followed in Emrich's footsteps as an increasing number of people realized that the Comstock represented a refuge, a last holdout, a place where freedom had a slightly finer edge than elsewhere. Among the most flamboyant of Virginia City's new residents were Lucius Beebe and his friend Charles Clegg. Beebe had first visited Virginia City in 1940 as a reviewer of the movie. A columnist for the *New York Tribune*, Beebe hated the film but loved the place. The same year, he fell in love again, this time with Clegg, who was to become his lifelong companion. War interrupted destiny, however, as Clegg joined the navy to defend his country, leaving Beebe, who was too old to serve. Shortly after Clegg's return, the two set their sights on the Comstock. In 1948 they rode in their ornate gilded Pullman car, the *Gold Coast*, across the country to its final destination in Carson City. The two bons vivants took to the Comstock and assumed Wild West personae, blending in as well as anyone from the East could.[3] They arrived just in time to see the disappearance of many of the western attributes that they enjoyed. As railroad enthusiasts, they witnessed the last run of the famed Virginia and Truckee Railroad in 1950 and mourned as scrappers hauled off its rails.[4]

Beebe and Clegg met with a Comstock that was an odd collection of locals and flotsam from the rest of the nation. In retrospect, one could see a great deal of continuity with the nineteenth century, when Virginia City attracted the famous and the notorious from the world over. As Comstock author Andria Daley-Taylor (who lives in the Beebe house) notes: "[The two men] met kindred spirits on their strolls around Virginia City. There were the remittance kids, children of the rich who were paid to stay away. A Delaware du Pont was tending bar at the Sky Deck Saloon. An Eastern socialite was running a hotel. A chef from Maxim's was cooking at the Bonanza Inn. Numerous members of Cafe Society were on hand, waiting out their six-week residencies at divorce ranches near Reno. . . . Soon pals and colleagues from back East were migrating to Virginia City, and townsfolk took it in good stride while hobnobbing with Cole Porter and other celebrities."[5] Many came to the Comstock to sink into the luxurious morass, a thief of time and ambition that served as refuge for internationally prominent ne'er-do-wells fleeing a high-pressure world. Unlike others, however, Beebe and Clegg set to work furiously writing books on the West.

In 1952 they revived the *Territorial Enterprise*, drawing on their own energy as well as the talents of those who settled in Virginia City. They wrenched their colleagues from dormancy to produce a nationally quoted weekly. As Daley-Taylor points out, writers included "Nevada's own Walter Van Tilburg Clark, Pulitzer Prize–winning historian Bernard De Voto, and folklorist Duncan Emrich. The 'two Katies'—Katherine Hillyer and Katherine Best, both contributors to national magazines—wrote a column called 'Comstock Vignettes.'"[6] Again the Comstock attracted the spotlight, and as in the nineteenth century, tourists began to include Virginia City on their "must-see" lists. Local businessmen enjoyed increased revenues, looking the other way when confronted with Beebe's and Clegg's personal lives. As Mary Andreasen, a resident of the district, recalled, "Virginia City was such a tolerant town then. There were many of those who espoused the bohemian lifestyle, writers and artists, and that's the way it was."[7]

In the midst of this dynamic time, one of the new Comstockers created a document that provides information about the transformation of the Comstock. In 1949 Duncan Emrich began recording interviews and conversations with some of the old-timers as they sat and drank in the Delta Saloon. For decades the tapes sat unused on the shelves of the Library of Congress until the University of Nevada's Oral History Program obtained copies and transcribed them. The resulting text reveals a dramatic contrast between the interests of Emrich and those he interviewed.

Ignoring the customary practice of allowing the informant to define topics and direct conversation, Emrich repeatedly interrupted his guests and attempted to bring them back to topics they clearly felt were irrelevant. Chief among the subjects discussed was Julia Bulette, who by the 1940s had faded from local view as little more than a blip in the past. Bulette apparently had not survived in local oral tradition, and the Comstock old-timers, when asked about Comstock vices, preferred to talk about opium dens or prostitutes other than Bulette. Emrich, however, belonged to a new generation, and he apparently felt the need to document Virginia City's obligatory "whore with the golden heart." He refused to accept the apathy that his informants felt for Bulette and instead dragged details out of them where free-flowing stories did not exist. Emrich worked to create his own folklore for the Comstock when he found the existing one not to his liking.[8]

Over the next ten years, Bulette became a cornerstone of local myth. The Virginia and Truckee, before disbanding, named one of its cars after the murdered prostitute. Entrepreneurs established a fake grave for Bulette opposite the hill bearing most of Virginia City's cemeteries, then promoted the story that proper citizens had not allowed her burial in the dignified final resting place. This, of course, ignored the fact that both sides of the ravine had active cemeteries in the 1860s. The "location" of Bulette's grave also disregarded her actual burial plot, which was apparently on the other side of the ridge. The new memorial, however, had a gleaming white fence, making it visible through a coin-operated telescope at the large east window of the Bucket of Blood Saloon.[9]

In 1959 the Comstock celebrated the centennial of its founding. The U.S. Postal Service issued a commemorative stamp, local citizens built a monument, and for a short time people looked back to see a faint flicker of the brilliant flame that had been so clearly visible throughout the world a hundred years earlier. In June, Virginia City hosted a party to recognize that momentous day when McLaughlin and O'Reilly sank their shovels into a treasure trove and when Henry Comstock arrived on the scene to claim part of the Lode for himself. Vice President Richard M. Nixon and his wife, Pat, herself a native of Nevada's mining frontier, attended, furnishing national recognition of the importance that the mining district once claimed.[10]

Later that year NBC aired the first of 440 episodes of a television program that would capture the nation for the next fifteen years. Ben, Hoss, and Little Joe Cartwright became ambassadors for a new era of Comstock fame.[11] It mattered little that their Virginia City stood on level ground without even a hint of the dramatic inclines that dominate the actual community's streets. The convenience of scriptwriters and executive producers created a place that chiefly served ranchers who drove their wagons into town for supplies, a drink in the local saloon, and invariably a little trouble. People could ignore the fact that ranching seemed more important than mining in this latter-day Comstock created by entertainment moguls. What mattered most was that Virginia City once again captured the national and eventually the international spotlight.

Tourists answered the call more than ever, now hoping to see the place that had inspired a television hit. The town they visited bore little resemblance to the place they had come to know on the screen, partly because

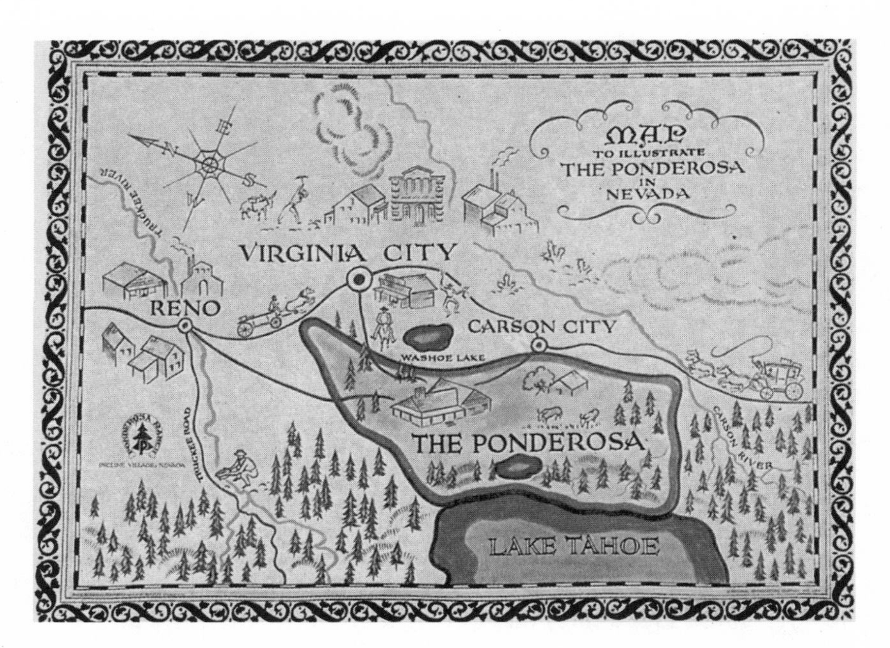

The map made internationally famous by the opening credits of *Bonanza*, the top-rated television show of the 1960s and 1970s. (Courtesy of the Ponderosa Ranch, Incline Village, Lake Tahoe, Nevada)

the imprint of the nineteenth century as it actually had been remained strong on the Comstock. If tourists left feeling they had not seen the real home of the Cartwrights, their complaints failed to find a clear place in the historical record. Nonetheless, many of the property owners on the Comstock felt the need to change their town to make it seem more like the mythic Virginia City of television. Soon vertical unpainted rough cedar boards covered finely built brick structures, disguising the real nineteenth-century industrial town in favor of myth born in the twentieth century.

Beebe and Clegg fumed over these changes. Moguls made millions, capitalizing on a history that they would corrupt even while exploiting it. The tourists who came by the thousands sounded the death knell for the old Comstock. The giant of the old mining West slumbered no more. It had awakened, and it carried an Instamatic and wore plaid Bermuda shorts. In 1960, Beebe and Clegg sold the *Territorial Enterprise* and spent an increasing amount of time living elsewhere. Beebe died six years later, but the connection with the Comstock ran deep: Clegg did not sell their Virginia City home until 1978, a year before he died.[12]

In spite of the objections of Beebe and Clegg, the television show *Bonanza* was the culmination of a process with deep roots in western history. For decades, historians and writers had tinkered with the image of the Comstock, inventing a myth to capture a place more the way it should have been than the way it actually was. Television continued the tradition while redefining its direction, and Virginia City was never the same.

At roughly the same time that television transformed the Comstock, a spark of genius from Beebe added yet another element to the twentieth-century renaissance. In 1959 the *Territorial Enterprise* ran a fictitious story about wild camel races in Virginia City. The *San Francisco Chronicle* fell for the hoax and reported the incident. Joke or not, it was an idea whose time had come. The following year the *Chronicle* sponsored a real race. Coincidentally, none other than John Huston was in the area, with Clark Gable and Marilyn Monroe, to film *The Misfits*, the last movie of the two entertainment legends. Never wanting to miss a novelty, the renowned director made a bet that he could win the race. For several days, Huston diligently trained his beast to expect food at the end of the track. On the day of the race, Huston and his camel rocketed to the finish line. Beebe crowned Huston the winner, and a tradition was born. Ever since then, Virginia City has been known for its camel races almost as much as for anything else.[13]

Only a few years later and while the popularity of *Bonanza* remained unchallenged, yet another new phenomenon struck the Comstock. Just as the artistic community of the 1940s and 1950s found a refuge in the Comstock, so too did their counterparts from the 1960s. Much of the feeling of freedom that had been part of the Comstock remained, and the environment proved ideal for the counterculture movement. Virginia City's Red Dog Saloon became a well-known hangout, featuring entertainment by Janis Joplin and Big Brother and the Holding Company. Ken Kesey and his Merry Pranksters came through the Comstock before Tom Wolfe wrote of their exploits in his *Electric Kool-Aid Acid Test.*

During the nineteenth century there had been constant communication between Virginia City and San Francisco, and this relationship found its echo a hundred years later. The Bay Area rock movement discovered that the Comstock was a refreshing retreat where anything was possible and experimentation could occur in a secluded environment. The seminal group, the Charlatans, became something of the house band at Virginia City's Red Dog, an odd combination of western saloon and San Francisco nightclub. Coming from the City on the Bay and having little or no experience, the performers and their leader (who did not play an instrument but simply hung around) developed a musical identity on the

Virginia City's Camel Races, with obligatory parade, have become an important part of the Comstock Calendar. (Courtesy of the Nevada Historical Society)

Comstock. When the Charlatans returned to California, they brought back the "San Francisco sound," as the region's particular brand of rock became known.[14]

Most of the others who had retreated to the Comstock also eventually returned to San Francisco or wherever, but like the earlier bohemians, a few lingered, finding the charm of the Comstock irresistible. For example, Lynn Hughes, who sang with the Charlatans and who helped found Stone Ground, another important early West Coast band, eventually forsook the life of rock and roll to settle in Silver City. Hughes died in 1993, long after she had played a pivotal role in the evolution of music, but the nation had not forgotten her contribution: while not featuring Hughes on its cover, *Rolling Stone* magazine did give her obituary several inches inside.[15] She was part of yet another dynamic period when the Comstock redefined itself. Echoing this movement, several geodesic domes, part of a back-to-nature rejection of conventional square architecture, also survive on the Comstock as testimony to a mentality and a time.[16]

In spite of the dramatic success that it represented, the television phase of Comstock history, like all the preceding periods, would have its day and its end. With the relegation of *Bonanza* to the world of reruns and an occasional, nostalgic made-for-TV movie, the influence of the Cartwrights waned. The doldrums again threatened the Comstock, but as had happened repeatedly before, some irrepressible part of its original heritage rose to the surface.

Occasionally visitors to the Comstock write the governor or other state officials to complain about the problems they perceive with the historic district. Criticism often focuses on the crass nature of signs and businesses, the telephone and electrical wires, and what visitors often see as twentieth-century side effects of tourism and gaming. The comments are naïve, revealing a poor understanding of the Comstock. Virginia City during its heyday roared boisterously, with garish signs, loudmouthed hawkers, and a freewheeling carnival atmosphere, exhibiting a degree of commercialism that would conceivably embarrass most of today's merchants. When confronted with exotic signs, Joe Page, veteran Comstocker and member of the Comstock Historic District Commission, is fond of recalling an early twentieth-century advertisement for a tobacco shop, shaped like a large cigar. The novelty was that it puffed real smoke.[17] Other aspects of the district, which critics perceive as inappropriate, date to an even earlier period: telegraph wires hung along the city's main street during the early 1860s. Long before the close of the nineteenth century, electrical and telephone wires added to the chaotic visual corridors.

Ironically, even the reputation of tourist trap strikes a common chord with the nineteenth-century heritage. The mining industry exploits nonrenewable resources. Unlike properly executed agriculture and timbering, which seek to conserve natural resources so that the land can repeatedly yield profits, mining is usually a one-time proposition. The industry tries to extract valuable minerals efficiently, usually with the expectation of eventually moving on to the next ore discovery.[18]

Similarly, many Comstock residents initially viewed tourists much like bodies of ore: they were to be exploited efficiently, and there was no expectation of reworking the resource. A remarkable assemblage of historic structures combined with the fame that television brought to the area, creating this new source of wealth. Like its counterpart of the nineteenth century, the bonanza was amply rich and large to justify wasteful practices. If visitors left unhappy because of a shabby experience, it did not matter. That they had spent their money was sufficient. Few conceived of return visits or worried about the bad public relations of word of mouth. This was a second big bonanza, and all that mattered was that the flush times had returned. Other historic mining districts have had their tourists, often depending on visitors for a steady income, but the Comstock had hit the mother lode. Its premier status in the nineteenth century had little to do with its more recent success. The thousands of visitors were a consequence of television, a phenomenon that could have focused just as easily on Telluride, Colorado, or Virginia City, Montana. Chance brought a boom to the Comstock now, just as it had a hundred years earlier. Fortunately, local entrepreneurs knew how to exploit an opportunity for as long as it lasted, and many remembered that following every bonanza is a *borrasca*.

With the end of the Cartwrights, it became increasingly possible to see the withering of Virginia City's twentieth-century bonanza. The growth experienced in the 1960s and early 1970s ceased in subsequent years. Slowly, many business owners came to realize that tourists could be a renewable resource and that public relations did matter. It was a radical change for a community geared for exploitation rather than cultivation.

Unfortunately, much of the historic district disappeared after years of neglect. For example, almost nothing remains today of the several blocks of the old Barbary Coast that lined the west side of South C Street, as illustrated in map 4. The stone ruins of the old tin shop still stand, as do a few back walls here and there. About ten feet to the south of the tin shop stands a row of old locust trees. They mark the border of Flowery Street that on the west side of C Street became an alley. As it ascended to B Street, the steep incline required stairs. The area is now choked with trees and undergrowth, but an old stone retaining wall and a scattering of lumber mark the spot where people used the passageway. Nearby plum and apple trees stand next to small level spots, sites of houses, ghosts of homes. The trees are descendants of nineteenth-century predecessors planted to make the neighborhood a pleasant place of blossoms in the spring, shade in the summer, and pies and preserves in the autumn. The house at 136 South C Street, two buildings south of the tin shop, was gone by the turn of the century, replaced with a new house built with scavenged lumber from the area. Significantly, the new house stands several feet further to the south, the residents at the time not wishing to live only a foot from their neighbors in nineteenth-century urban fashion. It is a small example of a subtle shift in the mining district's setting and use of the land.[19]

The creation of the Virginia City Historic District Commission in 1969 was a start at preserving what remained of the nineteenth century. When the Nevada legislature supplanted this organization with the Comstock Historic District Commission, placing a state agency in charge of local resources, it took action unprecedented in the nation to ensure proper management of a resource that was valuable to a sense of heritage and the regional economy. Since that time—especially since the early 1980s—business owners and others have used quality preservation, merchandise, and presentation to shed the darker aspects of Virginia City. Still, the mentality of exploitation runs deep. With the next television series or with the next ore strike, the Comstock could easily fall back into old patterns, relying on traditional, easier forms of exploitation instead of the more labor-intensive conservation.

The rehabilitation of Virginia City's Fourth Ward School serves as a good example of the process of deterioration followed by preservation on the Comstock. The county closed the school in 1936, and it remained unoccupied for the next fifty years. The harsh sun and weather caused the structure to deteriorate until it seemed a hopeless white elephant. Still, its alumni were prominent enough in Nevada history and politics that a rush to save the structure resulted in a state appropriation in 1964, the state's centennial year. It was a sentimental time, and it was possible for such a project to win support. Twenty years later, however, the building had again fallen into disrepair. The roof leaked, and a Washoe zephyr had knocked out a panel of windows. The Nevada State Historic Preservation Office provided the county with yet another grant to fix the building, but this step coincided with a growing realization that the Comstock should not be treated as a collection of standing ruins, propped up on occasion. Instead, all agreed that the Fourth Ward would need to be occupied if it were to have a secure future. In 1986, for the first time in fifty years, the venerated school opened once again to the public. Its local museum, created with funding from the Nevada Humanities Committee, now receives thousands of visitors a year.

Work on the statue of Justice that decorates the exterior of the 1877 Storey County Courthouse was yet another a benchmark of the 1980s. The seven-foot zinc statue cost $236, including shipping from its New York foundry. It is the only statue of Justice to grace the exterior of a Nevada courthouse, and it has long attracted attention because it lacks a blindfold. Originally the statue had a veneer of gold leaf, but by the 1940s much of the gold had weathered away, exposing the zinc underneath it. Locals painted the statue with silver skin and maroon robes, feeling that gold was less appropriate for this fine symbol of the Silver

State. By the 1980s the new layer of paint was falling off, again exposing the gold and zinc and making this dignified expression of an ideal look something like a leper. Locals began to agitate for its rehabilitation, and when one of the pans of Justice's scales dropped off, the situation reached a crisis. The Comstock Historic District Commission stepped forward with $3,000 and a plan: they decided to have Justice taken to Greg Melton, a sculptor and metal expert in Sutro. Melton stripped the goddess, soldered her joints, reattached the pan, and gave her a coat of automotive acrylic gold paint (all with the approval of the National Park Service, which agreed this would be the best course of action).

In 1988 Justice returned to her place overlooking Virginia City, in a celebration commemorating the Comstock. Honored guests included Nevada first lady Bonnie Bryan and Speaker of the Assembly Joe Dini, who gave an eloquent oration on the heritage of the old mining district and the importance of Lady Justice.[20] The statue, long a community symbol of nineteenth-century law and order, now exemplifies a new age for the Comstock. The 1980s and 1990s have been a time that saw several nationally approved tax credit projects, repairing structures scattered throughout the Comstock. One of the most dramatic rehabilitations was the main street project that returned the famed Delta Saloon and Sawdust Corner to an elegance in keeping with the nineteenth-century tenor of Virginia City's commercial corridor.[21] By the 1990s, only one Cartwright-era, cedar-veneer facade remained on C Street. The Bonanza Casino still boasts unpainted vertical boards, but a brick-and-iron building looms up from behind, waiting to take its place again as a reminder of the nineteenth century.

The Comstock's business community has also changed itself, as it adapts to evolving circumstance. Merchants no longer rely exclusively on hot dogs and T-shirts for profits. Many stores now carry quality jewelry, food, antiques, crystal, and other merchandise. To entice visitors, Virginia City offers parades for all occasions. It is unclear whether this is a conscious reference to a similar nineteenth-century tradition or merely coincidental. Nonetheless, these twenty-minute spectacles have become a Comstock custom, allowing the community to show off both the bizarre and the charming in its own unique blend. Many special events, scattered throughout the year, are led off with a thematic parade. Holidays and celebrations that inspire such processions include Saint Patrick's Day, Preservation Week, the Fourth of July, the camel races, Veterans Day, and the beginning of the Christmas season. Virginia City offers everything from a chili cook-off and a Rocky Mountain oyster fry to outhouse races. Founders Day, initiated in the early 1990s (with a parade), commemorates the first strike, in June 1859. The celebration includes the election of official "Town Characters." The post-television Comstock has relied on creativity as it continues to excavate the new ore body it discovered in tourism.

At the same time that the Comstock struggled with its image and sought to save visual reminders of its past, the role of mining within the historic district came into sharp focus. The increasing sophistication of cyanide extraction of microscopic gold allowed companies to mine low-grade ores at a profit. When the price of gold headed toward $600 an ounce in the late 1970s, much of the Lode became a candidate for profitable mining.[22] Houston Oil and Mineral proposed a large open pit south of Virginia City at the western edge of Gold Hill. Sentiment within the district was split: some rejoiced at the opportunity to return the Comstock to mining, its more honorable and natural economic base, but some saw large open-pit mines as a threat to tourism and a quality of life that had come to depend on the more sedate aspects of a former mining camp. The company won the first round. In 1979, it began excavating thousands of tons of earth, ultimately digging a hole approximately 450 feet deep, 800 feet wide, and 1,600 feet long.[23] The excavation threatened Gold Hill, and finally ended only a few feet from the original Comstock Highway that linked Virginia City with its sister communities to the south. Slumping to this day occasionally produces rifts in the road as the side walls slide into the pit, which remains without stabilization or reclamation.

Mining proponents correctly point out that the Comstock began with open pits, and that miners used this process throughout the 140 years of the district's development. While that is true, the Houston Oil and Mineral pit remains as testimony to the difference in scale that late twentieth-century machinery can effect. Nineteenth-century miners never dreamed of excavating a hole of this size, nor would they have found profit in such a marginal resource. It is one of the ironies of the Comstock that mining, upon which the district was founded, motivated Secretary of the Interior James Watt to list the historic area as an endangered landmark. Similarly, the project inspired the state legislature to alter the eminent domain law in 1981 to give local governments the authority to limit mining in historic districts if such an action was in the best interests of the community.[24] Although never implemented to date, the law testifies to the diminished relative importance of mining in Nevada since the days of the Big Bonanza. With the decline of the price of gold, mining on this scale ceased, but the 1970s and early 1980s have left their mark. The threat to other historic resources simply awaits the next price increase in a fickle market.

An aerial view taken November 30, 1979, shows the Houston Oil and Mineral Pit as Gold Hill at its very beginning. It already dwarfs the historical community at its edge. (Courtesy of Nevada State Historic Preservation Office)

No part of the Comstock has grappled with the nature of mining more than Silver City. This community has changed considerably since its inception in 1860. Some residents claim that it has more advanced degrees per capita that any other town in the nation. Given the cluster of archaeologists and others who have come to regard Silver City as a haven, this assertion may be true. At the same time, several of its roughly 160 residents have Comstock roots reaching back generations. Some of Silver City's residents feel that mining should comply with standards that do not infringe on the quality of life they have come to expect of the Comstock. Others maintain that the industry was first on the scene and has certain inherent rights to carry on unhindered. Mining companies have often proposed projects on the Storey County side of the border, just north of Silver City. Having a favorable hearing at the hands of a county generally more sympathetic to mining only heightens the concerns of tiny Silver City, which nonetheless appeals its case to the Bureau of Land Management and to court when necessary. These incidents have served to accentuate the contrast between Virginia City and Silver City, a difference that some locals maintain goes back to the Civil War, when the former was staunchly Union and many saw the latter with a Confederate cast. The nineteenth-century animosities are now gone, but on occasion they seem revived or replaced with new ones.[25] Ironically, it is the role of mining within this premier mining district that is often at the heart of the controversy.

Three recent incidents or issues go a long way toward demonstrating that although the Comstock has seen considerable change, there is at least some continuity. The first revolves around land ownership. During the late 1980s and early 1990s, it became apparent that nineteenth-century local governments created a severe problem as a consequence of a procedural error: Virginia City and Gold Hill were never formally platted with the federal government. Many, including the Bureau of Land Management, initially assumed that this meant that most of the communities remained on federal land. Since those who assumed that they were property owners (together with their predecessors) had paid local property taxes for more than a century, it was clear that the federal government would not wish to be in the position of chasing off the residents as squatters.

Subsequent research, however, demonstrated that those who owned original mining patents in the area may actually be the landowners. This line of reasoning maintained that since the communities were not platted as town sites, the only alternative way of gaining possession of the land would have been through mining claims, and since these were filed, the mining companies might have legitimate claim to much of the real estate in Virginia City and Gold Hill. The matter is complicated by the fact that the local government collected taxes on mining claims that apparently coincided with what it assumed to be valid private property, upon which homeowners also paid taxes. The controversy echoes J. Ross Browne's 1860 observation, quoted in the introduction, that no one has "title to property." His following assertion that there is "no property worth having" is hardly accurate, as now the issue of real estate ownership has millions of dollars at stake.[26] All parties have maintained that an amicable solution is possible, but the fine Comstock tradition of filing lawsuits over mining claims could easily become the alternative pursued.[27]

The year 1995 provided yet another indication of the way the Comstock's heritage survives. In January a controversy erupted over the election of Mrs. Virginia City. Jessi Winchester was the only candidate and won by default. The woman, who was in her fifties, listed her occupation as entertainer. With a background that included a longtime marriage and several children, she appeared to be a reasonable representative to the state contest. The *Comstock Chronicle*, however, disclosed that Winchester was a prostitute who apparently made her career choice in the face of family destitution because of her husband's incurring a disability. Sanctimonious outcries resounded in the area, and Winchester began to question whether the entrance fee into the state contest was worth it, since with the negative publicity, she was almost certain to lose. Almost at once, supporters emerged and drowned out the protests. Her family wrote letters to the *Comstock Chronicle* testifying to her upstanding character. Of even more interest were letters defending her career choice. These local authors insisted that Winchester was continuing a fine western tradition, and that, as one letter writer put it, "such simpering 'moralists' show no respect for our often colorful past by attempting to deny our colorful present." Some residents further advocated yet another popular regional practice of minding one's own business, echoing those who had no interest in objecting to the personal lives of Beebe and Clegg more than forty years earlier. "Given a choice between the Hooker and the Hypocrite," wrote former Storey County commissioner Karl Larson, "I'll go with the more honorable of the two (the Hooker) every time!"[28] Clearly these assertions were grounded in sometimes erroneous assumptions about the past. Still, enough people rallied to Winchester's defense to encourage her to continue her pursuit of a statewide title. She did not win the state contest and the moral protestations against her candidacy echoed in local papers for weeks, but Virginia City made a firm declaration about its character based on community perception of its heritage.

Yet another expression of the continuity of the Comstock can be found in the story of a local underground miner named Billy Varga. For decades, Varga had been a legend for his courage and hard work in the mines beneath the Comstock. In August 1993, Varga's microwave apparently malfunctioned, igniting several nearby sticks of dynamite. His house instantaneously turned into a mass of floating splinters. The blast blew out windows blocks away, and a thirty-pound mining jack soared up several streets, landing squarely on a Cadillac, the top of which crumpled. Varga, who was sitting in his living room at the time, walked away from the explosion with only a few scratches. The incident echoed the dynamite blast that killed Van Bokkelen and several others, presumably set off by the mining entrepreneur's pet monkey in 1873. Varga, however, was more fortunate than his nineteenth-century counterpart. Many locals persuasively pointed out that hard-rock mining was not for weaklings and that it had made a rugged man even tougher. When Varga died several months later of an unrelated illness, many mourned the passing of an age, suggesting that the Comstock would not see his kind again.[29] Right or not, it is remarkable enough that at least one formidable underground miner called the Comstock his home up to the 1990s. In all likelihood, others will follow in his footsteps and down his ladders.

Times might change, but the Comstock does not have to acknowledge that fact. The district could and will live on with its own brand of western and mining ethic, blithely ignoring its own evolution and the world around it, continually insisting that it maintains a firm anchor in the nineteenth century. Some aspects of the Comstock today seem clearly to be holdovers from its past. The occasional and unexpected opening of a deep nineteenth-century shaft serves to remind everyone of why Virginia City was founded in the first place. Still, that is only one of the many echoes from its history. After all, where else but in a mining town could a house blow up because of stored dynamite? In the same way, it seems that only the Comstock could carry over into the next millennium legal disputes over land ownership and mining claims that were initiated in the 1850s. Other aspects of the Comstock are holdovers exclusively in the minds of the residents, drawing on the image they have of their own heritage. During the heyday of the Comstock, the right of a prostitute to hold a title that made her an exemplary expression of marriage and womanhood would have been inconceivable. Still, there is nothing wrong with the process that has the local population recalling their heritage and modifying it to suit themselves and the changing times. From the 1850s to the present, the mining district has had to invent and reinvent itself to survive. Ultimately, who is better able to evaluate and shape the soul of a place that once captured the attention of the world than those who continue to claim the title "Comstocker"?

AFTERWORD

On May Day 1995, a sequence of events echoing two of the themes presented here left me convinced that the Comstock will survive for a long time to come. I traveled to Virginia City to videotape the house of Carol and Joe Page, who wish to leave their property to the Comstock Historic District Commission to serve as a museum. Their gift required them to describe the hundreds of furnishings and ornaments that make the house a remarkable portal into the past. The winter had been unusually wet, and on my way up the mountain, storm clouds spit rain. Small wildflowers cloaked much of the land, giving it a purple cast here and there. Cheatgrass and young mustard plants made the hills a vivid green, an unusual color for a Nevada landscape sure to grow yellow and brown with the heat of summer.

Carol has a fantastic memory for details and provided a history for almost all the objects that she had collected. She and Joe had worked lovingly for years to restore their house, which was built in 1864. In the same spirit, they selected furnishings that were precisely appropriate for the house. Whenever possible, they purchased Comstock relics, saving them from slipping away from the district. Indeed, preserving the Comstock is a decades-old ambition with them, and it has not stopped at their property. Joe is a long-term member of the Comstock Historic District Commission, and Carol helped found and promote Historic Preservation Weekend on the Comstock, making their house available for numerous house tours.

During the course of videotaping as Carol described her home, it became clear that the physical reminders of the Comstock were not the only things she had preserved. Often she would stop at a piece of furniture and point out, "This folding table belonged to John Mackay, and he used it in his box at Piper's Opera House" or "I bought this chair from Aileen Jacobsen, who assured me that it was from the Comstock." Without these recollections, the Pages' antiques would have been merely old objects from any number of places. Instead, their oral tradition preserves yet another aspect of the mining district's heritage. For more than a century and continuing into the present, the people of the Comstock have shaped their past with the stories they tell. The reality of the place, including the houses and other buildings that dot the landscape and the facts of its history, provides parameters to ensure that those who tell the stories do not stray too far. Nonetheless, the community's folklore about itself will persist as Comstockers continue to define their own past and their image.

As I walked out of the Page house that afternoon, I noticed the faint glimmer of a rainbow down Six Mile Canyon. It touched Sugarloaf, a volcanic plug at the base of the ravine. In a few seconds the rainbow grew increasingly bright, showing its entire span. Here, too, was a clear definition of the Comstock, since certainly there are few places in the world where it is possible to stand on solid ground and look down at a rainbow. Once again, the Comstock is a product of its place, and nothing is likely to change its dramatic setting.

Carol is fighting cancer. I hope that she will live long enough to see these words in print, but her prediction is that this is not likely, and I have learned not to challenge her intuition. The Comstock she helped preserve—with paint and with nail and with the retelling of Comstock stories—will be her legacy. And the rainbow, like some biblical contract, promises that the land embracing this remarkable place as well as the spirit of Carol and all the other wonderful people who call or have called themselves Comstockers will always be there.

Ronald M. James
May 2, 1995

In memory of Carol Page (1935–1996), who taught all associated with the Comstock a great deal about life.

NOTES

Acknowledgments

1. Christopher Lloyd, *Explanation in Social History* (Oxford: Basil Blackwell, 1986).

2. See especially E. J. Hobsbawm, *The Age of Capital, 1848–1875* (New York: Charles Scribner's Sons, 1975).

Introduction

1. J. Ross Browne, *A Peep at Washoe and Washoe Revisited* (1860 and 1863; reprint, Balboa Island, Calif.: Paisano Press, 1959), 57.

2. Mark Twain, *Roughing It* (1871; reprint, New York: Harper and Brothers Publishers, 1913), 2:43. See also Wilbur S. Shepperson, *Restless Strangers: Nevada's Immigrants and Their Interpreters* (Reno: University of Nevada Press, 1970), 1 ff., for a discussion of the international character of Nevada in the nineteenth century.

3. These terms are employed by Immanuel Wallerstein in *The Modern World-System: Capitalist Agriculture and the Origins of the European World-Economy in the Sixteenth Century* (New York: Academic Press, 1974); *The Capitalist World-Economy* (New York: Cambridge University Press, 1979); *World-Systems Analysis: Theory and Methodology* (Beverly Hills: Sage Publications, 1982); and *Politics of the World Economy: The States, the Movements, and the Civilizations* (New York: Cambridge University Press, 1984). Much of his world-system theory is problematic, but the overall message, of an increasingly united world framework, is a useful concept. E. J. Hobsbawm, in *The Age of Capital,* echoes much of this position without employing some of the methodology that has drawn criticism to Wallerstein. See also Thomas J. McCormick, "World Systems," *Journal of American History* 77 (1990): 125–32.

1. A Glimmer of Opportunity: The Setting

1. Eliot Lord, *Comstock Mining and Miners* (1883; reprint, San Diego: Howell-North, 1959), 11–14; Leonard J. Arrington and Davis Bitton, *The Mormon Experience: A History of the Latter-day Saints* (New York: Alfred A. Knopf, 1979), 174. See also James W. Hulse, *Silver State: Nevada's Heritage Reinterpreted* (Reno: University of Nevada Press, 1991), 50–66; Russell R. Elliott, *History of Nevada*, 2nd ed. (Lincoln: University of Nebraska Press, 1987), 61–68. For a firsthand account, claiming to have witnessed the discovery of gold in Gold Canyon in July 1850, see *Gold Hill Daily News*, 24 February 1880, 2:2.

2. Frequently ascribed to Samuel "Mark Twain" Clemens. Exact source unknown.

3. *Territorial Enterprise*, 5 February 1859, 2:4. Emphasis added.

4. See J. S. Holliday, *The World Rushed In: The California Gold Rush Experience* (New York: Simon and Schuster, 1981).

5. William Wright [Dan De Quille, pseud.], *The Big Bonanza* (1876; reprint, New York: Alfred A. Knopf, 1953), 14 ff.; Lord, *Comstock Mining*, 24–31; Henry DeGroot, *The Comstock Papers* (Reno: Grace Dangberg Foundation, 1985), 5; J. Wells Kelly, *First Directory of the Nevada Territory* (1862; reprint, Los Gatos, Calif.: Talisman Books, 1962), 196–97; Charles Howard Shinn, *The Story of the Mine as Illustrated by the Great Comstock Lode of Nevada* (1896, 1910; reprint, Reno: University of Nevada Press, 1980), 26–34. The most important source regarding the Grosh story is recently discovered letters from the brothers. See the *Mountain Democrat*, 15 August 1997.

6. Lord, *Comstock Mining*, 16, 19–21. See also Wright, *Big Bonanza*, 9–10.

7. Lord, *Comstock Mining*, 16, 20; Wright, *Big Bonanza*, 11.

8. Lord, *Comstock Mining*, 13–16; Wright, *Big Bonanza*, 10–11; *Territorial Enterprise*, 16 April 1859; Shepperson, *Restless Strangers;* Ronald M. James, Richard D. Adkins, and Rachel J. Hartigan, "A Plan for the Archeological Investigation of the Virginia City Landmark District" (Carson City: Nevada State Historic Preservation Office, 1993).

9. Lord, *Comstock Mining*, 33; see also extant issues of *Territorial Enterprise*, throughout the spring of 1859.

10. Lord, *Comstock Mining*, 33–34.

11. DeGroot, *Comstock Papers*, 4.

12. Ibid., 35–36; Wright, *Big Bonanza*, 21–23; *Territorial Enterprise*, 29 January 1859, 2:5.

13. Lord, *Comstock Mining*, 36; Wright, *Big Bonanza*, 21–23; *Territorial Enterprise*, 29 January 1859, 2:5.

14. *Territorial Enterprise*, 16 April 1859, 2:5; 21 April 1859, 1:2; 28 April 1859, 1:2.

15. Ibid., 21 May 1859, 2:2, and 4 June 1859, 2:4.

16. Lord, *Comstock Mining*, 37, 38; Wright, *Big Bonanza*, 24–25; *Territorial Enterprise*, 2 July 1859, 2:2. And see the *Enterprise* article of 4 June 1859, 2:4,

which apparently, though not conclusively, alludes to an earlier strike in the area. In addition, see Robert E. Kendall, "Henry Comstock's Offer Refused," *Comstock Chronicle*, 16 December 1994, 6, for an eyewitness account of the earliest events surrounding the McLaughlin-O'Riley strike. The account is taken from an article in the *Mining and Scientific Press.*

17. *Territorial Enterprise*, 28 April 1859, 1:2.

18. Lord, *Comstock Mining*, 38–39; Wright, *Big Bonanza*, 26–27; DeGroot, *Comstock Papers*, 7–8.

19. *Territorial Enterprise*, 25 June 1859, 2:3–4.

20. See, for example, Kelly, *First Directory*, 105. And see the *Territorial Enterprise*, 2 July 1859, 2:2.

21. *Territorial Enterprise*, 16 July 1859, 2:4; J. Ross Browne, *Resources of the Pacific Slope with a Sketch of the Settlement and Exploration of Lower California* (1869; reprint, New York: D. Appleton and Company, 1969), 2–14; *Territorial Enterprise*, 9 July 1859, 2:5.

22. Ibid., 5 February 1859, 2:4.

23. Ibid., 16 July 1859, 2:4. Houseworth, the district recorder, diligently made note of each claim and transaction, and although specific locations are often missing, his record book, Gold Hill Record Book A, on file in the Storey County Courthouse, remains an exceedingly useful source on the pivotal year of 1859. For an evaluation of the mines as worthless in early 1859, see Kendall, "Comstock's Offer Refused."

24. An arrastra is a device of Spanish-Mexican origin in which heavy stones are rolled and dragged over ore by an animal hitched to a turnstile. The pulverized rock can then be washed or treated with mercury to extract gold. See Donald L. Hardesty, *The Archaeology of Mining and Miners: A View from the Silver State*, Special Publication Series, no. 6 (Ann Arbor: Society of Historical Archeology, 1988), 9, 10, 39; Otis Young, *Western Mining* (Norman: University of Oklahoma Press, 1970), 69–71; Gold Hill Record Book A, p. 7, recorded 25 June 1859; Lord, *Comstock Mining*, 54.

25. Wright, *Big Bonanza*, 33; DeGroot, *Comstock Papers*, 11–12; Lord, *Comstock Mining*, 54–55; for a look at the early days of Aurora in the Sierra Nevada and its similar rush only a little later, see Roger D. McGrath, *Gunfights, Highwaymen, and Vigilantes: Violence on the Frontier* (Berkeley: University of California Press, 1984), 1–16.

26. See Gold Hill Record Book A; Lord, *Comstock Mining*, 45–49.

27. Browne, *Peep at Washoe*, 8. And see Genesis 10:29, 1 Kings 9:28, Job 28:16.

28. The Gold Hill Record Book A, which Wright claimed to have read, provides a chronicle of the names used. See Wright's *Big Bonanza*, 32; *Territorial Enterprise*, 24 December 1859, 2:4, and 31 December 1859, 2:2. See also the *Alta California*, 25 September 1859, reprinted from the *Territorial Enterprise*, 24 September 1859; Grant H. Smith, *The History of the Comstock Lode: 1850–1920*, 6th ed. rev. (1943; reprint, Reno: Nevada Bureau of Mines and the University of Nevada, 1966), 13–16.

29. DeGroot, *Comstock Papers*, 17, and see p. i; Lord, *Comstock Mining*, 57; Twain, *Roughing It*, 2:93; George D. Lyman, *The Saga of the Comstock Lode: Boom Days in Virginia City* (New York: Charles Scribner's Sons, 1951), 361 n. 3:4.

30. Lord, *Comstock Mining*, 65.

31. Wright, *Big Bonanza*, 64.

32. See, for example, *Gold Hill Daily News*, 23 June 1865, 2:2.

33. *Territorial Enterprise*, 24 December 1859, 2:4.

34. Some sources suggest that Finney's last name was in fact Fennimore. He appears in the 8th U.S. Manuscript Census of 1860 as Finney. See, however, Effie Mona Mack, *Nevada: A History of the State from the Earliest Times Through the Civil War* (Glendale, Calif.: Arthur H. Clark, 1936), 204–207.

35. See, for example, Gold Hill Record Book A. On p. 78, dated September 3, 1859, there is a record of a transaction signed by James Finney with an X, labeled "his mark."

36. "Notes and Sketches of the Washoe County," *Hutchings' California Magazine* 4, no. 10 (April 1860): 310. This article also underscores Finney's association with whiskey. *Alta California*, 22 July 1861, 1:4. See also the *Marysville Daily Appeal*, 17 July 1861, 4:1. Thanks to Eugene M. Hattori for assistance with these sources.

37. Kelly, *First Directory*, 105.

38. DeGroot, *Comstock Papers*, 9, 30; *Alta California*, 22 July 1861, 1:4.

39. Kelly, *First Directory*, 105.

40. See, for example, *Territorial Enterprise*, 29 January 1859, 2:5; 24 December 1859, 3:1; 21 May 1859, 2:2.

41. *Virginia Evening Bulletin*, 3 August 1863, 2:4. Comstock asserted that he had inherited claims from the Grosh brothers that remained valid and had never been purchased by the mine operators of the time.

42. Ibid., 1 October 1863, 3:2; *Gold Hill Daily News*, 21 November 1864, 2:6; 13 January 1865, 2:1; 9 August 1865, 3:1.

43. *Territorial Enterprise*, 20 February 1868, 2:2; 22 December 1868, 3:1; 13 January 1875, 2:5; 10 July 1875, 2:3.

44. DeGroot, *Comstock Papers*, 8, 10, 15–16, 29–30, 34, 38, 50.

45. *Territorial Enterprise*, 13 January 1875, 2:5.

46. Wright, *Big Bonanza*, 10, 20, 27, 29, 52–53, 54–57.

47. Ibid., 32; an article dating to April 1860 mentions that the town was named after Finney, but it says nothing of the supposed baptism. See "Notes and Sketches," *Hutchings' California Magazine*, 311.

48. An early version of Finney's death by a fall from a horse dates to a May 15, 1873, *Territorial Enterprise* article. The newspaper purports to preserve the text of an interview with a Paiute who witnessed the death. It is possible that the piece is fanciful. The text was reproduced in Wright's *Big Bonanza*, 53.

49. Wright, *Big Bonanza*, 28–29, 55–57. The *Territorial Enterprise*, 13 June 1875, 3:2, echoes the story of Finney's finding the original claim but mentions nothing of his drinking.

50. See Wright, *Big Bonanza*, 10, 20, 23, 25–28, 30–31, 42, 45–52, 62.

51. Ibid., 30–31, 49, 45–47. The *Territorial Enterprise*, 25 May 1875, 2:2, perpetuates the legend of Comstock's generosity by stating that Comstock was like all prospectors, representing a "generous, open-hearted race" who give treasure away "for a trifle."

52. Lord, *Comstock Mining*, 34, 411.

53. See George J. Young, "History of Mining in Nevada," in *The History of Nevada*, ed. Sam P. Davis (Reno: Elms Publishing, 1913), 316; Mack, *Nevada*, 204–205, 207, 438; Elliott, *History of Nevada*, 63–64; Hulse, *Silver State*, 67.

2. *The First Boom: Building the Community*

1. Residents of Chinatown later called their community Nevada City and then finally Dayton, a name it retains to this day. For Silver City, see *Territorial Enterprise*, 31 December 1859, 2:2. And see "Notes and Sketches of the Washoe County," *Hutchings' California Magazine* 4, no. 10 (April 1860): 312.

2. For observations about woodcutters in the mid-1860s, see Hemmann Hoffmann, *Californien, Nevada, und Mexico: Wanderungen eines Polytechnikers* (Basel: Schweighauserische, 1871), 131–35. The *Nevada State Journal* published a translation of the Nevada material on Sundays, April through May 1949, in eight installments, but the translation is inadequate. Thanks to Brian Galloway for his assistance with the German.

3. *Territorial Enterprise*, 10 December 1859, 1:4 (and see 17 December 1859, 1:6; 24 December 1859, 2:4; 18 February 1860, 2:4); Wright, *Big Bonanza*, 76.

4. Browne, *Peep at Washoe*, 64. John W. Reps sees a more orderly planned community documented in the historical record. There is some evidence of community planning in 1860, but much was clearly left to random choice. See John W. Reps, "Bonanza Towns: Urban Planning on the Western Frontier," in *Pattern and Process: Research in Historical Geography*, ed. Ralph E. Ehrenberg (Washington, D.C.: Howard University Press, 1975), 276–77. For an evaluation dating to 1864 of community planning, see Hoffmann, *Californien, Nevada, und Mexico*, 104–106. Hoffmann regarded community planning favorably, especially in comparison with his Swiss homeland. For a turn-of-the-century account of housing during this period, see Clifton Johnson, *American Highways and Byways: The Pacific Coast* (New York: Macmillan, 1908), 189–90.

5. Browne, *Peep at Washoe*, 24, 16.

6. Lord, *Comstock Mining*, 77–78; see also Twain, *Roughing It*, 2:16–23.

7. Lord, *Comstock Mining*, 73.

8. Ibid., 66, 67; see also 94–95. Clarence King, *Report of the Geological Exploration of the Fortieth Parallel*, Vol. 3, *Mining* (Washington, D.C.: Government Printing Office, 1870), 147–66, discusses the amount of supplies needed for mining. On the rush, see *Sacramento Union*, 22 March 1860, 3:1; 11 April 1860, 2:2; 27 April 1860, 3:2; 4 May 1860, 2:1; 9 May 1860, 5:2.

9. Browne, *Peep at Washoe*, 32.

10. All citations to the 1860 census are references to the 8th U.S. Manuscript Census for Utah Territory. Professor Kenneth H. Fliess of the University of Nevada, Reno, Anthropology Department entered the data into a computer and made it available for research on this project. For early doctors, see Browne, *Peep at Washoe*, 77–80; for the pottery shop, see the *Virginia Evening Bulletin*, 3 August 1863, 3:1. On establishing Comstock as an urban community, see Eugene Moehring, "The Comstock Urban Network," *Pacific Historical Review* 66 (1997).

11. For the history of the *Territorial Enterprise*, see Richard E. Lingenfelter and Karen Rix Gash, *The Newspapers of Nevada: A History and Bibliography, 1854–1979* (Reno: University of Nevada Press, 1984), 253–54. On Twain in Nevada, see his *Roughing It* and Katherine Hillyer, *Young Reporter: Mark Twain in Virginia City* (Sparks, Nev.: Western Printing, 1964); Effie Mona Mack, *Mark Twain in Nevada* (New York: Charles Scribner's Sons, 1947); George Williams III, *Mark Twain: His Life in Virginia City, Nevada* (Riverside, Calif.: Trees by the River, 1986).

12. Lord, *Comstock Mining*, 65; Shinn, *Story of the Mine*, 107–13; Hoffmann, *Californien, Nevada, und Mexico*, 110–11.

13. *Territorial Enterprise*, 3 March 1860, 1:6.

14. Ibid., 2:2; see also *Sacramento Union*, 17 March 1860, 2:3; 25 April 1860, 2:2; 5 September 1860, 1:7.

15. Twain, *Roughing It*, 2:91; Lord, *Comstock Mining*, 193; Browne, *Peep at Washoe*, 24–26, 182–84.

16. Wright, *Big Bonanza*, 99.

17. Ibid., 161–62, maintains that loads of up to 80,000 pounds were the norm, but his assertions can never be taken at face value. Nonetheless, Lord, *Comstock Mining*, 255, confirms loads of up to 75,050 pounds—90,690, including the wagon. See also Shinn, *Story of the Mine*, 105–107; John Debo Galloway, *Early Engineering Works Contributory to the Comstock* (Reno: Nevada State Bureau of Mines, 1947), 29–41; *Sacramento Union*, 8 May 1860, 3:1; 6 June 1860, 2:2.

18. Wells Drury, *An Editor on the Comstock Lode* (Palo Alto: Pacific Books, 1948), 138–42; on stages, see Shinn, *Story of the Mine*, 113–17; Browne, *Peep at Washoe*, 25, 60–62, 149–66, also comments on stage drivers.

19. Wright, *Big Bonanza*, 320–21; Twain, *Roughing It*, 1:137–43.

20. *Harper's Weekly*, 30 June 1877, 501–502; Douglas McDonald, *Camels in Nevada* (Las Vegas: Nevada Publications, 1983); Mary McNair Mathews, *Ten Years in Nevada, or Life on the Pacific Coast* (1880; reprint, Lincoln: University of Nebraska Press, 1985), 95–96; Hoffmann, *Californien, Nevada, und Mexico*, 111–12. On salt, see Jake Highton, "Salt: The Unsung 'Hero' of Nevada Mining," *Nevada Historical Society Quarterly* 26 (1983): 172–86. For a later reference to sightings of Comstock camels, see the *Silver Bow Standard*, 2 September 1905, 1:4.

21. Lord, *Comstock Mining*, 94; "Notes and Sketches," *Hutchings' California Magazine*, 311.

22. Wright, *Big Bonanza*, 85; Lord, *Comstock Mining*, 74.

23. *Territorial Enterprise*, 10 March 1860, 1:3. Wages apparently increased by late 1860, but it remained too early for the Comstock to take on the appearance of permanence. Congress did not create Nevada Territory until 1861. Lord, *Comstock Mining*, 96. Most Cornish have distinctive last names derived from Brythonic roots. Identification of ethnicity by name is problematic, but in the case of the Cornish, it is one of the few means available. See Ronald M. James, "Defining the Group: Nineteenth-Century Cornish on the Mining Frontier," in *Cornish Studies: 2*, ed. Philip Payton (Exeter: University of Exeter, 1994), and G. Pawley White, *A Handbook of Cornish Surnames*, 2nd ed. (Exeter: A. Wheaton, 1981).

24. These women either professed occupations or appear to have been living in situations that suggest professions other than prostitution.

25. See Ronald M. James, "Women of the Mining West: Virginia City Revisited." *Nevada Historical Society Quarterly* 36 (1993): 153–77. See also Holliday, *The World Rushed In*, 354; Sally Springmeyer Zanjani, *Goldfield: The Last Gold Rush on the Western Frontier* (Athens, Ohio: Swallow Press, 1992), 102–103, 104–108; Ralph Mann, *After the Gold Rush: Society in Grass Valley and Nevada City, California, 1849–1870* (Stanford: Stanford University Press, 1982), 201; Lord, *Comstock Mining*, 93.

26. Twain, *Roughing It*, 2:54; Lord, *Comstock Mining*, 75; Sally Springmeyer Zanjani, *"Ghost Dance Winter" and Other Tales of the Frontier* (Reno: Nevada Historical Society, 1994), 118–35.

27. Wright, *Big Bonanza*, 87–88; Lord, *Comstock Mining*, 75–77. Myron Angel (ed.), *History of Nevada: 1881* (Oakland, Thompson and West, 1881), 344, records these as the details of separate murders.

28. Wright, *Big Bonanza*, 88; Angel, *History*, 356–57.

29. Hoffmann, *Californien, Nevada, und Mexico*, 116. All this is not to say that there was not violence. Of course, Virginia City, like any western boomtown, had its share. For comparative purposes, see McGrath, *Gunfights, Highwaymen, and Vigilantes*.

30. Twain, *Roughing It*, 2:12.

31. See Angel (ed.), *History*, 206–208; Rachel J. Hartigan, "Looking for a Friend Among Strangers: Virginia City's Religious Institutions as Purveyors of Community" (undergraduate thesis, Yale University, 1993), 1–12; Kevin Rafferty, "Catholics in Nevada" (draft chapter for the *Nevada Comprehensive Preservation Plan*, 1992, on file in the Nevada State Historic Preservation Office); Rose Marian Shade, "Virginia City's Ill-Fated Methodist Church," *Journal of the West* 8 (1969): 447–53; Vincent A. Lapomarda, S.J., "Saint Mary's in the Mountains: The Cradle of Catholicism in Western Nevada," *Nevada Historical Society Quarterly* 35 (1992): 58–62; Drury, *Editor*, 29–30. For a nearly contemporary reluctance for religion to arrive in a mining boomtown, consider the case of Aurora in the Sierra Nevada; see McGrath, *Gunfights, Highwaymen, and Vigilantes*, 10–11. Compare Laurie F. Maffly-Kipp, *Religion and Society in Frontier California* (New Haven: Yale University Press, 1994). On Washoe zephyrs, see Twain, *Roughing It*, 1:147–48; Shinn, *Story of the Mine*, 64–65.

32. Browne, *Peep at Washoe*, 68.

33. For the dominance of U.S.–born citizens in newly found western mining districts, compare Mann, *After the Gold Rush*, 197, and Rodman Paul, *Mining Frontiers of the Far West, 1848–1880* (New York: Holt, Rinehart, and Winston, 1963), 25.

34. For the raw numbers of foreign-born in 1860, see the 9th Census of the United States, *Statistics of Population*, 299, table 4. The summary of the 9th U.S. Census (1870) is more valuable for Nevada in 1860, since the summary of the 8th U. S. Census reports information only for Utah Territory. The summary ten years later compares data with 1860 but also reports data for the Nevada part of Utah Territory separately.

35. Almost all the Hispanics lived in Virginia City: only four born in Spain and the one from Panama were listed in Gold Hill and the Virginia Mining District outside Virginia City during the 1860 census. Some people born in Mexico appear with the place of nativity as Sonora.

36. The census enumerator listed the Hispanic packers together. He probably recorded people consecutively, but this was not necessarily the case and cannot be the basis of firm conclusions. See also Browne, *Peep at Washoe*, 16, 22, 25–27; McGrath, *Gunfights, Highwaymen, and Vigilantes*, 139–40.

37. See the 8th U.S. Manuscript Census of 1860. Only one Hispanic woman, a milliner, declared an occupation in Virginia City and Gold Hill.

38. *Territorial Enterprise*, 3 March 1860, 1:6; Virginia Mining Record Book C, 5 March 1860, p. 1, Storey County Recorder's Office. See also Angel (ed.), *History*, 56. See the 1861 bird's-eye view of African American artist Grafton T. Brown in Douglas McDonald, *Virginia City and the Silver Region of the Comstock Lode* (Las Vegas: Nevada Publications, 1982), 18–19. The Maldonado residence is one of the few houses that Brown illustrated with an enlargement on the border.

39. Lord, *Comstock Mining*, 90 (see also the *Virginia Evening Bulletin*, 8 August 1863, 3:1); Young, *Western Mining*, 79, 155–56.

40. *Territorial Enterprise*, 3 March 1860, 1:5, 2:2.

41. Wright, *Big Bonanza*, 85.

42. *Territorial Enterprise*, 18 February 1860, 2:2.

43. Thanks for this insight to Michael Brodhead, professor emeritus of the University of Nevada, Reno.

44. *Territorial Enterprise*, 18 February 1860, 2:2; 3 March 1860, 1:2–3.

45. Angel (ed.), *History*, 151.

46. Lord, *Comstock Mining*, 59.

47. Nevada City Pioneer Cemetery; for the career of Meredith, see Mann, *After the Gold Rush*, 77; and see William M. Stewart, *Reminiscences of Senator William M. Stewart of Nevada*, ed. George Rothwell Brown (New York: Neale Publishing, 1908),123–25, for a personal recollection by a friend of Meredith, later to serve as a U.S. senator from Nevada.

48. Wright, *Big Bonanza*, 79. Letter from Bryant to his father, dated 31 May 1860, on file at the Nevada State Historic Preservation Office and in the Storey

County Recorder's Office, Virginia City, Nevada. Bryant was the husband of Louise Bryant, who later became Mrs. John Mackay, the wife of one of the richest men on the Comstock.

49. Three excellent overviews of this conflict and its resolution are Alice M. Baldrica, "Lander and the Settlement of the Pyramid Lake War," in *Frederick West Lander: A Biographical Sketch (1822–1862)*, ed. Joy Leland (Reno: Desert Research Institute), 151–90; McGrath, *Gunfights, Highwaymen, and Vigilantes*, 17–54; and Ferol Egan, *Sand in a Whirlwind: The Paiute Indian War of 1860* (Reno: University of Nevada Press, 1985). See also the *Sacramento Union*, 12 May 1860, 2:1; 14 May 1860, 2:2; 15 May 1860, 2:3–4; 19 May 1860, 2:3–4; 21 May 1860, 1:4; 2 June 1860, 1:7; 4 June 1860, 1:4; 4 June 1860, 2:2.

50. Roy S. Bloss, *The Pony Express: The Great Gamble* (Berkeley: Howell-North, 1959). See also LeRoy R. Hafen, *The Overland Mail: 1849–1869* (Cleveland: Arthur H. Clark, 1926), 165–91; Sam P. Davis, *The History of Nevada* (Reno: Elms Publishing, 1913), 260–65.

51. *Territorial Enterprise*, 17 December 1859, 1:6. See also Frankie Sue Del Papa, *Political History of Nevada*, 9th ed. (Carson City: Nevada State Printer, 1990); David Alan Johnson, *Founding the Far West: California, Oregon, and Nevada, 1840–1890* (Berkeley: University of California Press, 1992), 71–78.

52. Richard F. Burton, *City of the Saints, and Across the Rocky Mountains to California* (1862; reprint, New York: Alfred A. Knopf, 1963), 556–58. Burton's illustration is the most detailed record of his Comstock sojourn. He visited the area after an extensive tour of the West and was perhaps tired of recording the exhaustive observations for which he is noted. On the early growth of infrastructure, see Browne, *Peep at Washoe*, 178–79, 184; for the role of the Comstock as an international tourist attraction in the nineteenth century, see Wilbur S. Shepperson and Ann Harvey, *Mirage-Land: Images of Nevada* (Reno: University of Nevada Press, 1992), 33–35, 38. Thanks to Eugene M. Hattori, Nevada State Historic Preservation Office, for assistance in locating the place where Burton sat to make his drawing of Virginia City.

53. Carson County Records, Plats and Surveys, Storey County Courthouse. The survey was recorded on September 6, 1860.

3. The First Boom: Building the Mines

1. See Hardesty, *Archaeology of Mining and Miners*, 38 ff.; Angel (ed.), *History*, 583; "The Patio Amalgamation," *Mining and Scientific Press* 8, No. 20 (14 May 1864): 323, and (28 May 1864): 361; *Frank Leslie's Illustrated Newspaper*, 27 November 1875, 193–94; Young, *Western Mining*, 69–75. See also Highton, "Salt."

2. Lord, *Comstock Mining*, 80–88, maintains that Paul's first mill had twenty-four stamps. The figure appears as thirty-two in Kelly's 1862 *First Directory*, 100, where the facility is described as employing fifteen men and having a capacity of thirty tons per day.

3. "Treatment of Silver Ores," *Mining and Scientific Press* 8, no. 12 (19 March

1864): 178; also see King, *Geological Exploration*, 197–205, 206–72; and for the chemistry of the process, see 273–93. See also Lynn R. Bailey, *Supplying the Mining World: The Mining Equipment Manufacturers of San Francisco, 1850–1900* (Tucson: Westernlore Press, 1996), 11–14.

4. See Lord, *Comstock Mining*, 80–88, for a discussion of the growth and processes of the early Comstock mills, and see Kelly, *First Directory*, 100 ff., 108 ff., 170 ff., 199 ff., and 215 ff.; Angel (ed.), *History*, 60, 67, 584; *Sacramento Union*, 24 March 1860, 2:4; Lord, *Comstock Mining*, 113.

5. Lord, *Comstock Mining*, 117 ff.

6. Ibid., 116; the Nevada territorial manuscript census, 1861–1864, reveals an increasingly complex population as compared to the 1860 federal document.

7. Kelly, *First Directory*, 100; DeGroot, *Comstock Papers*, 84–85.

8. Smith, *History of the Comstock*, 83. See also DeGroot's "Comstock Papers," *Mining and Scientific Press* 33 (12 August 1876): 112. The latter does not agree completely with the former. Unfortunately, both sources are problematic. A Maldonado, presumably one of the brothers, died from gunshot wounds incurred during a dispute in 1863. See *Virginia Evening Bulletin*, 23 September 1863, 3:2; 24 September 1863, 3:1; 25 September 1863, 3:1.

9. Kelly, *First Directory*, 110. See also Lord, *Comstock Mining*, 124–28.

10. Smith, *History of the Comstock*, 25, citing *Alta California*, 8 April 1862, 1:5. See also Lord, *Comstock Mining*, 114, 116; and Henry DeGroot, "Pioneer Mills and Millers," *Mining and Scientific Press* 34, no. 6 (10 February 1877), reprinted in *Comstock Papers*, 71 ff.

11. Kelly, *First Directory*, 100.

12. Historical Cultural Resources Survey (HCRS), U.S. Department of the Interior, "Greiner's Bend: A Case Study" (Carson City: Comstock Project and Nevada State Historic Preservation Office, 1980).

13. *Territorial Enterprise*, 20 July 1861, 2:4.

14. Precise classification of the Washoe pine on Geiger Grade, distinct from other closely related pines, remains to be done. Washoe pine is known to exist in similar nearby andesite habitats. Thanks to James D. Morefield and Glenn H. Clemmer of the Nevada Natural Heritage Program for insight concerning Geiger Grade and its andesite ecosystem.

15. William G. White and Ronald M. James, "Little Rathole on the Big Bonanza: Historical and Archaeological Assessment of an Underground Resource," Survey Report (Carson City: State Historic Preservation Office, 1991).

16. Lord, *Comstock Mining*, 217–18; *Mining and Scientific Press* 6, no. 39 (27 July 1863): 1; Browne, *Peep at Washoe*, 213, 216; Shinn, *Story of the Mine*, 98; Smith, *History of the Comstock*, 83.

17. King, *Geological Exploration*, 50, 61, 103–16. See also Smith, *History of the Comstock*, 24.

18. Angel (ed.), *History*, 573–74; Lord, *Comstock Mining*, 90; Shinn, *Story of the Mine*, 94–97; Young, *Western Mining*, 244–47.

19. Douglas H. Strong, *Tahoe: An Environmental History* (Lincoln: University of Nebraska Press, 1984); Ronald H. Limbaugh, "John Muir and the Mining Industry," *Mining History Association Annual* (1996): 61–66; Mann, *After the Gold Rush*, 79; Twain, *Roughing It*, 2:93; Shinn, *Story of the Mine*, 118–20.

20. Lord, *Comstock Mining*, 314–16; see also Otis Young, "Philipp Deidesheimer, 1832–1916, Engineer of the Comstock," *Historical Society of Southern California* 57 (1975): 361–69.

21. Lord, *Comstock Mining*, 89.

22. Ibid., 232; King, *Geological Exploration*, 124–31; Shinn, *Story of the Mine*, 98–99.

23. Smith, *History of the Comstock*, 278 ff.; *Frank Leslie's Illustrated Newspaper*, 6 April 1878, 77; Young, *Western Mining*, 168 ff.; Robert E. Kendall, "Pitfalls and Perils of Deep Mining on the Comstock," *Nevada Historical Society Quarterly* 39 (1996): 216–31; Harry M. Gorham, *My Memories of the Comstock* (Los Angeles: Suttonhouse, 1939), 160–61.

24. *Alta California*, 13 April 1860, 1:9.

25. Lord, *Comstock Mining*, 297–300; Robert E. Stewart, Jr., and Mary Frances Stewart, *Adolph Sutro: A Biography* (Berkeley: Howell-North, 1962). For a thorough treatment of the water problem, discussing in part Sutro, see G. W. Dickie, "The Men and Machinery of the Comstock," *Engineering and Mining Journal* 98 (29 August–26 December 1914).

26. Browne, *Peep at Washoe*, 77–80; Kelly, *First Directory*, 108; Shinn, *Story of the Mine*, 100–101.

27. Virginia Mining Record Book A, Storey County Recorder's Office, Storey County Courthouse, 3 November 1859, 59–60, and 24 February 1860, 382, respectively. The Storey County Recorder's Office has several nineteenth-century maps that attempt to untangle the myriad mining claims. See also George Ferdinand Becker, *Atlas to Accompany the Monograph on the Geology of the Comstock Lode and the Washoe District* (Washington, D.C.: Government Printing Office, 1882).

28. Carson County Records: Plats and Surveys, 12–13.

29. Lord, *Comstock Mining*, 136–37; *Gold Hill Daily News*, 23 April 1864, 3:1; see also Browne, *Peep at Washoe*, 80–83, 215–19, 220–25.

30. In 1860 twenty-two lawyers lived in the part of Utah Territory that would become Nevada. See the 8th U.S. Manuscript Census of 1860. For a survey of several lawyers who worked on the Comstock, see George Thomas Marye, Jr., *From '49 to '83 in California and Nevada: Chapters from the Life of George Thomas Marye, a Pioneer of '49* (San Francisco: A. M. Robertson, 1923); for an early lawsuit, see *Sacramento Union*, 22 September 1860, 1:7.

31. Lord, *Comstock Mining*, 137–39; Virginia Mining Record Book E, 101.

32. Stewart, *Reminiscences*, 152–63; Russell R. Elliott, *Servant of Power: A Political Biography of Senator William M. Stewart* (Reno: University of Nevada Press, 1983), provides a full treatment of Stewart's life; see especially 26–33. See also

David Alan Johnson, "A Case of Mistaken Identity: William M. Stewart and the Rejection of Nevada's First Constitution," *Nevada Historical Society Quarterly* 22 (1979): 186–98; and see *Gold Hill Daily News*, 12 October 1863, 2:2, and 26 July 1864, 2:2.

33. Elliott, *History of Nevada*, 77 ff., has an excellent discussion of the move to statehood. See also Leslie Burns Gray, *The Source and the Vision: Nevada's Role in the Civil War* (Sparks, Nev.: Gray Trust, 1989), for a discussion of Nevada and the Civil War; McGrath, *Gunfights, Highwaymen, and Vigilantes*, 55–69. Also see Davis, *History*, 266–72; Alvin M. Josephy, Jr., *The Civil War in the American West* (New York: Alfred A. Knopf, 1991), 237, 262–63; Hoffmann, *Californien, Nevada, und Mexico*, 114–18.

34. See David Alan Johnson, "The Courts and the Comstock Lode: The Travail of John Wesley North," *Pacific Historian* 27 (1983): 31–46, especially 43 ff. For the depression and a broad treatment of its effects, see David Alan Johnson, "Industry and the Individual on the Far Western Frontier: A Case Study of Politics and Social Change in Early Nevada," *Pacific Historical Review* 51 (1982): 243–64; and Johnson, *Founding the Far West*, 71–97, 189–230, 313–15. See Lord, *Comstock Mining*, 131–80, for excellent overviews of the unfolding of litigation and politics in the period; and see William H. Brewer, *Up and Down California in 1860–1864* (New Haven: Yale University Press, 1930), 554, for a brief firsthand account of the 1864 depression. The need to define the complicated geology of the Comstock, for legal as well as exploratory reasons, was by no means settled, and many reports followed. See, for example, Baron Ferdinand von Richthofen, *The Comstock Lode: Its Character, and the Probable Mode of Its Continuance in Depth* (San Francisco: Sutro Tunnel Company, 1866); John Adams Church, *The Comstock Lode: Its Formation and History* (New York: J. Wiley and Sons, 1879); and Becker, *Atlas*.

35. Johnson, "Courts," 44; Lord, *Comstock Mining*, 163; and see Merlin Stonehouse, *John Wesley North and the Reform Frontier* (Minneapolis: University of Minnesota Press, 1965), 176.

36. Stewart, *Reminiscences*, 164–67; Elliott, *Servant of Power*.

37. Lord, *Comstock Mining*, 173; see also Smith, *History of the Comstock*, 64–74, and Shinn, *Story of the Mine*, 130–35, for overviews of the single-ledge controversy.

38. That is not to say that legal contests as well as armed conflict over claims did not continue. For an 1874 example, see Drury, *Editor*, 73–75.

39. Lord, *Comstock Mining*, 171–77; Elliott, *Servant of Power*, 21.

40. Lord, *Comstock Mining*, 146.

41. Alice B. Addenbrooke, *The Mistress of the Mansion* (Palo Alto: Pacific Books, 1959); Swift Paine, *Eilley Orrum: Queen of the Comstock* (Indianapolis: Bobbs-Merrill, 1929); see also Drury, *Editor*, 26–29; Brewer, *Up and Down California*, 557–58; Davis, *History*, 724–27; and Bernadette S. Francke, "Divination on Mount Davidson: An Overview of Women Spiritualists and Fortunetellers on the Comstock," in *Comstock Women: The Making of a Mining Community*, ed.

Ronald M. James and C. Elizabeth Raymond (Reno: University of Nevada Press, 1997), 165–78.

42. See Twain, *Roughing It;* Hillyer, *Young Reporter;* Mack, *Mark Twain;* and Williams, *Mark Twain.*

43. Judith Robinson, *The Hearsts: An American Dynasty* (New York: Avon Books, 1992), 41–44, 50, 53–55, 61, 63, 65–66.

4. Grief, Depression, and Disasters: Successes in the Midst of Failures

1. *Reese River Reveille,* 21 April 1864, 2:2, and 23 April 1864, 2:2; Angel (ed.), *History,* 268–70; Lord, *Comstock Mining,* 207–208; *Gold Hill Daily News,* 16 May 1864, 3:1, and 17 May 1864, 2:1, 3:2; J. Ross Browne, *Mining Adventures: California and Nevada, 1863–1865* (1863, 1865, 1869; reprint, Balboa Island, Calif.: Paisano Press, 1961), 114–20. In contrast with Angel, the *Gold Hill Daily News* asserts that Austin raised $5,335.

2. *Gold Hill Daily News,* 10 April 1865, 2:3; Alfred Doten, *The Journals of Alfred Doten: 1849–1903,* ed. Walter Van Tilburg Clark (Reno: University of Nevada Press, 1973), 830; Hoffmann, *Californien, Nevada, und Mexico,* 146; see also Smith, *History of the Comstock,* 55–56.

3. *Gold Hill Daily News,* 15 April 1965, 2:1.

4. Doten, *Journals,* 831; see also *Gold Hill Daily News,* 15 April 1865, 2:3.

5. Doten, *Journals,* 831; *Gold Hill Daily News,* 16 April 1865, 2:1; Angel (ed.), *History,* 271. See also *Virginia Daily Union,* 20 April 1865, 2:6, and 21 April 1865, 3:1.

6. *Gold Hill Daily News,* 20 April 1865, 3:1; *Virginia Daily Union,* 20 April 1865, 2:2. See also Mathews, *Ten Years in Nevada,* 52; Hoffmann, *Californien, Nevada, und Mexico,* 146–48; Smith, *History of the Comstock,* 56–57.

7. Smith, *History of the Comstock,* 59–60; Shinn, *Story of the Mine,* 136–53. For a general discussion of the Comstock and its depression of the mid-1860s, see Hoffmann, *Californien, Nevada, und Mexico,* 113–15, 149–50, 157–58.

8. Angel (ed.), *History,* 594; Smith, *History of the Comstock,* 60. Smith may have based his estimate on figures such as those that Allen C. Bragg gave, citing Virginia City's population as 30,000 in 1864. It is an unlikely figure that was based on a forty-five-year-old recollection. See Allen C. Bragg, "Pioneer Days in Nevada," *Second Biennial Report of the Nevada Historical Society, 1909–1910* (1911), 77. Brewer, *Up and Down California,* 554, estimated the 1863 population as 16,000 to 24,000 people, although it is unclear whether he referred to the Comstock in general or only to Virginia City. Hemmann Hoffmann, a Swiss traveler, estimated Virginia City's population in August 1864 to be more than 18,000; see Hoffmann, *Californien, Nevada, und Mexico,* 102. The territorial census has Storey County figures at fewer than 5,000 in both 1861 and 1862. It seems highly improbable that the population could have increased from 5,000 to anywhere from 16,000 to 24,000 in one year.

9. Anne M. Butler, "Mission in the Mountains: The Daughters of Charity

in Virginia City," in *Comstock Women: The Making of a Mining Community*, ed. Ronald M. James and C. Elizabeth Raymond (Reno: University of Nevada Press, 1997), 142–64; Smith, *History of the Comstock*, 59–60, 80.

10. This was the conclusion of preliminary surveys conducted by the Nevada State Historic Preservation Office as research for the preparation of James, Adkins, and Hartigan, "A Plan for the Archeological Investigation of the Virginia City Landmark District"; see also Ronald M. James, "On the Edge of Bonanza: Declining Fortunes and the Comstock Lode," *Mining History Association Annual* (1996): 101–108.

11. Drury, *Editor*; Lingenfelter and Gash, *Newspapers of Nevada;* Jake Highton, *Nevada Newspaper Days: A History of Journalism in the Silver State* (Stockton, Calif.: Heritage West Books, 1990).

12. See Smith, *History of the Comstock*, 49–50; David Lavender, *Nothing Seemed Impossible: William C. Ralston and Early San Francisco* (Palo Alto: American West Publishing, 1975), 165–82; Cecil Gage Tilton, *William Chapman Ralston: Courageous Builder* (Boston: Christopher Publishing House, 1935), 90–115.

13. Wright, *Big Bonanza*, 401–402; Tilton, *William Chapman Ralston*, 138–40; Lord, *Comstock Mining*, 244.

14. Lavender, *Nothing Seemed Impossible*, 186–87; Tilton, *William Chapman Ralston*, 137–48.

15. Lord, *Comstock Mining*, 244–48; Shinn, *Story of the Mine*, 162–66; Smith, *History of the Comstock*, 49–51.

16. Jackson, *Treasure Hill*, 5–33.

17. *Territorial Enterprise*, 9 April 1869, 3:2–3; Lord, *Comstock Mining*, 252. The exact amount of subscriptions is cited differently in the various sources and is probably clouded by the Bank of California control of many of the mines.

18. David Myrick, *Railroads of Nevada and Eastern California* (Berkeley: Howell-North, 1963), 155, and, for a general discussion, 136–62; see also Mark Wurm and Harry Demoro, *The Silver Short Line: A History of the Virginia and Truckee Railroad* (Virginia City: Virginia and Truckee Railroad, 1983), 36; Shinn, *Story of the Mine*, 166–68. Lucius Beebe and Charles Clegg in their fanciful history, *Virginia and Truckee: A Story of Virginia City and Comstock Times* (Oakland: Grahame H. Hardy, 1949), 15, maintain that the twists and turns amounted to twenty-two full circles.

19. *Territorial Enterprise*, 23 May 1869, 3:2; Lord, *Comstock Mining*, 250–53; Galloway, *Early Engineering Works*, 51–56.

20. Angel (ed.), *History*, 280–83; Lord, *Comstock Mining*, 250–56; Doten, *Journals*, 1061–1063; *Nevada Appeal*, 1 October 1869, 2:1–2.

21. *Gold Hill Daily News*, 13 November 1869, 3:2; Lord, *Comstock Mining*, 254; Wurm and Demoro, *Silver Short Line*, 30–55.

22. *Territorial Enterprise*, 24 June 1869, 3:2; Beebe and Clegg, *Virginia and Truckee*, 54–55.

23. *Territorial Enterprise*, 24 June 1873, 3:2; Wurm and Demoro, *Silver Short Line*, 82.

24. Beebe and Clegg, *Virginia and Truckee*, 54–55; Del Papa, *Political History*, 9–12; Guy Louis Rocha and Dennis Myers, "Myth #8: The Trestle of the State Seal," *Sierra Sage*, 16 August 1996, 16.

25. *Gold Hill Daily News*, 19 November 1869, 3:1–2; Lord, *Comstock Mining*, 255.

26. Lord, *Comstock Mining*, 255; 8th, 9th, and 10th U.S. Manuscript Censuses of 1860, 1870, and 1880, respectively.

27. Death records and information on tombstones frequently allude to mining accidents that do not appear in the local papers.

28. *Gold Hill Daily News*, 26 April 1869, 3:1–2; Lord, *Comstock Mining*, 270; Shinn, *Story of the Mine*, 231–34. For a full treatment of the disaster, see also *Territorial Enterprise*, 8 April 1869, 2:1 and 3:2–3; 9 April 1869, 3:4; *Gold Hill Daily News*, 7 April 1869, 2:2 and 3:1–2; 8 April 1869, 3:1–2; 9 April 1869, 3:1; 10 April 1869, 3:1–2; Wright, *Big Bonanza*, 126–31; Doten, *Journals*, 1041.

29. *Territorial Enterprise*, 8 April 1869, 3:2–3; 9 April 1869, 3:4; *Gold Hill Daily News*, 7 April 1869, 3:1. The newspaper accounts vary from Lord, *Comstock Mining*, 270–71.

30. *Territorial Enterprise*, 9 April 1869, 3:4.

31. Ibid., 8 April 1869, 3:2; Lord, *Comstock Mining*, 272; Wright, *Big Bonanza*, 126. The version presented here is from the contemporary newspaper, which varies from the later histories of Lord and Wright.

32. *Territorial Enterprise*, 8 April 1869, 3:2; Drury, *Editor*, 69–70; Wright, *Big Bonanza*, 129; Lord, *Comstock Mining*, 272; *Gold Hill Daily News*, 8 April 1869, 2:1. The *Gold Hill Daily News* estimated the loss at thirty-five; see 10 April 1869, 3:1.

33. Lord, *Comstock Mining*, 273; Wright, *Big Bonanza*, 129.

34. *Gold Hill Daily News*, 26 April 1869, 3:1–2.

35. For the relationship between the Yellow Jacket disaster and the Sutro Tunnel, see Stewart and Stewart, *Adolph Sutro*, 70–80. See also Adolph Sutro, *The Bank of California Against the Sutro Tunnel* (Washington, D.C.: M'Gill and Witherow, 1874), and his *Closing Argument of Adolph Sutro, the Sutro Tunnel* (Washington, D.C.: M'Gill and Witherow, 1872); United States Congress, House, *Report . . . in Regard to the Sutro Tunnel* (Washington, D.C.: M'Gill and Witherow, 1872).

36. *Territorial Enterprise*, 23 October 1869, 3:2. For an overview of the tunnel and its construction, see Stewart and Stewart, *Adolph Sutro*, 81–160; and see Shinn, *Story of the Mine*, 194–208.

37. *Territorial Enterprise*, 24 July 1869, 3:2.

5. A Time of Bonanza

1. There were 2,785 males fifteen years or older in 1860 in what would be the Storey County portion of the district.

2. Butler, "Mission in the Mountains."

3. Marion S. Goldman, *Gold Diggers and Silver Miners: Prostitution and Social Life on the Comstock Lode* (Ann Arbor: University of Michigan Press, 1981), remains

the best overview of Comstock prostitution. See, however, the comments in James, "Women of the Mining West." See also Anne M. Butler, *Daughters of Joy, Sisters of Misery: Prostitution in the American West, 1865–1909* (Urbana: University of Illinois Press, 1985), for one of the best overviews of nineteenth-century prostitution in the West.

4. Mathews, *Ten Years in Nevada*. See also James, "Women of the Mining West." For prostitutes five years later, see, for example, Nevada, *Census of the Inhabitants*, 2:251, 253, 337, 339, 340, 344, 347, 367, 369–75, 395–96, 401–402.

5. See James, Adkins, and Hartigan, "Competition and Coexistence." Location can be surmised from the 9th U.S. Manuscript Census of 1870, but the enumerator recorded most of the Chinese in several large blocks, leaving little question, particularly when compared to the patterns in the 1880 census, that they lived in segregated neighborhoods. Some of the Chinese living as domestics in Euro-American households appear in the census as cooks. The 1875 state census reports that the Chinese population had increased to 1,254 men and 87 women. See Nevada, *Census of the Inhabitants*, 2:615.

6. Sue Fawn Chung, "The Chinese Experience in Nevada: Success Despite Discrimination," *Nevada Public Affairs Review* 2 (1987): 43–51; and Sue Fawn Chung, "Their Changing World: Chinese Women on the Comstock, 1860–1910," in *Comstock Women: The Making of a Mining Community*, ed. Ronald M. James and C. Elizabeth Raymond (Reno: University of Nevada Press, 1997).

7. Chung, "Their Changing World." For the Chinese population in 1875, see Nevada, *Census of the Inhabitants*, 2:369–75, 395–96. That document is suspect, however, particularly when dealing with Asians. The 1875 census provides virtually no information except "Chinaman" or "Chinawoman" and an age for hundreds of people. It is not possible to regard the total number of Asians as an accurate count, and the information provided allows for no in-depth comparative analysis.

8. Table 2.1 in chapter 2 lists 17 percent of the population in 1860 as female, but this figure includes children. The 1870 census lists a handful of people as of Mexican nativity but bearing Irish or British surnames and Anglo first names. They are removed from the category of Hispanics here, although it is impossible to precisely determine their actual associations.

9. Mathews, *Ten Years in Nevada*, 168. Besides the 9th U.S. Manuscript Census of 1870, see Nevada, *Census of the Inhabitants*, which appears to document many of the same trends within the Hispanic population.

10. See Kelly, *First Directory*, and J. Wells Kelly, *Second Directory of the Nevada Territory* (Virginia City: Valentine and Company, 1863). Teamsters often brought wives, and the fact that he had boarders suggests that Jones may have had a wife with him. See James, "Women of the Mining West."

11. Kelly, *Second Directory*.

12. These figures are far removed from those published in the summaries. The census figures here take into account duplications, since many moved during a census and were documented twice or more, and also remove Hispanics and

Portuguese, since they were often inappropriately listed as mulatto. The summaries, particularly for the 1875 state census, differ widely from the published lists and are not to be trusted.

13. When directories are compared with census manuscripts, families are reported as included in the former when they appear in the latter.

14. John F. Uhlhorn, *Virginia and Truckee Railroad Directory, 1873–74* (Sacramento: H. S. Crocker and Company, 1873). The map in James and Raymond (eds.), *Comstock Women*, 220, showing Chinese neighborhoods, inadvertently added a third neighborhood on the north end of Virginia City. This is a mistake and was not a Chinese enclave.

15. Smith, *History of the Comstock*, 14–15. See also Lyman, *Saga of the Comstock Lode*, 68–69; and for a general overview, see Ethel Manter, *Rocket of the Comstock: The Story of John William Mackay* (Caldwell, Id.: Caxton Printers, 1950).

16. Smith, *History of the Comstock*, quoting D. O. Mills, *New York Herald*, 21 July 1902; *San Francisco Chronicle*, 21 July 1902. Lord, *Comstock Mining*, 301–302.

17. White and James, "Little Rathole."

18. Smith, *History of the Comstock*, 106.

19. Ellin Mackay Berlin, *The Silver Platter* (Garden City, N.Y.: Doubleday, 1957), 121–66.

20. Lord, *Comstock Mining*, 303–304; Angel (ed.), *History*, between 48 and 49, 97–98. See also Shinn, *Story of the Mine*, 173–93; and Smith, *History of the Comstock*, 116–18. Most early histories of Fair and Mackay rely on the recollections of the two men for their rise on the Comstock. Smith produced a more accurate overview with a greater reliance on newspapers and annual reports of the mine. He points out that it is difficult to determine whether Fair was foreman or assistant superintendent of the Hale and Norcross.

21. *Territorial Enterprise*, 18 October 1868, 3:1, and 15 November 1868, 3:3; C. C. Goodwin, *As I Remember Them* (Salt Lake City: Salt Lake Commercial Club, 1913), 161.

22. Lord, *Comstock Mining*, 302–303.

23. *Territorial Enterprise*, 26 February 1869, 3:2, and 11 March 1869, 3:1. See also Smith, *History of the Comstock*, 116–18.

24. Smith, *History of the Comstock*, 126–28; Lord, *Comstock Mining*, 281–82; Wright, *Big Bonanza*, 360–61.

25. Angel (ed.), *History*, 91–92; Smith, *History of the Comstock*, 129–35; Lord, *Comstock Mining*, 282–84. Sharon managed to secure his own Senate seat two years later, making him a junior legislator to his old adversary. He subsequently lost his seat to Fair in 1881. Jones remained in office for thirty years. See also Johnson, *Founding the Far West*, 318–19.

26. Lord, *Comstock Mining*, 308–309.

27. Ibid., 309–10. Smith, *History of the Comstock*, 152, points out that the story cited by Lord is largely the fanciful, self-serving invention of Fair.

28. Lord, *Comstock Mining*, 311.

29. Bertrand F. Couch and Jay A. Carpenter, *Nevada's Metal and Mineral Production (1859–1940, Inclusive)*, Geology and Mining Series, no. 37 (Reno: Nevada State Bureau of Mines and Mackay School of Mines, 1943), 132–38. Smith, *History of the Comstock*, provides an excellent overview of technical aspects of the big bonanza, including the subsequent development of the Bonanza Firm's mines.

30. Part of the silver is on display at the Silver Legacy Hotel Casino in Reno, on loan from the University of Nevada, Reno. The display includes text that provides some of the details here.

31. Galloway, *Early Engineering Works*, 63–74, provides an excellent overview of Schussler's plan; see also Shinn, *Story of the Mine*, 102–104; and Hugh Shamberger, *The Story of the Water Supply for the Comstock*, Geological Survey Professional Paper 779 (Washington, D.C.: U.S. Geological Survey, Government Printing Office, 1965). On the Risdon Iron Works, see Bailey, *Supplying the Mining World*, 75–81.

32. Hoffmann, *Californien, Nevada, und Mexico*, 106, describes fires based on his stay in Virginia City during 1864 and 1865, but these tended to destroy only individual structures, one at a time.

33. *Territorial Enterprise*, 27 October 1875, 1:1. See also *Gold Hill Daily News*, 26 October 1875, 2:1, 3:2–3, and 27 October 1875, 3:1–3; Lord, *Comstock Mining*, 325–29; Wright, *Big Bonanza*, 428–36; Doten, *Journals*, 1261–62, and Steven R. Frady, *Red Shirts and Leather Helmets: Volunteer Fire Fighting on the Comstock Lode* (Reno: University of Nevada Press, 1984), 170–83. Thanks to Steven R. Frady, Nevada Division of Forestry, for discussing the fire and answering questions about its development and the efforts of the firefighters.

34. *Territorial Enterprise*, 27 October 1875, 1:2.

35. Ibid., 28 October 1875, 1:4; Alice McCully Crane, transcription of letter dated 31 October 1875, on file in the Nevada State Historic Preservation Office and the Comstock Historic District Commission Office.

36. *Territorial Enterprise*, 27 October 1875, 1:3. See also Lord, *Comstock Mining*, 327; Wright, *Big Bonanza*, 429.

37. Goodwin, *As I Remember Them*, 161; Smith, *History of the Comstock*, 192, oddly reverses the roles. See also William Breault, S.J., *The Miner Was a Bishop: The Pioneer Years of Patrick Manogue, California—Nevada, 1854–1895* (Rancho Cordova, Calif.: Landmark, 1988), 87–88.

38. James Woods, *Recollections of Pioneer Work* (San Francisco: Joseph Winterburn, 1878), 225; Frady, *Red Shirts and Leather Helmets*, 183. Thanks to Storey County's sheriff, Robert Del Carlo, for pointing out the sealed hole and for insight regarding its discovery and possible use.

39. Frady, *Red Shirts and Leather Helmets*, 176–77, points out that Father Manogue's diary of contributions does not list Mackay. On the other hand, legend attributes a special role to Mackay, that of allowing the father to draw on his account whenever needed, which may have produced an economic relationship different from those recorded in the pledge book. See, for example, Goodwin, *As I Remember Them*, 161. Nonetheless, Mackay's role remains problematic.

40. Eliza Buckland, 31 October 1875, letter to sister and brother, on file in the Nevada Historic Preservation Office. Thanks to Bernadette Francke for assistance with this material.

41. *Territorial Enterprise*, 27 October 1875, 1:2–3, and 29 October 1875, 1:3; Lord, *Comstock Mining*, 327–28; see also Mathews, *Ten Years in Nevada*, 163–66.

42. *Territorial Enterprise*, 28 October 1875, 1:4.

43. Ibid., 27 October 1875, 1:3; 28 October 1875, 1:3; 29 October 1875, 1:5; Wright claims only two fatalities, but newspaper accounts suggest three. Wright, *Big Bonanza*, 432, also notes that "after the fire two or three men were killed by falling walls." Perhaps this is the source of the confusion.

44. *Territorial Enterprise*, 27 October 1875, 1:1; Wright, *Big Bonanza*, 434.

45. *Territorial Enterprise*, 27 October 1875, 1:1, and 29 October 1875, 1:4. See also Francke, "Divination on Mount Davidson"; Margaret Marks, interview by Ann Harvey, University of Nevada Oral History Program, Reno, 1988,14–15.

46. Alice McCully Crane, transcription of letter dated 31 October 1875, on file in the Nevada State Historic Preservation Office and the Comstock Historic District Commission Office; *Territorial Enterprise*, 28 October 1875, 1:3; John Taylor Waldorf, *A Kid on the Comstock: Reminiscences of a Virginia City Childhood* (1968; reprint, Reno: University of Nevada Press, 1991), 92.

47. *Territorial Enterprise*, 28 October 1875, 1:3.

48. Ibid., 29 October 1875, 1:3, 4, and 30 October 1875, 1:1, 5, 6; Lord, *Comstock Mining*, 328; Wright, *Big Bonanza*, 434; Woods, *Pioneer Work*, 227.

49. *Territorial Enterprise*, 29 October 1875, 1:5.

50. Ibid., 28 October 1875, 1:1, and 30 October 1875, 1:2; Lord, *Comstock Mining*, 328, 329.

51. Lord, *Comstock Mining*, 328–29.

52. Ronald M. James, *Temples of Justice: County Courthouses of Nevada* (Reno: University of Nevada Press, 1994), 133–42. Competition for the title of grandest courthouse is provided by the federal government with its Carson City federal courthouse, built in 1889, but nothing on the state or local level in Nevada comes close.

53. Richard C. Datin, *Elegance on C Street: Virginia City's International Hotel* (Reno: Privately published, 1977); Drury, *Editor*, 121.

54. Wright, *Big Bonanza*, 432.

55. See, for example, Duncan Emrich, "In the Delta Saloon: Conversations with Residents of Virginia City, Nevada," tape recordings, 1949, 1950, transcription in University of Nevada Oral History Program, Reno, 1991, 74–76.

6. The Workers: Labor in an Industrialized Community

1. Cedric E. Gregory, *Concise History of Mining* (New York: Pergamon Press, 1980), 221–22.

2. Register of Death, Storey County Recorder's Office, Virginia City, Nevada. See p. 112 for the example cited. A copy of the document is on file in the State

Historic Preservation Office, Carson City, Nevada. See also Lord, *Comstock Mining*, 404; Smith, *History of the Comstock*, 243–44; Kendall, "Pitfalls and Perils," 227–31.

3. Hugh Gallagher, longtime Comstock resident, recalls how such an incident claimed the life of his grandfather and several other miners. See Hugh James Gallagher, interview by Lucy Scheid, University of Nevada Oral History Program, Reno, 1984, 2.

4. Lord, *Comstock Mining*, 389–406; King, *Geological Exploration*, 84–87, and on ventilation, see 131–33.

5. *Territorial Enterprise*, 3 September 1876, 3:4.

6. Ibid., 1 July 1868, 3:1.

7. See, however, Jack Flanagan, interview by Ann Harvey, University of Nevada Oral History Program, 1984, 4.

8. Wright, *Big Bonanza*, 250–53; Ronald M. James, "Knockers, Knackers, and Ghosts: Immigrant Folklore in the Western Mines," *Western Folklore* 51 (1992): 153–77.

9. The profile of the occupants is based in part on documents found within the building. Furnishings remained intact and undisturbed until a rehabilitation of the structure in 1995. Rooms on the north side were 9½ feet wide, but on the south side they were only 7½ wide. The length of the rooms varied greatly, measuring from 7 to 11½ feet. Thanks to property owner Debra Thomas for supplying a floor plan of the property, on file at the Nevada State Historic Preservation Office. For a later view of the life of the miner in his apartments, see Joyce Hart Harper, *Rejoice: Comstock Memoirs* (Sparks, Nev.: Maverick, 1991), 9–14; see also Browne, *Peep at Washoe*, 202–205.

10. Otis Young, *Black Powder and Hand Steel: Miners and Machines on the Old Western Frontier* (Norman: University of Oklahoma Press, 1978), 95.

11. Lord, *Comstock Mining*, 224; see also Young, *Black Powder*, 97–99; Phillip I. Earl, *This Was Nevada* (Reno: Nevada Historical Society, 1986), 113–15; Smith, *History of the Comstock*, 47; Bailey, *Supplying the Mining World*, 115–26. Thanks to Guy L. Rocha, State Archives and Records Administrator, Nevada State Library and Archives, for assistance with Hallidie.

12. King, *Geological Exploration*, 117–24, 133–46; Young, *Black Powder*, 105–107, and see 108–109 for yet another Comstock innovation.

13. *Frank Leslie's Illustrated Newspaper*, 9 March 1878, 5; see also Twain, *Roughing It*, 2:94–95; Browne, *Peep at Washoe*, 208–15.

14. Shinn, *Story of the Mine*, 222–23; compare also Brewer, *Up and Down California*, 555–56.

15. The longest drill at a Comstock mine recorded by archaeologists was 3 feet, 6½ inches; the shortest was 11 inches; see White and James, "Little Rathole," 28.

16. Angel (ed.), *History*, 610; *Frank Leslie's Illustrated Newspaper*, 9 March 1878, 6; Lord, *Comstock Mining*, 390, 392, 395, 400; Flanagan, interview, 38–39.

17. The process of drilling was actually more complicated than the overview presented here. See Young, *Black Powder*, 17–18; Young, *Western Mining*, 182–86.

18. Waldorf, *Kid on the Comstock*, 114–19.

19. Doten, *Journals*, 1305, 1688, 2083. Experiments with electricity date to as early as 1868 with wiring for electric bells for communication. See the *Territorial Enterprise*, 28 April 1868, 3:1; Lord, *Comstock Mining*, 366. See, however, Flanagan, interview, 55, and H. L. Slosson, *Deep Mining on the Comstock* (San Francisco: H. L. Slosson, 1910).

20. Lord, *Comstock Mining*, 335–37; Young, *Black Powder*, 35–36; Young, *Western Mining*, 204–11; Hardesty, *Archaeology of Mining and Miners*, 21; Smith, *History of the Comstock*, 246–47; Bailey, *Supplying the Mining World*, 16–17.

21. Mann, *After the Gold Rush*, 183–85; *Territorial Enterprise*, 2 December 1868, 2:3; 9 January 1868, 1:2; 7 April 1868, 3:2; 18 August 1868, 3:1; 2 December 1868, 2:3; 9 February 1870, 3:1; Lord, *Comstock Mining*, 366–67; Young, *Western Mining*, 212–14; Hardesty, *Archaeology of Mining and Miners*, 22; Smith, *History of the Comstock*, 245–46.

22. Wright, *Big Bonanza*, 148.

23. *Territorial Enterprise*, 1 July 1873, 2:1–2, 3:1–3; 2 July, 3:2; 3 July, 3:2–3; 4 July, 3:1; Wright, *Big Bonanza*, 292; Angel (ed.), *History*, 598.

24. White and James, "Little Rathole," 45–50; Young, *Western Mining*, 79–89. For a description of this process, see *Frank Leslie's Illustrated Newspaper*, 2 March 1878, 445–46. See also Twain, *Roughing It*, 2:16–19.

25. *Territorial Enterprise*, 8 April 1871, 3:2, and 14 April 1871, 3:2. And see White and James, "Little Rathole." Thanks to Jan Loverin of the Nevada State Museum for identification of the clothing. Samuel Clemens records his attempts at small-time mining in *Roughing It*, 1:243–44. Journalist Alfred Doten also worked his own claim of a rathole mine, recording the effort in his private journal. See Doten, *Journals*, 717–99.

26. Information on workers appears in the 9th and 10th U.S. Manuscript Censuses, 1870 and 1880, respectively; Wurm and Demoro, *Silver Short Line*, 87–88.

27. It is difficult to associate some people—such as engineers, laborers, mechanics, and even carpenters—with any one industry, since workers with such generic titles could have served a mine, a mill, the railroad, or in some cases a foundry. Timothy Francis McCarthy, a blacksmith, worked for several of these industries throughout his Comstock career. See "Diaries of Timothy Francis McCarthy," on file at the Comstock Historic District Commission. All five of the men who appear under foundries for 1860 were machinists, a trade that this survey groups with foundries in later census records. The problem with such groupings is underscored by the fact that since there were no foundries on the Comstock in 1860, these machinists were certainly working for other facilities: the same is likely for at least some of their later counterparts who also appear under foundries. See table 6.1.

28. The 2 December 1879 issue of the *Territorial Enterprise*, for example, includes an advertisement for four of the foundries (1:5). Angel (ed.), *History*, 588, 589, 602–604, identifies seven foundries as operating at one time or another. Others probably came and went in the fluid economy of the first twenty years. Lord, *Comstock Mining*, 201; Angel (ed.), *History*, 603–604. On the original Fulton Foundry in San Francisco and its McCone–owned branch, see Bailey, *Supplying the Mining World*, 64–75.

29. See Burial Records for Storey County, Storey County Recorder's Office, Virginia City, Nevada. A copy is available in the Nevada State Historic Preservation Office, Carson City, Nevada.

30. Doten, *Journals*, 1068; see Browne, *Peep at Washoe*, 208–15, for a discussion of the dangers of tours.

31. Wright, *Big Bonanza*, 98; see also *Frank Leslie's Illustrated Newspaper*, 13 April 1878, 94; Smith, *History of the Comstock*, 256–57.

32. Hoffmann, *Californien, Nevada, und Mexico*, 108–109, describes the network established for meat and dairy products in 1864 and 1865.

33. Strong, *Tahoe;* Kelly J. Dixon, "The Frontier Lumbering Industry of Henness Pass Road: 1860s–1880s" (paper presented at twenty-fifth Great Basin Anthropological Conference, Kings Beach, Calif., October 1996; on file at the Nevada Historic Preservation Office); and see Kelly J. Dixon, Erika Johnson, and Juanita L. A. Spencer, "Sagehen Basin Analysis Area Historic Site Evaluation, Volume 2" (Tahoe National Forest Report #05-17-816, 1997), 7.

34. See Howard Hickson, *Mint Mark: "CC" The Story of the United States Mint at Carson City, Nevada* (Carson City: Nevada State Museum, 1972), and see S. Allen Chambers, Jr., *The Architecture of Carson City, Nevada* (Washington, D.C.: The Historic American Buildings Survey, National Park Service, n.d.), 21–26.

35. This is according to a hand count of the 1875 state census. Unlike the federal censuses, the state census was not computerized, because it is unreliable. The hand count may result in some inaccuracies that would not occur with the computerized document.

36. Lord, *Comstock Mining*, 387; Shinn, *Story of the Mine*, 226; *Frank Leslie's Illustrated Newspaper*, 9 March 1878, 5; Miriam Florence Squire Leslie, *California: A Pleasure Trip from Gotham to the Golden State* (New York: G. W. Carleton, 1877), 281; see also Drury, *Editor*, 19; Smith, *History of the Comstock*, 241.

37. See Guy Louis Rocha, "The Many Images of the Comstock Miners' Unions," *Nevada Historical Society Quarterly* 39 (1996): 163–81, for an excellent overview of the development and historiography of mining unions on the Comstock and in the West. See also Stewart, *Reminiscences*, 164–65.

38. Elliott, *History of Nevada*, 143.

39. Elliott's overview of the Comstock union activity is one of the best; see ibid., 141–44; and see Smith, *History of the Comstock*, 241–43. Other groups also organized, but their unions were less important to the community, dominated as it was by the mines. There was, for example, a mechanics' union. See *Territorial Enterprise*, 2 December 1879, 3:5.

40. See Rocha, "Images"; and see Gunther Peck, "Manly Gambles: The Politics of Risk on the Comstock Lode, 1860–1880," *Journal of Social History* 26 (1993): 701–23. Thanks to Guy Louis Rocha, administrator of the Nevada State Archives and Records, for his assistance with the topic of unions on the Comstock.

41. John Rule, "Some Social Aspects of the Cornish Industrial Revolution," in *Industry and Society in the South-West*, ed. Roger Burt (Exeter: University of Exeter Press, 1970), 71–106; John Rule, *The Experience of Labour in Eighteenth-Century English Industry* (New York: St. Martin's Press, 1981); John Rule, *The Labouring Classes in Early Industrial England, 1750–1850* (New York: Longman, 1986). Also see White and James, "Little Rathole," 47–48.

7. *The International Community: Ethnicity Celebrated*

1. Shepperson, *Restless Strangers*, provides an excellent overview of immigrants in Nevada, although examples and citations are problematic. For a nineteenth-century observation of diversity, see Mathews, *Ten Years in Nevada*, 165; Hoffmann, *Californien, Nevada, und Mexico*, 153.

2. See, for example, Fredrik Barth (ed.), *Ethnic Groups and Boundaries: The Social Organization of Cultural Difference* (Boston: Little, Brown), 9–38; G. Carter Bentley, "Ethnicity and Practice," *Comparative Studies in Society and History* 29 (1987): 24–55; Stanford M. Lyman and William A. Douglas, "Ethnicity: Strategies of Collective and Individual Impression Management," *Social Research* 40 (1973): 344–65; Mary C. Waters, *Ethnic Options: Choosing Identities in America* (Berkeley: University of California Press, 1990).

3. See Ronald M. James, "Erin's Daughters on the Comstock: Building Community," in *Comstock Women: The Making of a Mining Community*, ed. Ronald M. James and C. Elizabeth Raymond (Reno: University of Nevada Press, 1997), for an overview of the literature. See also Robert Arthur Burchell, *San Francisco Irish, 1848–1880* (Berkeley: University of California Press, 1980); David M. Emmons, *The Butte Irish: Class and Ethnicity in an American Mining Town, 1875–1925* (Urbana: University of Illinois Press, 1989); James P. Walsh, "The Irish in the New America: 'Way Out West," in *America and Ireland, 1776–1976: The American Identity and the Irish Connection*, ed. David Noel Doyle and Owen Dudley Edwards (Westport, Conn.: Greenwood Press, 1980); and see the special issue of the *Journal of the West* 31, no. 2 (April 1992) devoted to the western Irish. Johnson, *Founding the Far West*, 323, asserts that the Irish were not skilled when they came west, contradicting what Emmons has found in Butte. And see Ronald M. James, "Timothy Francis McCarthy: An Irish Immigrant Life on the Comstock," *Nevada Historical Society Quarterly* 39 (1996): 300–308.

4. Joseph Wickenden, "History of the Nevada Militia, 1862–1912" (compiled under the direction of Brigadier General Jay H. White, Adjutant General, 1941; microfilm copy in University of Nevada, Reno, Library); Angel (ed.), *History*, 587–88; *Territorial Enterprise*, 26 February 1867, 3:2, and 27 February 1867, 3:2; William D'Arcy, *The Fenian Movement in the United States: 1858–1886* (Washing-

ton, D.C.: Catholic University of America Press, 1947); Maurice Harmon (ed.), *Fenians and Fenianism* (Seattle: University of Washington Press, 1970); R. F. Foster, *Modern Ireland: 1600–1972* (London: Allen Lane, Penguin Press, 1988), 390–405.

5. See, for example, Angel (ed.), *History*, 262; *Territorial Enterprise*, 4 December 1874, 3:2; 2 October 1872, 2:6; 1 April 1873, 3:2; 10 February 1875, 2:8; 4 April 1875, 3:1; 7 April 1875, 2:6; 30 September 1880, 3:4–5; 12 October 1880, 2:8; 14 October 1880, 3:3; Foster, *Modern Ireland*, 403–19, 432–40.

6. Hasia R. Diner, *Erin's Daughters in America: Irish Immigrant Women in the Nineteenth Century* (Baltimore: Johns Hopkins University Press, 1983), 30–34; James, "Erin's Daughters."

7. Diner, *Erin's Daughters in America*. See also James, "Erin's Daughters"; James, Adkins, and Hartigan, "Competition and Coexistence."

8. James, "Defining the Group." For an example of the Cornish hiring their own and, in this case, dismissing German competitive labor, see Hoffmann, *Californien, Nevada, und Mexico*, 179.

9. Doten, *Journals*, 1193; James, "Defining the Group."

10. *Territorial Enterprise*, 3 August 1870, 3:1; 4 August 1870, 3:1.

11. Ibid., 6 August 1878, 3:2; 16 March 1876, 3:2; see also *Gold Hill Daily News*, 21 March 1864, 3:1; *Territorial Enterprise*, 22 March 1864; and Lord, *Comstock Mining*, 183. See also Doten, *Journals*, 899, for another incident in 1866.

12. *Territorial Enterprise*, 27 March 1879, 3:4, and 10 April 1879, 3:2–3; for an overview of the Chinese on the Comstock, see Judy Anne Thompson, "Historical Archaeology in Virginia City, Nevada: A Case Study of the 90-H Block" (master's thesis, University of Nevada, Reno, 1992); and see Loren B. Chan, "The Chinese in Nevada: An Historical Survey, 1856–1970," *Nevada Historical Society Quarterly* 25 (1982): 266–314; for additional analysis of the Chinese on the Comstock, see Sue Edwards, "Chinese Prostitution on the Comstock Lode, 1860–1880" (unpublished manuscript, 1983, in University of Nevada, Las Vegas, Library, Special Collections), and Sue Edwards, "Statistical Analysis of the Chinese on the Comstock Lode, 1870–1880" (unpublished manuscript, 1985, in University of Nevada, Las Vegas, Library, Special Collections). See Nevada, *Census of the Inhabitants*, 2:369–75, for the flourishing Chinatown of 1875, at which time Storey County had well over 1,000 Chinese residents; see also Alice Mildred Byrne, interview by Ann Harvey, Reno, 1984, University of Nevada Oral History Program, 12–14, 33; William Leslie Marks, interview by Ann Harvey, Reno, 1984, University of Nevada Oral History Program, 33–34.

13. Mathews, *Ten Years in Nevada*, 252.

14. For an overview of Comstock laundries, see James, Adkins, and Hartigan, "Competition and Coexistence." For a discussion on miners versus the Chinese, see Peck, "Manly Gambles," 701–23.

15. James, Adkins, and Hartigan, "Competition and Coexistence."

16. See, for example, Sharon Lowe, "The 'Secret Friend': Opium in Comstock Society, 1860–1887," in *Comstock Women: The Making of a Mining Commu-*

nity, ed. Ronald M. James and C. Elizabeth Raymond (Reno: University of Nevada Press, 1997). For occupations, see the 9th and 10th Manuscript Censuses of 1870 and 1880, respectively.

17. See Chung, "Their Changing World." See also Drury, *Editor*, 38; Anton P. Sohn, "Chinese Doctors in Nevada and the Great Basin," *Greasewood Tablettes* 8 (1997): 1–2.

18. Gray, *The Source and the Vision;* Elmer R. Rusco, *"Good Times Coming?": Black Nevadans in the Nineteenth Century* (Westport, Conn.: Greenwood Press, 1975).

19. Brown was the artist responsible for one of the earliest bird's-eye views of Virginia City.

20. The conclusion about the location of the boardinghouse is based on the probability that a smaller group of African Americans lived at 4 South D Street, an address indicated by comparison of directories and names listed together in 1870. The larger boardinghouse was only forty entries away in the manuscript census, suggesting that the enumerator had not traveled far between the two places. For an early discussion of this part of the 9th Manuscript Census of 1870, see Rusco, *"Good Times Coming?"*, 128.

21. In most cases these were not servants. See the 9th and 10th U.S. Manuscript Censuses of 1870 and 1880, respectively, and the 1875 state census. See also Drury, *Editor*, 37.

22. *Territorial Enterprise*, 5 April 1870, 3:1; see also Rusco, *"Good Times Coming?"*, 58.

23. Wright, *Big Bonanza*, 312. Thanks to Kelly J. Dixon for assistance with the literature on African Americans.

24. Doten, *Journals*, 1213; 9th U.S. Manuscript Census of 1870; 1875 state census; Kelly, *Second Directory;* Henry G. Langley, *The Pacific Coast Business Directory for 1867* (San Francisco: Henry G. Langley, 1867); M. D. Carr, *The Nevada Directory for 1868–69* (San Francisco: M. D. Carr and Company, 1868); see also Rusco, *"Good Times Coming?"*, 58, 73–80.

25. Brown appears in the 1875 state census, but then his presence on the Comstock is no longer indicated. His name appears in a Book of Deeds in the Storey County Recorder's Office; Book 49, page 1, dated 1885 and titled "Conveyance of Real Estate for Delinquent Taxes of 1885." The county apparently confiscated two lots he owned after he failed to pay $9.50 in taxes. Thanks to Kelly J. Dixon for her help with this material. See also Kelly, *Second Directory;* Charles Collins, *Mercantile Guide and Directory for Virginia City and Gold Hill* (Virginia City: Agnew and Deffebach, 1864–65); L. M. McKenney and Company, *McKenney's Business Directory of the Principal Towns of California, Nevada, Utah, Wyoming, Colorado and Nebraska, 1880–81* (Sacramento: H. S. Crocker and Company, 1882); Carr, *Nevada Directory for 1868–69;* Rusco, *"Good Times Coming?"*, 56; Doten, *Journals*, 889. For African Americans and saloons in Denver, see Thomas J. Noel, *The City and the Saloon: Denver, 1858–1916* (1982; reprint, Niwot, Colo.: University of Colorado Press, 1996), 26–28, 46–47. It appears that

Comstock African Americans fared better in the business than did those of Denver.

26. *Territorial Enterprise*, 6 June 1886, 3:1; see also the 9th and 10th U.S. Manuscript Censuses (1870 and 1880), which list her birthplace as Massachusetts and Kentucky, respectively. Her parents were born in Virginia. She too may have been a slave. The census lists her as keeping house, but the obituary supplies more details. Henry G. Langley, *The Pacific Coast Business Directory for 1871–73*, 2nd ed. (San Francisco: Henry G. Langley, 1871), identifies Payne as having a saloon at 11 South D Street. Uhlhorn, *Virginia and Truckee Railroad Directory*, 120, lists her residence at 78 North C Street. In addition, D. M. Bishop and Company, *Bishop's Directory: Virginia City, Gold Hill, Silver City, Carson City, and Reno, 1878–79* (San Francisco: B. C. Vandall, 1878), 199, lists her as owning a saloon at 94-B North C Street. The 10th U.S. Manuscript Census of 1880 shows Payne as living at 112 North C Street.

27. Rusco, *"Good Times Coming?"*, 75, 76, 99, 114, 116, 177, 182, 184; 10th U.S. Manuscript Census of 1880; 1875 state census; *Territorial Enterprise*, 9 April 1870, 1:5; 16 November 1876, 2:6.

28. Mathews, *Ten Years in Nevada*, 110–14; unfortunately, the incident turned out negatively, but the wish for literacy serves well here.

29. Angel (ed.), *History*, 217. For information on the African American churches on the Comstock, see Rusco, *"Good Times Coming?"*, 174–79. For the African American community in general, see Rusco and see also Michael S. Coray, "Influences on Black Family Household Organization in the West, 1850–1860," *Nevada Historical Society Quarterly* 31 (1988): 1–31; and Michael S. Coray, "African-Americans in Nevada," *Nevada Historical Society Quarterly* 35 (1992): 239–57.

30. Wright, *Big Bonanza*, 116–19; DeGroot, *Comstock Papers*, 103–105; Lord, *Comstock Mining*, 93, 389; Twain, *Roughing It*, 2:35; David Arnold D'Ancona, *A California-Nevada Travel Diary of 1876: The Delightful Account of a Ben B'rith*, ed. William M. Kramer (Santa Monica: N. B. Stern, 1975), 51.

31. *Gold Hill Daily News*, 20 July 1865, 3:2. The newspaper refers to a meeting a few months later in the headquarters of the "Mexican Patriotic Club." It is likely that this was the same organization as the Liberal Mexican Club. See ibid., 29 September 1865, 3:1.

32. *Territorial Enterprise*, 5 December 1867, 2:6; Doten, *Journals*, 863, 897, and see 1020; Shepperson, *Restless Strangers*, 21–23, discusses these aspects of the calendar.

33. Two excellent sources provide the basis for the discussion of Paiutes on the Comstock. See Eugene M. Hattori, *Northern Paiutes on the Comstock: Archaeology and Ethnohistory of an American Indian Population in Virginia City, Nevada* (Carson City: Nevada State Museum, 1975); and Eugene M. Hattori, "'And Some of Them Swear Like Pirates': American Indian Women in Nineteenth-Century Virginia City," in *Comstock Women: The Making of a Mining Community*,

ed. Ronald M. James and C. Elizabeth Raymond (Reno: University of Nevada Press, 1997). See also Drury, *Editor*, 38–39; Alice Mildred Byrne interview, 14; Carroll Dolve, interview by Ann Harvey, Reno, 1984, University of Nevada Oral History Program, 10–12.

34. Angel (ed.), *History*, 264; *Territorial Enterprise*, 7 October 1873, 3:2; 25 October 1873, 3:1; 9 December 1876, 3:3; 10 January 1880, 3:3; 11 December 1880, 3:1; *Gold Hill Daily News*, 4 January 1865, 3:2; 24 January 1865, 2:3; 26 January 1865, 3:1; Doten, *Journals*, 1035, 1617; *Territorial Enterprise*, 29 December 1874, 2:8; 28 December 1876, 2:6; 26 January 1877, 3:3–4; 3 January 1880, 2:6.

35. *Territorial Enterprise*, 16 July 1876, 2:4; 13 August 1876, 3:3; 9 July 1880, 2:7; 8 August 1880, 3:3–4. And see Doten, *Journals*, 1323, 1739: Doten mentions going to a Caledonian picnic as late as 1889.

36. *Territorial Enterprise*, 7 October 1873, 3:2.

37. Angel (ed.), *History*, 584, 264.

38. Doten, *Journals*, 798, 828, 933, 1201; Angel (ed.), *History*, 339, 423, 548, 598. Thanks to Robert Nylen of the Nevada State Museum for his help with van Bokkelen. Shepperson, *Restless Strangers*, 20, misspells his name as von Bokkelen, thereby implying a confused ethnicity, "von" being characteristically German and "van" being associated with the Netherlands. See the 9th U.S. Manuscript Census of 1870 for his New York nativity. That document lists van Bokkelen as a miner. *Territorial Enterprise*, 1 July 1873, 2:1–2, 3:1–3; 2 July, 3:2; 3 July, 3:2–3; 4 July, 3:1.

39. 9th U.S. Manuscript Census of 1870. Those listing nativity as other than one of the German states included two Englishmen, an Italian, a Frenchman, three additional people listing French nativity but having German names, and a man born in New York but having a German name. The percentage of Germans involved in the liquor trade relative to the total number of men employed in the industry in the various census manuscripts is 1860, 23 percent; 1870, 27 percent; and 1880, 21 percent. Compare Noel, *The City and the Saloon*, 53–55.

40. Lingenfelter and Gash, *Newspapers of Nevada*, 257, 260.

41. D'Ancona, *California-Nevada Travel Diary*, 38–55; Israel ben Joseph Benjamin, *Three Years in America, 1859–1862* (New York: Arno Press, 1975), 210; *Territorial Enterprise* 1 October 1872, 3:2, and 4 October 1872, 3:1. Mathews, always able to furnish an example of prejudice and narrow-mindedness, comments on Jews in Virginia City. See *Ten Years in Nevada*, 52–53; for comparison, see Noel, *The City and the Saloon*, 58–59.

42. Harriet Rochlin and Fred Rochlin, *Pioneer Jews: A New Life in the Far West* (Boston: Houghton Mifflin, 1984), 32–34, 190–93; John P. Marshall, "Jews in Nevada, 1850–1900," *Journal of the West* 23 (1984): 62–64; Norton B. Stern, "Notes on a Virginia Police Chief," *Western States Jewish Historical Quarterly* 21 (1979): 89–91.

43. See, for example, *Territorial Enterprise*, 11 March 1868, 2:5; 14 March 1868 3:1; 15 March 1868, 2:5; 6 July 1877, 3:4; 29 July 1873, 3:1; 12 September

1873, 2:4; 26 September 1873, 2:4; 14 September 1880, 2:7 and 3:4; 29 October 1880, 3:4. And see *Virginia Evening Bulletin,* 1 August 1863, 3:2; 3 August 1863, 3:1; 4 August 1863, 2:1 and 3:1; 15 August 1863, 3:2.

44. *Territorial Enterprise,* 26 October 1871, 3:2.

45. Lord, *Comstock Mining,* 93, citing *Sacramento Union,* 1 October 1860, 1:6. See also Mathews, *Ten Years in Nevada,* 58.

46. Twain, *Roughing It,* 2:43; Rossiter W. Raymond, *A Glossary of Mining and Metallurgical Terms* (Easton, Penn.: American Institute of Mining Engineers, 1881).

47. *Territorial Enterprise,* 27 March 1879, 3:4, and 10 April 1879, 3:2.

48. Ibid., 18 January 1872, 2:4; 20 January 1872, 3:1; 14 May 1872, 3:2. Compare the effect of the Chinese inspiring unity among Irish and Cornish miners in Grass Valley and Nevada City, California. See Mann, *After the Gold Rush,* 187–89, 216.

49. *Territorial Enterprise,* 1 September 1876, 2:3, and 5 September 1976 3:5.

50. James, "Defining the Group," 38–40.

51. James, "Women of the Mining West."

52. James, "Erin's Daughters."

53. James, Adkins, and Hartigan, "Competition and Coexistence."

8. The Moral Options: Sinners

1. Doten, *Journals,* 994–95.

2. William R. Jones (ed.), *Murder of Julia Bulette: Virginia City, Nevada, 1867* (Golden, Colo.: Outbooks, 1980); Douglas McDonald, *The Legend of Julia Bulette* (Las Vegas: Nevada Publications, 1980); Susan A. James, "Queen of Tarts," *Nevada Magazine* 44 (1984): 51–53.

3. Twain, *Roughing It,* 2:54.

4. Ibid., 2:76; Fred Lockley, *Vigilante Days at Virginia City* (Portland: Fred Lockley, 1924), 4–5, 6.

5. *Gold Hill Daily News,* 4 March 1871, 3:2; *Nevada State Journal,* 5 September 1969, 4, and 10 June 1959, 21; *Reno Gazette-Journal,* 7 April 1991, 13C; Drury, *Editor,* 143–57; Twain, *Roughing It,* 2:58.

6. Angel (ed.), *History,* 568.

7. Lord, *Comstock Mining,* 111.

8. See, for example, the argument presented by Richard Maxwell Brown, *Strain of Violence: Historical Studies of American Violence and Vigilantism* (New York: Oxford University Press, 1975), 91–179; McGrath, *Gunfighters, Highwaymen, and Vigilantes,* 225–46.

9. Lockley, *Vigilante Days,* 6.

10. Doten, *Journals,* 2190.

11. Wright, *Big Bonanza,* 181–85.

12. *Territorial Enterprise,* 26 March 1871, 2:1, 3:1–2.

13. Ibid., 28 March 1871, 2:2, 3:1; 21 July 1871, 3:2.

14. Wright, *Big Bonanza*, 185–86.

15. Twain, *Roughing It*, 2:62–64.

16. Drury, *Editor*, 17; see also 158–68; *Virginia Evening Bulletin*, 23 September 1863, 3:2; 24 September 1863, 3:1; 25 September 1863, 3:1.

17. Lord, *Comstock Mining*, 210–12, 378; see also Mathews, *Ten Years in Nevada*, 233–34.

18. This assertion is based on the assumption that vagrancy and "sleeping on the sidewalk" were basically the same, making a difference of 12 to 7, respectively, between 1863 and 1880.

19. Leslie, *California*, 280.

20. *Territorial Enterprise*, 2 July 1871, 3:2; see also 22 July 1877, 2:1, for an editorial on wife beating and a Nevada Revised Statute that called for the pillorying of wife beaters. The editor, who cited a case of wife beating, called for the enforcement of the law.

21. Mathews, *Ten Years in Nevada*, 109, 110–14, and see 242.

22. *Territorial Enterprise*, 4 July 1873, 3:2.

23. Mathews, *Ten Years in Nevada*, 242. Compare Drury's discussion of Black Billy the goat in Drury, *Editor*, 18–19.

24. *Territorial Enterprise*, 10 April 1879, 3:3. Hoffmann, *Californien, Nevada, und Mexico*, 105, describes these wooden boardwalks as early as 1864 and 1865.

25. Lord, *Comstock Mining*, 288–90; *Territorial Enterprise*, 11 January 1868, 3:1; 19 January 1871, 2:1; *Gold Hill Daily News*, 11 January 1868, 3:2; 15 February 1869, 3:1; see also Mathews, *Ten Years in Nevada*, 175–76.

26. For an early expression of this idea, see Mathews, *Ten Years in Nevada*, 193.

27. The figure for 1880 is adjusted above the number of prostitutes actually listed as such in the manuscript census. The increased count is possible by using criteria based on address, age, and living situation. Goldman, *Gold Diggers*, employed a similar system, but her use of the sources was problematic, causing her count of prostitutes in 1880 to be beyond what the data support. No such adjustments are possible for the 1870 census because that document does not include addresses, and yet even with this adjustment, prostitutes are much more common in 1870 than in 1880. See James, "Women of the Mining West." For a treatment of prostitutes in 1870, see George M. Blackburn and Sherman L. Ricards, "The Prostitutes and Gamblers of Virginia City, Nevada: 1870," *Pacific Historical Review* 48 (1979): 235–51.

28. This conclusion is suggested by the few addresses provided for Chinese prostitutes during the 10th U.S. Manuscript Census of 1880.

29. Chung, "Their Changing World."

30. James, "Queen of Tarts."

31. Doten, *Journals*, 817. See also *Virginia Daily Union*, 28 December 1864, 3:2, and 24 January 1865, 3:3.

32. Doten, *Journals*, 820; Goldman, *Gold Diggers*, 80–81.

33. Doten, *Journals*, 862, 863, 871, 873, 905, 910, 934, 976; Goldman, *Gold Diggers*, 51–52, 62–63, 79–80.

34. Wright, *Big Bonanza*, 276–77. Much of the eastern downhill side of South C Street in the vicinity of the Barbary Coast is too steep for development. Newspaper accounts on Comstock activities consistently provide even-numbered addresses for the Coast, giving the basis for concluding a western location on South C Street.

35. *Territorial Enterprise*, 7 August 1875, 3:5, and 20 January 1877, 3:2–3; *Gold Hill Daily News*, 19 January 1877, 3:2–3.

36. See, for example, *Territorial Enterprise*, 20 November 1875, 3:1; 5 April 1877, 3:2; 6 April 1877, 3:2.

37. Ibid., 7 April 1877, 3:4. Three months later the *Enterprise* felt it could herald the closing of yet another "Barbary-coast den of iniquity." See 21 June 1877, 3:3.

38. Ibid., 7 June 1877, 3:4.

39. Ibid., 21 July 1877, 3:3, and 22 July 1877, 3:4.

40. See the Sanborn-Perris Fire Insurance Map of 1890. In addition, the 10th U.S. Manuscript Census of 1880 confirms that the former Barbary Coast had become more respectable by that time, as indicated by noncontroversial occupations and the existence of families in the neighborhood.

41. See, for example, Blackburn and Ricards, "Prostitutes and Gamblers," 249.

42. Angel (ed.), *History*, 572.

43. Browne, *Peep at Washoe*, 186–88.

44. Wright, *Big Bonanza*, 268; Lord, *Comstock Mining*, 377 (Lord cites "Records of the United States Census Agents, 1880"); Drury, *Editor*, 122. Mathews also discusses the abundance of saloons; see *Ten Years in Nevada*, 192–93.

45. Ron Rothbart, "The Ethnic Saloon as a Form of Immigrant Enterprise," *International Migration Review* 27 (1993): 247; Perry R. Duis, *The Saloon: Public Drinking in Chicago and Boston, 1880–1920* (Urbana: University of Illinois Press, 1983), 28. Looking at the ratio of saloons to adult men probably would bring the Comstock into line with Chicago and the Shenandoah Valley. Roger McGrath suggests that Aurora in the Sierra Nevada in the early 1860s had about one saloon for every 200 men, a statistic expressed in a format different from saloon to citizens in general. It is difficult, therefore, to say how it compares with the other communities presented here, but Aurora appears to have had generally the same ratio. In Bodie it was about 1:100. See McGrath, *Gunfights, Highwaymen, and Vigilantes*, 11, 111–13. Noel's determination that the ratio of saloons to people in Denver in 1880 was 1:360 is out of line with the other communities. See Noel, *The City and the Saloon*, 23–24.

46. Wright, *Big Bonanza*, 268.

47. Ibid.; for mention of Gold Hill's only two-bit saloon, see Drury, *Editor*, 237–38; see also Jack Flanagan interview, 48–50. Denver's segregation of saloons into nickel, short-bit, long-bit, and two-bit establishments seems to be a finer gradation than was found on the Comstock. See Noel, *The City and the Saloon*, 23.

48. Lord, *Comstock Mining*, 93, citing *Sacramento Union*, 1 October 1860, 1:6; *Territorial Enterprise*, 12 September 1867, 3:1. Even-numbered addresses for

northern streets indicated the downhill side. The numbers switched at the east-west Union Street, to the south of which even-numbered addresses indicated an uphill, or western location.

49. Wright, *Big Bonanza*, 272–73.

50. *Territorial Enterprise*, 1 January 1875, 1:4.

51. Ibid.

52. For ethnic differences in drinking habits, see Rothbart, "Ethnic Saloon," 332–58; Duis, *Saloon*, 143–45; Noel, *The City and the Saloon*, 53–66.

53. Lord, *Comstock Mining*, 377. Lord cites "Records of the United States Census Agents, 1880." The figures in Lord's estimate seem to coincide with an 1872 estimate in the *Territorial Enterprise* that lists monthly consumption from local breweries at 1,000 barrels per month, or $20,000 (see 8 June 1872, 3:1). According to Lord's estimate, 225,000 gallons cost about $300,000, or $1.33 per gallon. The 1872 estimate of $20,000 per month for local beer would convert to 15,000 gallons per month or 180,000 gallons per year, which is only slightly more than the 147,996 that Lord maintained local breweries produced in 1880.

54. Twain, *Roughing It*, 2:54; Drury, *Editor*, 124–27. Drury mourned the loss of the dignified customs associated with the free lunch and proper saloon conduct. The free lunch was one of the Comstock's oldest saloon traditions. See, for example, the *Territorial Enterprise*, 12 September 1867, 3:1. Duis argues that Chicago saloonkeepers invented the free lunch after that city's great fire of 1871, but clearly earlier examples from the Comstock indicate that this was not the case. See Duis, *Saloon*, 52.

55. See especially the 10th U.S. Manuscript Census of 1880, which, unlike those before, has street addresses. Compare Noel, *The City and the Saloon*, 41–52.

56. Drury, *Editor*, 122–27.

57. *Gold Hill Daily News*, 2 June 1865, 3:1.

58. Elliott West, *The Saloon on the Rocky Mountain Mining Frontier* (Lincoln: University of Nebraska Press, 1979), 51–72, 97–129; Duis, *Saloon*, 48, 83.

59. See, for example, *Territorial Enterprise*, 12 September 1867, 3:1.

60. Browne, *Peep at Washoe*, 187; Hoffmann, *Californien, Nevada, und Mexico*, 144; Leslie, *California*, 280.

61. Outlets for local beer manufacturers were called breweries even when they only sold the product, but the title indicated that they had fresh beer on tap.

62. *The Lariat*, August 1877, 1:1, 2:2.

63. For diversity of functions among saloons, see Noel, *The City and the Saloon*, 11–20, 53–66.

64. See, for example, *Territorial Enterprise*, 8 June 1872, 3:1; and see Hattori, *Northern Paiutes*, and Hattori, "'And Some of Them Swear.'"

65. Lord, *Comstock Mining*, 73; and see *Sacramento Union*, 25 April 1860, 2:2.

66. Bishop and Company, *Bishop's Directory*, 218; McKenney and Company, *McKenney's Business Directory*, 201.

67. Donald L. Hardesty, "Public Archaeology in the Virginia City Landmark District: The 1993 and 1994 Field Seasons" (report prepared for the Nevada State

Historic Preservation Office, 1994); and Donald L. Hardesty, "Public Archaeology on the Comstock" (report prepared for the State Historic Preservation Office, 1996). Thanks to Dr. Hardesty for his work with the material and for making his conclusions and insights available. For a twentieth-century—though still pertinent—view of Comstock saloon life, see Harper, *Rejoice*, 9–14.

68. See the 9th U.S. Manuscript Census of 1870 for Storey County; and see Blackburn and Ricards, "Prostitutes and Gamblers," 252–58; Hoffmann, *Californien, Nevada, und Mexico*, 144.

69. Mathews, *Ten Years in Nevada*, 185.

70. Drury, *Editor*, also lists such exotic games as euchre and pedro; see 128–37. See also, for example, Doten, *Journals*, 1243; John M. Findlay, *People of Chance: Gambling in American Society from Jamestown to Las Vegas* (New York: Oxford University Press, 1986), 79–109.

71. Louise M. Palmer, "How We Live in Nevada," in *So Much to Be Done: Women Settlers on the Mining and Ranching Frontier*, ed. Ruth B. Moynihan, Susan Armitage, and Christiane Fischer Dichamp (Lincoln: University of Nebraska Press, 1990), 116.

72. Mathews, *Ten Years in Nevada*, 186–87, 116.

73. Drury, *Editor*, 106–16, and see 128–37.

74. Eric Moody and Robert A. Nylen, *Brewed in Nevada: A History of the Silver State's Beers and Breweries* (Carson City: Nevada State Museum, 1980), 5–7; and see Eric Moody and Robert A. Nylen, "The Comstock Brewing Industry," *OAH Newsletter* (February 1988): 4–5. For insight into the Comstock breweries of the 1860s, see Hoffmann, *Californien, Nevada, und Mexico*, 119, 128, 149–50.

75. See *Territorial Enterprise*, 8 June 1872, 3:1. Other breweries included the Pacific, the California, and the Washington. See ibid., 8 June 1872, 3:1; 6 July 1866, 2:5; 3 January 1869, 1:5; 30 November 1879, 2:8 and 3:5; 15 April 1880, 2:6; 3 June 1880, 3:5; and the *Gold Hill Daily News*, 28 January 1865, 3:1. Hemman Hoffmann, a Swiss "chemist," worked at the Pacific Brewery in 1864 and 1865. He names the Virginia, the Bavaria, the Washington, and the Philadelphia as competitors. See Hoffmann, *Californien, Nevada, und Mexico*, 128.

76. See the manuscript census records for 1860, 1870, and 1880.

77. *Territorial Enterprise*, 7 April 1877, 3:4; Lowe, "The 'Secret Friend.'"

9. The Moral Options: Saints

1. Leslie's summation of Virginia City in 1877 as a place of "very few women except of the worst class, and as few children" appears to be ill-informed and superficial. See her *California*, 278. See also Mathews, *Ten Years in Nevada*, 237–39.

2. For a discussion of image and reality, see Elizabeth Jameson, "Women as Workers, Women as Civilizers: True Womanhood in the American West," in *The Women's West*, ed. Susan Armitage and Elizabeth Jameson (Norman: University of Oklahoma Press, 1987), 145–64. Compare Mann's assertion in *After the Gold Rush*, 197–98.

3. Lowe, "The 'Secret Friend.'"

4. Wright, *Big Bonanza*, 156.

5. James, "Erin's Daughters"; Diner, *Erin's Daughters in America*, notes that the Irish were unique among immigrants in sending almost equal numbers of men and women to North America. Once established as an ethnic group, the Irish represented about one third of the Comstock population, but because their number included so many women and because women overall were in the minority, the Irish were particularly important.

6. Mathews, *Ten Years in Nevada*, 132; see also Kathryn D. Totton, "'They Are Doing So to a Liberal Extent Here Now': Women and Divorce on the Comstock, 1859–1880," in *Comstock Women: The Making of a Mining Community*, ed. Ronald M. James and C. Elizabeth Raymond (Reno: University of Nevada Press, 1997), for an excellent overview of divorce on the Comstock. See, however, Drury, *Editor*, 36.

7. Palmer, "How We Live in Nevada," 113.

8. Wright, *Big Bonanza*, 157.

9. "First Biennial Report of the Superintendent of Public Instruction." See also Lord, *Comstock Mining*, 206–207, 375–76, for citations from the previous year's report; and see the 10th U.S. Manuscript Census of 1880.

10. See the 9th and 10th U.S. Manuscript Censuses of 1870 and 1880, respectively; Storey County did not, of course, directly employ all of these teachers. Some, like Mary McNair Mathews, were self-employed, running private schools. These numbers also do not include the Daughters of Charity, who are discussed below.

11. Mathews, *Ten Years in Nevada*.

12. Waldorf, *Kid on the Comstock*, 43.

13. See the manuscript census reports for the various years and the 1875 state census.

14. See Rocha, "Images."

15. Elliott, *History of Nevada*, 354–55.

16. Angel (ed.), *History*, 75–99, 572, 576–77.

17. Elliott, *Servant of Power*, 47–53, 66–69.

18. See the manuscript censuses for the respective years and the 1875 state census. The 1875 census documents the transition described here by the 1870 and 1880 manuscript census records. Again, these statistics do not include the Daughters of Charity, who are discussed below and who opened a hospital between 1870 and 1880, providing even more for the community in the way of health care. See also Anton P. Sohn, "The Acceptance of Women Physicians in Nineteenth-Century Nevada," *Greasewood Tablettes* 6 (1995): 1–2. For general observations about medicine on the Comstock, see Duane A. Smith, "Comstock Miseries: Medicine and Mining in the 1860s," *Nevada Historical Society Quarterly* 36 (1993): 1–12.

19. Gorham, *My Memories of the Comstock*, 20.

20. Mathews, *Ten Years in Nevada*, 55; Smith, "Comstock Miseries," 9.

21. See the census manuscripts for the respective years: some of the firefighters probably worked for mining companies; indeed, some listed their profession as "fireman in mine." And see the 1875 state census. These figures do not include those listed specifically as such.

22. See Frady, *Red Shirts and Leather Helmets*, for the definitive treatment of Comstock firefighters; see also Mathews, *Ten Years in Nevada*, 184.

23. Butler, "Mission in the Mountains."

24. *Territorial Enterprise*, 16 March 1876, 3:2; see also Lee Lukes Pickering, *The Story of St. Mary's Art Center—Now—and St. Mary Louise Hospital—Then* (Carson City: N.p., 1986).

25. Angel (ed.), *History*, 208; Shade, "Virginia City's Ill-Fated Methodist Church," 448–49. For an example of a Methodist Church benefit, see Hoffmann, *Californien, Nevada, und Mexico*, 158–59.

26. Thomas K. Gorman, *Seventy-five Years of Catholic Life in Nevada* (Reno: Journal Press, 1935), 17, 41–43; for an excellent overview of religion on the Comstock, see Rachel J. Hartigan, "Looking for a Friend Among Strangers: Virginia City's Religious Institutions as Purveyors of Community" (undergraduate thesis, Yale University, 1993); and see Francis P. Weisenburger, "God and Man in a Secular City," *Nevada Historical Society Quarterly* 14 (1971): 3–24.

27. John B. McGloin, S.J., "Patrick Manogue: Gold Miner and Bishop," *Nevada Historical Society Quarterly* 14 (1971): 25–32; Lapomarda, "Saint Mary's in the Mountains," 58–62; Breault, *The Miner Was a Bishop;* John T. Dwyer, *Condemned to the Mines: The Life of Eugene O'Connell* (New York: Vantage Press, 1976). The name of the Catholic church in Virginia City appears several ways, two of which are shown on signs on the front of the building.

28. William Pugh, "The History of the Baptist Church in Virginia City," *Comstock Chronicle*, 20 June, 14 July, and 28 July 1995; Charles Jeffrey Garrison, "How the Devil Tempts Us to Go Aside from Christ: The History of the First Presbyterian Church of Virginia City, 1862–1867," *Nevada Historical Society Quarterly* 36 (1993): 13–34; Charles Jeffrey Garrison, "Over 125 Years of Struggle at the Comstock's Oldest Church," *Touring the Lode: Comstock Chronicle Summer Tour Guide*, 4 April 1989.

29. See the 9th and 10th U.S. manuscript census records for 1870 and 1880, respectively, and the 1875 state census.

30. McCarthy letters (1882), on file at the Nevada Historic Preservation Office. Thanks to John McCarthy for access to this material.

31. Mathews, *Ten Years in Nevada*, 195. The 1872 Comstock diary of an Irish blacksmith, Timothy Francis McCarthy, records many Sundays when he noted, apparently with regret, that he was unable to attend mass because he had to work. See the files of the Nevada State Historic Preservation Office. Thanks to John McCarthy for access to this material.

32. Hoffmann, *Californien, Nevada, und Mexico*, 99; and see the *Nevada State Journal*, 17 April 1949, 6:1, for a translation of the text, although the quote presented here is based on the original and deviates from this translation.

33. See, for example, *Gold Hill Daily News*, 1 March 1864, 3:1. The *Virginia Evening Bulletin*, 11 July 1863, 2:2, advocated temperance and called for "success to the Sons as well as Daughters of Temperance," but it is unclear whether the newspaper was referring to a local group or merely to the nationwide movement. Hoffmann, who lived in Virginia City in 1864, describes the temperance movement; see *Californien, Nevada, und Mexico*, 98–100. See also McGrath, *Gunfights, Highwaymen, and Vigilantes*, 113.

34. *Gold Hill Daily News*, 1 March 1864, 3:1.

35. Ibid., 15 July 1865, 3:1.

36. Ibid., 27 July 1865, 3:1; and see 19 and 22 July 1865, 3:1 and 2:5, respectively.

37. Mathews, *Ten Years in Nevada*, 116; and see 115.

38. *Virginia Evening Chronicle*, 26 January 1875, 3:1; see also Anita Ernst Watson, Jean E. Ford, and Linda White, "'The Advantages of Ladies' Society': The Public Sphere of Women on the Comstock," in *Comstock Women: The Making of a Mining Community*, ed. Ronald M. James and C. Elizabeth Raymond (Reno: University of Nevada Press, 1997).

39. Hardesty, "Public Archaeology."

40. Drury, *Editor*, 47; and see 32–34, 41–53, 212–13.

41. On the positive role of the unions and other groups, see Mathews, *Ten Years in Nevada*, 180–82; Wickenden, "History of the Nevada Militia"; and see the discussion of militias in chapter 7. See also Drury, *Editor*, 90.

42. Drury, *Editor*, 39.

43. Mathews, *Ten Years in Nevada*, 251–60; Twain, *Roughing It*, 2:105.

44. *Harper's Weekly*, 29 December 1877, 1025. Under the laundry table one man is smoking. The artist probably intended this as a depiction of opium use. The laundry worker is spitting starch on clothes, a practice that became the focus of western anti-Chinese folklore. In addition, the sausage makers are probably depicted to instill revulsion. Nonetheless, these negative aspects do not dominate the scenes. See also James, Adkins, and Hartigan, "Competition and Coexistence."

45. 10th U.S. Manuscript Census of 1880 for Storey County. There was only one man in the city jail. All the rest were in the county facility. An African American janitor is listed with his place of residence as the county jail, but it appears that he was not incarcerated but rather worked there. The per capita incarceration ratio for females according to this document was 1:1357, far below that for males.

46. Thompson, "Historical Archaeology." For comparative purposes see Eugene M. Hattori, Mary K. Rusco, and Donald R. Touhy, "Archaeological and Historical Studies at Ninth and Amherst, Lovelock, Nevada" (Carson City: Nevada State Museum, 1979).

47. Waldorf, *Kid on the Comstock*, 60–63; see also Drury, *Editor*, 64–66; Oscar Lewis, *Silver Kings: The Lives and Times of Mackay, Fair, Flood, and O'Brien, Lords of the Nevada Comstock* (1947; reprint, Reno: University of Nevada Press, 1986), 47–114.

48. Waldorf, *Kid on the Comstock*, 61–62; Miriam Michelson, *The Wonderlode of Silver and Gold* (Boston: Stratford, 1934), 278–85; Berlin, *Silver Platter*, 140, 169–70. Drury, *Editor*, 66, offers a more generous portrait of Fair; see also Lewis, *Silver Kings*, 115–216. Whether the story is based on an actual incident is unclear. Of more importance is its significance in local folklore.

49. Lewis, *Silver Kings*, 169–71, 178; Andria Daley-Taylor, "Girls of the Golden West," in *Comstock Women: The Making of a Mining Community*, ed. Ronald M. James and C. Elizabeth Raymond (Reno: University of Nevada Press, 1977).

50. For a treatment of theaters on the Comstock, see Andria Daley-Taylor, "A Feasibility Study for Piper's Opera House: Virginia City, Nevada" (report submitted to the Storey County Commissioners, 1995).

51. Besides the census, see Drury, *Editor*, 34–35. On theater, see Margaret G. Watson, *Silver Theatre: Amusements of Nevada's Mining Frontier, 1850–1864* (Glendale, Calif.: Arthur H. Clark, 1964).

52. Drury, *Editor*, 89–90; Hoffmann, *Californien, Nevada, und Mexico*, 142–43; Doten, *Journals*, 900, 1138. See also Drury, *Editor*, 90–93. The theater could also be a place of practical jokes. See ibid., 41–46. For animal fights in Aurora of the Sierra Nevada, see McGrath, *Gunfighters, Highwaymen, and Vigilantes*, 12

53. Waldorf, *Kid on the Comstock*, 140–43; McCarthy letters (1882), on file at the Nevada Historic Preservation Office.

54. Doten, *Journals*. See the index for numerous citations.

55. 9th U.S. Manuscript Census of 1870. See the commentary associated with Waldorf's recollection of the circus, *Kid on the Comstock*, 143–45. After circuses found that they could transport their show on the railroad all the way to Virginia City, they set up below town on Washington Street, but the earlier shows probably found the final haul uphill too formidable. Swiss immigrant Hemmann Hoffmann describes a circus visiting Virginia City in 1864. See Hoffmann, *Californien, Nevada, und Mexico*, 108.

56. See the 9th and 10th U.S. manuscript censuses of 1870 and 1880, respectively.

57. He appears in both the 9th and the 10th U.S. Manuscript Censuses of 1870 and 1880, respectively. See Earl, *This Was Nevada*, 104–106. See also Dave Basso, *The Works of C. B. McClellan* (Sparks, Nev.: Falcon Hill, 1987), and Dave Basso, *The Washoe Club: The Story of a Great Social Institution* (Sparks, Nev.: Falcon Hill, 1988). In addition, see *Territorial Enterprise*, 1 October 1876, 3:2.

58. Thanks to Bernadette S. Francke for assistance with this material.

59. Mathews, *Ten Years in Nevada*, 36, and see 54.

60. Thanks to Bernadette S. Francke for help with Muckle and the stonecarvers. See the files of the Comstock Historic District Commission.

61. *Gold Hill Daily News*, 19 April 1879, 3:3, and 21 April 1879, 3:5.

62. Doten, *Journals*, 1063–64. For a discussion of leisure activities, see Roy Rosenzweig, *Eight Hours for What We Will: Workers and Leisure in an Industrial City, 1870–1920* (Cambridge: Cambridge University Press, 1983). For Aurora of

the Sierra Nevada, compare McGrath, *Gunfighters, Highwaymen, and Vigilantes,* 12–13.

63. *Territorial Enterprise,* 1 May 1877, 3:3.

64. Doten, *Journals,* 905.

65. Waldorf, *Kid on the Comstock,* 68–69.

66. Gorham, *My Memories of the Comstock,* 91, and see 88–93.

67. Paul Simon, *Freedom's Champion: Elijah Lovejoy* (Carbondale: Southern Illinois University Press, 1994).

68. William G. Chrystal, "The 'Wabuska Mangler' as Martyr's Seed: The Strange Story of Edward P. Lovejoy," *Nevada Historical Society Quarterly* 37 (1994): 18–34. Thanks to Rev. Chrystal, First Congregational Church of Reno, for his assistance with this material.

10. Princes and Paupers: Contrasts in Class

1. See, for example, Richard H. Peterson, *The Bonanza Kings: The Social Origins and Business Behavior of Western Mining Entrepreneurs, 1870–1900,* 2nd ed. (Norman: University of Oklahoma Press, 1991); see also Richard H. Peterson, *Bonanza Rich: Lifestyles of the Western Mining Entrepreneurs* (Moscow: University of Idaho Press, 1991). In the preface to the second edition of *Bonanza Kings,* Peterson offers a good overview of the historiography of class in the mining West.

2. There are, of course, numerous studies on the lower end of social ladders, particularly when linked with ethnicity. Examinations of the full spectrum of economic options are less common, however.

3. Waldorf, *Kid on the Comstock,* 114.

4. Ibid., 114–19; Mathews, *Ten Years in Nevada,* 120–21.

5. Ibid., 225.

6. 10th U.S. Manuscript Census of 1880.

7. Lord, *Comstock Mining,* 204; Wright, *Big Bonanza,* 215, 291–92; Mathews, *Ten Years in Nevada,* 224–25. Mathews maintained that the Chinese even cut sagebrush.

8. Waldorf, *Kid on the Comstock,* 98–99, 123.

9. See Hattori, *Northern Paiutes,* and Hattori, "'And Some of Them Swore Like Pirates.'" In addition, see James, Adkins, and Hartigan, "Competition and Coexistence," for a treatment of the values of ethnic groups, including the Northern Paiutes.

10. This is based on analysis of the 10th U.S. Manuscript Census of 1880 for Storey County. See the files of the State Historic Preservation Office, Carson City, Nevada.

11. See, for example, Julie Nicoletta, "Redefining Domesticity: Women and Lodging Houses on the Comstock," in *Comstock Women: The Making of a Mining Community,* ed. Ronald M. James and C. Elizabeth Raymond (Reno: University of Nevada Press, 1997).

12. Mathews, *Ten Years in Nevada.*

13. James, Adkins, and Hartigan, "Competition and Coexistence."

14. 10th U.S. Manuscript Census of 1880. The 1890 Sanborn-Perris Fire Insurance Map of Virginia City indicates that there are "mountain cabins" on the uphill periphery of the community, but they are not included on the map, presumably because the cartographers regarded them as uninsurable. See James, "On the Edge of Bonanza."

15. Goldman, *Gold Diggers*, 92–93.

16. 1875 state census of Storey County; also see Goldman, *Gold Diggers*, 77.

17. Goldman, *Gold Diggers*, 124–35.

18. For an excellent, balanced overview of Chinese women on the Comstock, see Chung, "Their Changing World"; see also Goldman, *Gold Diggers*, 95–98, 112–13. Goldman, however, misunderstands aspects of the Chinese community, which Chung corrects. See also James, "Women of the Mining West," for corrections of Goldman's research techniques and some of her conclusions.

19. 10th U.S. Manuscript Census of 1880.

20. Waldorf, *Kid on the Comstock*, 69–70. This may be the James Anderson who appears in the 10th U.S. Manuscript Census of 1880 as a seventy-five-year-old illiterate blacksmith from North Carolina living alone on 7 South E Street.

21. *Territorial Enterprise*, 18 August 1878, 3:4, and 6 September 1878, 3:4; and see Phillip I. Earl, "Brutal Times for Early-Day Virginia City Woman," *Reno Gazette-Journal*, 5 October 1986, 2E, with appreciation to the author for help with related citations.

22. 9th and 10th U.S. Manuscript Censuses of 1870 and 1880, respectively; and see Butler, "Mission in the Mountains." The 1875 state census shows only 37 girls associated with the Daughters of Charity. Since the document does not include households, it is not possible to determine exactly how many girls were recorded at the orphanage. In addition, the 1875 state census is seriously flawed, and so it may not have included everyone who was living there, and it is not useful in this context.

23. Doten, *Journals*, 1568, 1674, 1678, 1679, 1792, 1825, 2165; Earl, *This Was Nevada*, 85–88. His voice was so high, people occasionally classified Jose as a mezzo tenor. Joe Pengelly, "Juan Ricardo José: Ricardo Jose, the Great Spanish Singer from Lanner," *Cornish World* 6 (1995): 27; Arthur Cecil Todd, *The Cornish Miner in America*, 2nd ed. (Spokane: Arthur H. Clark, 1995), 104–106.

24. Mathews, *Ten Years in Nevada*, 268–81, and see 38–39 for another example of charity. See Watson, Ford, and White, "'The Advantages of Ladies' Society,'" for a discussion of a variety of charity organizations.

25. There was a handful of children for whom the census enumerator declared wealth. In addition, 16 of 466 people from fifteen to nineteen years old declared wealth. Averages for men are affected slightly by the computer program for the 9th U.S. Manuscript Census, which did not allow for declarations of wealth that were equal to or greater than $100,000. Two declarations of real estate and seven of personal estate appear, therefore, as $99,999 for the sake of

computation. This has a minimal effect on averages and actually serves to smooth out the effects of just a few people who would have otherwise skewed the results with their remarkable financial success.

26. See also Nicoletta, "Redefining Domesticity."

27. Palmer, "How We Live in Nevada," 113. For an example of class distinction, however, see Mathews, *Ten Years in Nevada*, 39–41.

28. Mathews, *Ten Years in Nevada*, 130, 37; Palmer, "How We Live in Nevada," 115.

29. Janet I. Loverin and Robert A. Nylen, "Creating a Fashionable Society: Comstock Needleworkers from 1860 to 1880," in *Comstock Women: The Making of a Mining Community*, ed. Ronald M. James and C. Elizabeth Raymond (Reno: University of Nevada Press, 1997).

30. Bernadette S. Francke, "The Neighborhood and Nineteenth-Century Photographs: A Call to Locate Undocumented Historic Photographs of the Comstock Region," *Nevada Historical Society Quarterly* 35 (1992): 258–69.

31. James, "Defining the Group," 43; Angel (ed.), *History*, 238, 637; Doten, *Journals*, 1128, 1303, 1581, 2001, 2131.

32. 10th U.S. Manuscript Census, 1880; and see Mathews, *Ten Years in Nevada*, 249–50.

33. Leslie, *California*, 282; *Frank Leslie's Illustrated Newspaper*, 9 March 1878, 6.

34. Francke, "Divination on Mount Davidson"; Addenbrooke, *Mistress of the Mansion*, 36.

35. Peterson, *Bonanza Kings*, 1–14.

36. Brown's bird's-eye view poster of Virginia City.

37. Lewis, *Silver Kings*.

38. Mathews, *Ten Years in Nevada*, 106–109.

39. Stewart, *Reminiscences;* and see Elliott, *Servant of Power*.

40. Lewis, *Silver Kings*, 227–29. See also Peterson, *Bonanza Rich*, 63.

41. Peterson, *Bonanza Rich*, 73–106.

42. Waldorf, *Kid on the Comstock*, 132.

43. Datin, *Elegance on C Street*, 23.

44. *Frank Leslie's Illustrated Newspaper*, 2 March 1878, 446.

45. Peterson, *Bonanza Rich*, 119; Lewis, *Silver Kings*, 6, 26, 53; Basso, *The Washoe Club; Territorial Enterprise*, 21 February 1875, 3:2; 22 April 1875, 3:1; 13 January 1876, 3:2; 3 September 1876, 3:4.

46. A recent archaeological survey by the Nevada State Historic Preservation Office indicates that poorer domiciles existed above the community on the higher slopes of Mount Davidson. Although also possessing a remarkable view, these locations required their occupants to hike up and down the steep incline to reach places of employment and shopping.

47. Ronald M. James and John McCarthy, "McCarthy House" (National Register Nomination; Carson City: Nevada State Historic Preservation Office, 1995).

11. *Over Time: Bonanza and* Borrasca *(1877–1942)*

1. Lord, *Comstock Mining,* 342; see also Stewart and Stewart, *Adolph Sutro.*

2. Stewart and Stewart, *Adolph Sutro;* and see Carl Dahlgren, *Adolph Sutro: A Brief Story of a Brilliant Life* (San Francisco: Press of San Francisco, 1895); Theodore Sutro, *The Sutro Tunnel Company and the Sutro Tunnel* (New York: J. J. Little and Company, 1887).

3. The 10th U.S. Manuscript Census of 1880, for example, shows the construction trades as having some of the highest levels of unemployment.

4. McKenney and Company, *McKenney's Business Directory,* 663.

5. Smith, *History of the Comstock,* 217, and see 207–31. Smith gives this estimate. He also furnishes a good overview of the decline of the Comstock.

6. Todd, *The Cornish Miner in America.*

7. Letter from "Fatty" to his "Friend Joe," 28 July 1898, McCarthy Collection, on file at the Nevada State Historic Preservation Office.

8. William S. McFeely, *Grant: Biography* (New York: W. W. Norton, 1981), 450–77.

9. Doten, *Journals,* 1352–55; *Territorial Enterprise,* 26 October 1879, 2:2; 27 October 1879, 2:2, 3:3–5; 28 October 1879, 3:2–4; 29 October 1879, 2:2, 3:2–3. See also Smith, *History of the Comstock,* 226–28.

10. Doten, *Journals,* 1370–71; *Territorial Enterprise,* 5 September 1880, 3:3; 7 September 1880, 2:2; 8 September 1880, 2:2, 3:3–5; 9 September 1880, 3:3; Smith, *History of the Comstock,* 228–29.

11. Smith, *History of the Comstock,* 230.

12. Ibid., 231; and see Lord, *Comstock Mining,* 407.

13. Paula Petrik has observed this same phenomenon in Helena, Montana, as local mining interests declined. See Paula Petrik, *No Step Backward: Women and Family on the Rocky Mountain Mining Frontier, Helena, Montana, 1865–1900* (Helena: Montana Historical Society Press, 1987), 21. For statistics related to the Comstock, see the 10th U.S. Manuscript Census of 1880. See also Mathews, *Ten Years in Nevada,* 144.

14. *Virginia Evening Chronicle,* 13 July 1880, 3:2; and see Waldorf, *Kid on the Comstock,* 175.

15. Drury, *Editor,* 237–38.

16. McKenney and Company, *McKenney's Business Directory,* 663–707.

17. Lord, *Comstock Mining,* 355–60, 380–81; for an optimistic viewpoint from the late 1880s, see William Wright [Dan De Quille, pseud.], *A History of the Comstock Lode and Mines* (Virginia City: F. Boegle, 1889); on wages, see *Virginia Evening Chronicle,* 23 January 1883, 3:2.

18. Smith, *History of the Comstock,* 269–87; *Mining and Scientific Press* 96, no. 24 (13 June 1908): 804; Marye, *From '49 to '83 in California and Nevada,* 200–204; Bailey, *Supplying the Mining World,* 17.

19. Smith, *History of the Comstock,* 249–52, 256.

20. Ibid., 252; Hickson, *Mint Mark: "CC".*

21. Drury, *Editor*, 181–82. The issue referred to, in January 1893, is apparently no longer extant. The newspaper resumed printing in December 1893, but no longer enjoyed its former prestige.

22. St. Mary's Hospital ledger book, St. Mary in the Mountains Catholic Church archives, Virginia City, Nevada. Thanks to Caroline Beaupre, archivist for St. Mary in the Mountains, for her assistance. See also Butler, "Mission in the Mountains."

23. Letters and other documents associated with the final Comstock days of the Daughters of Charity are on file at the State Historic Preservation Office, Carson City, Nevada, and at St. Mary in the Mountains Catholic Church archives, Virginia City, Nevada. See also Butler, "Mission in the Mountains." Use of the term "Sisters of Charity" in the ledger book is technically incorrect; it should have been "Daughters of Charity."

24. *Territorial Enterprise*, 9 September 1897, 1:1; Basso, *Washoe Club*.

25. At first the process used potassium cyanide, but this was later replaced with sodium cyanide. See *Mining and Scientific Press* 105, no. 22 (30 November 1912): 703; Smith, *History of the Comstock*, 258; Hardesty, *Archaeology of Mining and Miners*, 51; Russell R. Elliott, *Nevada's Twentieth-Century Mining Boom: Tonopah, Goldfield, Ely* (Reno: University of Nevada Press, 1966), 157–58; Jack Flanagan, interview, 25–27.

26. Waldorf, *Kid on the Comstock*, 172.

27. Both quotes are from the McCarthy Collection, on file at the Nevada State Historic Preservation Office. The first dates from 7 September 1897 (Mary McCarthy to her brother Joseph); the second dates from 7 October 1898 (Frances McCarthy to her son Joseph).

28. *Mining and Scientific Press* 96, no. 24 (13 June 1908): 804–806.

29. Waldorf, *Kid on the Comstock*, 172; see also *Mining and Scientific Press* 90, no. 5 (4 February 1905): 1, 73–74; and 96, no. 24 (13 June 1908): 804–806.

30. Elliott, *Nevada's Twentieth-Century Mining Boom*, 157; for a good overview of prospects on the Comstock, see *Mining and Scientific Press* 99, no.1 (3 July 1909): 24–29; 102, no. 1 (14 January 1911): 121; 108, no. 16 (18 April 1914): 652–53.

31. Many of these traditions are extremely difficult to confirm, but there is reason to believe that at least some of the Comstock housing served elsewhere. See, however, Michael Hagen, "The Mason Valley—Comstock Connection: Exploring the Moving of Old Buildings" (unpublished manuscript, 1996, on file in Nevada State Historic Preservation Office).

32. Sanborn Fire Insurance Maps (1907, 1923).

33. For employment statistics, the term "adult men" is defined as men who are at least twenty years old.

34. Waldorf, *Kid on the Comstock*, 173; for another treatment of turn-of-the-century Northern Paiutes on the Comstock, see Johnson, *American Highways and Byways*, 200–201; see also 202–207.

35. Smith, *History of the Comstock*, 286–88. See also Whitman Symmes, "The

Comstock Lode," *Engineering and Mining Journal* 95 (1913): 129, and Whitman Symmes, *Report: The Situation on the Comstock* (San Francisco: Blair-Murdock, 1913), for critical looks at the management of the Comstock mines after the turn of the century. See also *Engineering and Mining Journal* 98, no. 14 (3 October 1914): 616–17, and 99, no. 3 (16 January 1915): 155. For a look at small-time mining on the Comstock, see R. T. King, "Verne Foster and the Nevada Mining Association" (unpublished manuscript, 1988, University of Nevada Oral History Program, Reno), 21–29.

36. Waldorf, *Kid on the Comstock*, 182; see also *Engineering and Mining Journal* 112, no. 10 (3 September 1921): 362.

37. Ty Cobb, "Tales of Famous General's Exploits in Virginia City," *Reno Gazette-Journal*, 1 April 1992.

38. *The Artemisia* (Reno: Yearbook of the University of Nevada; now University of Nevada, Reno, 1904), 173.

39. *Evening Gazette*, 3 July 1909; 6 July 1909.

40. Drury, *Editor*, 54–61. A list of players inside the Opera House museum boasts dozens of notables.

41. Daley-Taylor, "Feasibility Study," 22.

42. Datin, *Elegance on C Street*, 41–46; for a description of the International after the turn of the century, when it had lost much of its elegance, see Johnson, *American Highways and Byways*, 182.

43. Census enumerators probably overcounted Chinese women as prostitutes because of prejudice and misunderstanding about their living situations. Sue Fawn Chung suggests the number should be as low as 71. See Chung, "Their Changing World."

44. *Nevada State Journal*, 8 March 1933, 2:2–3. see also Chung, "Their Changing World." Johnson, *American Highways and Byways*, 182, refers to Ching as "landlord" of the International.

45. William Leslie Marks interview, 27–28; Edward Daniel Gladding, interview by Ann Harvey, 1984, University of Nevada Oral History Program; Ty Cobb, "'Rustling' Up Work on the Comstock," *Reno Gazette-Journal*, 14 July 1996, 2F.

46. Waldorf, *Kid on the Comstock*, noted the destruction of nineteenth-century houses after a visit to his hometown in 1924; see also Drury, *Editor*, 292–93.

47. Ty Cobb, "Do-It-Yourself Roller Coaster on the Comstock," *Reno Gazette-Journal*, 18 June 1995; see also Ty Cobb, *The Best of Cobbwebs* (Reno: Black Rock Press, 1997); and Kathryn M. Totton, "Comstock Memories: 1920s–1960s" (unpublished manuscript, 1988, University of Nevada Oral History Program), 18.

48. Ronald H. Limbaugh, "Making the Most of Experience: The Career of William J. Loring, Nevada Mining Engineer," *Mining History Association Annual* (1994): 11; see also Drury, *Editor*, 292–93; *Nevada State Journal*, 26 August 1935, 1:1–2, 2:1–4. This was a special issue of the *Nevada State Journal*, which included several articles on Comstock mining in the mid-1930s.

49. Wurm and Demoro, *Silver Short Line*, 135–62; John Giuffra, interview by

Lucy Scheid, 1984, University of Nevada Oral History Program, 13; for an excellent view of life in Virginia City during the 1930s, see Harper, *Rejoice.*

50. See U.S. Bureau of Mines, *Minerals Yearbook: 1942*, 80–85; authority for the War Production Board originated with Executive Order 9024, issued on 24 January 1942. Shortly after that, the board issued Preference Rating Order P-56, the first to place limits on gold and silver mining. Thanks to Mona L. Reno, Federal Publications Librarian, Nevada State Library and Archives, for her assistance with this material. See also Gerald D. Nash, *World War II and the West: Reshaping the Economy* (Lincoln: University of Nebraska Press, 1990), 18–40.

51. Dorothy Nichols, interview by Ann Harvey, 1984, University of Nevada Oral History Program, 16; Inez Solaga, interview by Lucy Scheid, 1984, University of Nevada Oral History Program, 1; Harper, *Rejoice*, 145–75.

52. This is based on oral testimony of numerous Comstock residents. It is also clear, however, that some prostitutes remained, although with their careers ended or made unofficial.

12. The Sequel to the Big Bonanza: Tourism and Television

1. Emrich, *In the Delta Saloon.* See the excellent introduction by R. T. King. See also *Virginia City News*, 23 February 1940, 3:5; *Territorial Enterprise*, 18 November 1955, 1:2–3, 4:3. Thanks to Andria Daley-Taylor for her assistance with information on the weddings.

2. David Barnett, "*Virginia City* the Movie," *Nevada Magazine* 50 (1990): 43–47; William J. Henley, "Old-Time Movie Houses and the 'Virginia City' Premiere Fiasco," *Comstock Chronicle*, 12 February 1993, 7.

3. Andria Daley-Taylor, "Boardwalk Bons Vivants," *Nevada Magazine* 52 (1992): 20–24. 35–37; Dorothy Nichols, interview, 17–19; David Toll, interview by Lucy Scheid, 1984, University of Nevada Oral History Program, 14–19; Alice Mildred Byrne interview, 72; Earl S. Pomeroy, *In Search of the Golden West: The Tourist in Western America* (1957; reprint, Lincoln: University of Nebraska Press, 1990), 190–92.

4. Wurm and Demoro, *Silver Short Line;* John Giuffra interview, 13.

5. Daley-Taylor, "Boardwalk Bons Vivants," 22.

6. Ibid., 23; and see Daley-Taylor's "Girls of the Golden West."

7. Daley-Taylor, "Boardwalk Bons Vivants," 35.

8. Emrich, *In the Delta Saloon*, see especially 99–101, 292–93.

9. Thanks to Don McBride for information related to his saloon, the Bucket of Blood. The telescope was removed during a remodeling in 1962.

10. *Nevada State Journal*, 7 June 1959, 1:1–8; 14 June 1959, 1:1, 12:1–8; 13 June 1959, 1:3, 4:1, 6:6–7.

11. Vincent Terrace, *The Complete Encyclopedia of Television Programs: 1947–1979* (New York: A. S. Barnes and Company, 1979), 1:132–34.

12. Daley-Taylor, "Boardwalk Bons Vivants." See also Shepperson and Harvey, *Mirage-Land*, 141–44, 164.

13. James Goode, "The Making of 'The Misfits,'" *Nevada Magazine* 46 (1986): 12–17; Michael Sion, "Camel Race in Virginia City," *Nevada Magazine* 52 (1992): E-17; *Comstock Chronicle*, 8 September 1995, 10–11; Alice Mildred Byrne interview, 76–78.

14. Anthony DeCurtis and James Henke (eds.), *The Rolling Stone Illustrated History of Rock and Roll*, 2nd ed. (New York: Random House, 1980), 362–63; see also the *Comstock Chronicle*, 1 April 1994, 8; 8 April 1994, 8; 28 June 1996, 6.

15. *Rolling Stone*, 13 May 1993. See also ibid., 26 February 1976; *Comstock Chronicle*, 18 March 1994, 2; 25 March 1994, 2; 1 and 8 April, 1994, 8; Chandler Laughlin, "Rockin' at the Red Dog," *Nevada Magazine* 55 (1995): 80–83.

16. Jim McCormick, "Silver City: Reminiscences, Facts, and a Little Gossip," *Nevada Historical Society Quarterly* 30 (1987): 37–55.

17. On the chaos of early signage, see Browne, *Peep at Washoe*, 186.

18. Later miners may rework mill tailings and mine dumps, but that is rarely planned for in the original designs of a mine. Of course, farmers and the lumber industry can also be wasteful and view the land as a shortcut to profit, with little regard for the future. The difference is that while slash-and-burn or extensive agriculture can be practiced with low population, intensive cultivation is more commonly the goal, if not always the practice. Mining is different because it intends to exhaust a resource that cannot grow back.

19. Willard Lowther, interview by author, Virginia City, 25 April 1995.

20. For information on the Statue of Justice and its rehabilitation, see the files of the State Historic Preservation Office (Carson City), and see James, *Temples of Justice*, 136–41.

21. For a discussion of change on the Comstock from another point of view, see Dorothy Stroup, "Renaissance in Virginia City," *San Francisco Examiner*, 18 June 1989.

22. U.S. Bureau of Mines, *Minerals Yearbook: 1978–79*, 392.

23. The company later partially backfilled the pit with about 100 feet of material. Gauging depth is problematic because of widely varying elevations of side walls, created during the digging of the pit into the side of a hill. Thanks to Doug Driesner of the Nevada Minerals Division for help with this information. Mr. Driesner worked on the Gold Hill project from January 2, 1979, until 1983, when the company ceased operation.

24. See Nevada Revised Statute 37.038, amended in 1981.

25. Compare Mann's observations of modern Nevada City and Grass Valley, California. See Mann, *After the Gold Rush*, 194.

26. Browne, *Peep at Washoe*, 57.

27. *San Francisco Chronicle*, 23 March 1994, 1; *Comstock Chronicle*, 26 November 1993, 1, and 2 November 1995, 1; *Nevada Appeal*, 28 November 1993, A:1, 8; *Nevada Weekly*, 26 January –1 February 1994, 8; *Reno Gazette-Journal*, 22 August 1996, 1:2–5; "Virginia City Townsite: A Prepared Statement by International Hotel and Land Company for Discussion Purposes Only" (unpublished, undated paper on file in the State Historic Preservation Office, Carson City, Nevada).

28. *Comstock Chronicle*, 27 January 1995, 7; and see 10 February 1995, 4, 5, for a reply from Winchester and for yet another letter of support.

29. See running commentaries throughout August, September, October, and November 1993 in the *Comstock Chronicle*. See especially 20 August 1993, 1; 15 October 1993, 1; and 5 November 1993, 6.

BIBLIOGRAPHY AND BIBLIOGRAPHICAL ESSAY

This volume is an attempt to provide an overview of Comstock history and to create a framework for future research. Although the federal government recognized the Comstock as one of the largest National Historic Landmarks, there has not been a scholarly, comprehensive examination of the subject since the 1883 publication of Eliot Lord's *Comstock Mining and Miners.* Lord's book will remain of critical importance because it is insightful and well researched. In addition, it incorporates material from interviews, as well as from primary sources that are no longer extant.

This is not to say that there have not been subsequent useful sources that take a broad look at Comstock history. Charles Shinn's *The Story of the Mine* appeared in 1896, but it does little more than borrow liberally from Lord. Grant Smith's *History of the Comstock Lode*, published in 1943, is well written and thoroughly documented. Smith offers corrections to Lord that are extremely helpful, and his work serves as an excellent companion to that of his nineteenth-century predecessor. A number of other publications have also documented the Comstock, relying heavily on primary photographs. Barbara Richnak's *Silver Hillside*, published in 1984, is particularly noteworthy for its presentation of many rarely seen images. The text is a good summation of Comstock history, but it is not intended to go beyond Lord. Much the same is true of Douglas McDonald's *Virginia City and the Silver Region of the Comstock Lode*, which appeared in 1982.

Several secondary sources have played important roles in developing specific topics important to the history of the Comstock. John Debo Galloway's *Early Engineering Works Contributory to the Comstock*, appearing in 1947, and Steven R. Frady's 1984 publication, *Red Shirts and Leather Helmets*, provide specialized views of important aspects of the area's past. The same can be said for Eugene M. Hattori's *Northern Paiutes on the Comstock*, published in 1975. Marion S. Goldman's *Gold Diggers and Silver Miners* was a groundbreaking study that appeared in 1981. With the computerized manuscript censuses it has been possible to check Goldman's research. Sadly, much of her work was flawed, conducted at a time when such a resource was not available. The need to assess the roles of specific individuals before drawing conclusions about groups requires a correction of Goldman's work. Nonetheless, her book will continue to serve as a key source of insight and analysis when dealing with the topic of western prostitution.

The Comstock has recently been the focus of increasing scholarship that will continue to rewrite the history of this important mining district. Numerous articles and theses have appeared during the past ten years that challenge previously held assumptions and add considerably to the historical image of the area. Noteworthy among these are the master's theses of Sharon Lowe and Judy Anne Thompson. Bernadette S. Francke, during her tenure with the Comstock Historic District Commission, helped rewrite Comstock history with her comprehensive organization of Comstock material and her articles on specific topics. Kelly J. Dixon is continuing this tradition, but in the direction of archaeology, which promises to be an increasingly important means of reassessing the Comstock. Kenneth Fliess is adding yet another important point of view with his demographic research. Although yet to be published on its own, this work has affected this volume as well as *Comstock Women: The Making of a Mining Community* (1998), a collection of articles that provides radically new perspectives on a wide variety of gender issues that will influence the district's historiography for decades.

There is every reason to believe that these efforts are only the beginning of a continuing trend to reevaluate the district's past. Subsequent studies could rely on the volumes of corporate records to arrive at conclusions about stock and other financial manipulations. Opportunities also exist to pursue analysis of mortality and fertility rates, property ownership, and small-business longevity and profitability. And of course there are hundreds of other possibilities. The comprehensive histories of Silver City and Dayton, yet to be written, promise to complement the Comstock's story with their own unique sagas. The newly founded Comstock Archaeology Center's ongoing research increases the potential for additional insight. Ten or twenty years of continued scholarship may require this publication to appear in a second edition featuring dramatically changed conclusions and points of emphasis. On the other hand, a new volume drawing on future research may replace it entirely.

Besides Lord and Smith, a number of primary sources are worthy of note. Samuel Clemens (Mark Twain) wrote *Roughing It*, one of the first comprehensive treatments of the Comstock. Clemens's book deals with his life in Virginia City and the West in the early 1860s. It is fanciful, but some parts of it provide unparalleled insight into this formative period. William Wright, using the nom de plume Dan De Quille, published his outstanding *Big Bonanza* in 1876. Though not a scholarly treatment, it is more thorough than *Roughing It*, which inspired Wright to a certain extent. Wright's book provides an intimate portrait of his Comstock in the mid-1870s. Along this line, J. Ross Browne's *A Peep at Washoe and Washoe Revisited* offers an insightful, comical treatment of Virginia City during the 1860s. Mary McNair Mathews, with far less humor, provides a detailed examination of her Comstock life during the 1870s in *Ten Years in Nevada.*

Other important primary sources include Alfred Doten's extensive and detailed *Journals.* Henry DeGroot's *The Comstock Papers* originally appeared in the *Mining and Scientific Press,* and the Grace Dangberg Foundation later reprinted them as a collection. His observations add considerably to the firsthand knowledge about the district's earliest years. Wells Drury's *An Editor on the Comstock Lode* offers insight into Virginia City during the late 1870s and the 1880s. Several other, later memoirs similarly complement our understanding of Comstock history and development.

This study relies heavily on the U.S. Census manuscripts, which have recently been entered into a computerized database. This format has proved to be an enormously powerful tool because of the ease with which it allows investigators to pose a variety of questions involving gender, ethnicity, marriage, employment, and geographic distribution. Clearly, it will continue to shape the way we visualize the Comstock. This effort did not include the territorial and state censuses, since their format differs significantly from that of the federal documents and they are undependable in profound ways. It was still possible to work with these documents using techniques that sidestep the issue of statistical reliability, but they can never be as valuable as the federal documents.

The records available in the Storey County Recorder's Office have also played a pivotal role in the research for this book. Until recently, these valuable sources of information were difficult to tap because of a lack of organization. The material that is now readily accessible could easily provide the basis for a dozen dissertations and many more articles. Similarly, numerous directories document the Comstock from the earliest years throughout its development, augmenting the census with the names of individuals and providing information about businesses and general community development.

The collection of material at the University of Nevada, Reno's Oral History Program office is extremely useful for the insight it offers Comstock research. Although most of this is relevant only to twentieth-century history, Duncan Emrich's *In the Delta Saloon,* recorded in 1949 and 1950, reached back to the Comstock's heyday. Without the efforts of the Oral History Program, that volume would not be available to researchers. The program is also credited with fourteen oral histories, completed in the 1980s, that will remain important to the development of Comstock history.

This book draws on a variety of historical sources, including oral histories, but it also relies on archaeological data. In addition, it attempts to integrate methodology from gender studies, folklore, demography, ethnicity studies, and geography, as well as from architectural and art history. Ultimately, it seems that only an interdisciplinary approach can capture a place as complex and diverse as the Comstock.

Primary Sources

Angel, Myron, ed. *History of Nevada: 1881.* Oakland, Calif.: Thompson and West, 1881.

Becker, George Ferdinand. *Atlas to Accompany the Monograph on the Geology of the Comstock Lode and the Washoe District.* Washington, D.C.: Government Printing Office, 1882.

———. *Geology of the Comstock Lode and the Washoe District.* Washington, D.C.: Government Printing Office, 1882.

Benjamin, Israel ben Joseph. *Three Years in America, 1859–1862.* New York: Arno Press, 1975.

Bishop, D. M., and Company. *Bishop's Directory: Virginia City, Gold Hill, Silver City, Carson City, and Reno, 1878–79.* San Francisco: B. C. Vandall, 1878.

Bragg, Allen C. "Pioneer Days in Nevada." *Second Biennial Report of the Nevada Historical Society, 1909–1910.* (1911): 72–81.

Brewer, William H. *Up and Down California in 1860–1864.* New Haven: Yale University Press, 1930.

Brown, Grafton T. "Bird's-Eye View of Virginia City." San Francisco: Britton and Company, 1861.

Browne, J. Ross. *Mining Adventures: California and Nevada, 1863–1865.* 1863, 1865, and 1869. Reprint, Balboa Island, Calif.: Paisano Press, 1961. Page citations are to the reprint edition.

———. *A Peep at Washoe and Washoe Revisited.* 1860 and 1863. Reprint, Balboa Island, Calif.: Paisano Press, 1959. Page citations are to the reprint edition.

———. *Resources of the Pacific Slope with a Sketch of the Settlement and Exploration of Lower California.* 1869. Reprint, New York: D. Appleton and Company, 1969. Page citations are to the first series in the volume.

Burton, Richard F. *City of the Saints and Across the Rocky Mountains to California.* 1862. Reprint, New York: Alfred A. Knopf, 1963. Page citations are to the first series in the volume.

Carr, M. D. *The Nevada Directory for 1868–69.* San Francisco: M. D. Carr and Company, 1868.

Church, John Adams. *The Comstock Lode; Its Formation and History.* New York: J. Wiley and Sons, 1879.

———. *The Heat of the Comstock Mines.* Reno: University of Nevada, printed by author [ca. 1878?].

Cobb, Ty. *The Best of Cobbwebs.* Reno: Black Rock Press, 1997.

———. "Do-It-Yourself Roller Coaster on the Comstock," *Reno Gazette-Journal,* June 18, 1995.

———. "'Rustling' Up Work on the Comstock," *Reno Gazette-Journal,* 14 July 1996, 2F.

———. "Tales of Famous General's Exploits in Virginia City," *Reno Gazette-Journal,* April 1, 1992.

Collins, Charles. *Mercantile Guide and Directory for Virginia City and Gold Hill.* Virginia City: Agnew and Deffebach, 1864–65.

D'Ancona, David Arnold. *A California-Nevada Travel Diary of 1876: The Delightful Account of a Ben B'rith.* Edited by William M. Kramer. Santa Monica: N. B. Stern, 1975.

DeGroot, Henry. *The Comstock Papers.* Reno: Grace Dangberg Foundation, 1985.

Doten, Alfred. *The Journals of Alfred Doten: 1849–1903.* Edited by Walter Van Tilburg Clark. Reno: University of Nevada Press, 1973.

Drury, Wells. *An Editor on the Comstock Lode.* Palo Alto: Pacific Books, 1948.

Goodwin, C. C. *As I Remember Them.* Salt Lake City: Salt Lake Commercial Club, 1913.

Gorham, Harry M. *My Memories of the Comstock.* Los Angeles: Suttonhouse, 1939.

Harper, Joyce Hart. *Rejoice: Comstock Memoirs.* Sparks, Nev.: Maverick, 1991.

Kelly, J. Wells. *First Directory of the Nevada Territory.* 1862. Reprint, Los Gatos, Calif.: Talisman Books, 1962.

———. *Second Directory of the Nevada Territory.* Virginia City: Valentine and Company, 1863.

King, Clarence. *Report of the Geological Exploration of the Fortieth Parallel.* Vol. 3. *Mining.* Washington, D.C.: Government Printing Office, 1870.

Langley, Henry G. *The Pacific Coast Business Directory for 1867.* San Francisco: Henry G. Langley, 1867.

———. *The Pacific Coast Business Directory for 1871–73.* 2d ed. San Francisco: Henry G. Langley, 1871.

Leslie, Miriam Florence Squire. *California: A Pleasure Trip from Gotham to the Golden Gate.* New York: G. W. Carleton, 1877.

Lockley, Fred. *Vigilante Days at Virginia City.* Portland: Fred Lockley, 1924.

Lord, Eliot. *Comstock Mining and Miners.* 1883. Reprint, San Diego: Howell-North, 1959.

Marye, George Thomas, Jr. *From '49 to '83 in California and Nevada: Chapters from the Life of George Thomas Marye, a Pioneer of '49.* San Francisco: A. M. Robertson, 1923.

Mathews, Mary McNair. *Ten Years in Nevada or Life on the Pacific Coast.* 1880. Reprint, Lincoln: University of Nebraska Press, 1985. Page citations are to the reprint edition.

McKenney, L. M., and Company. *McKenney's Business Directory of the Principal Towns of California, Nevada, Utah, Wyoming, Colorado and Nebraska, 1880–81.* Sacramento: H. S. Crocker and Company, 1882.

Michelson, Miriam. *The Wonderlode of Silver and Gold.* Boston: Stratford, 1934.

Nevada, State of. *Census of the Inhabitants of the State of Nevada, 1875. Appendix to the Journals of the Senate and Assembly: 8th Session.* Carson City: State Printer, 1877.

Ninth Census of the United States. *Statistics of Population: Tables I to VIII, Inclusive.* Washington, D.C.: Government Printing Office, 1872.

Palmer, Louise M. "How We Live in Nevada." In *So Much to Be Done: Women Settlers on the Mining and Ranching Frontier*, edited by Ruth B. Moynihan, Susan Armitage, and Christiane Fischer Dichamp. Lincoln: University of Nebraska Press, 1990. Originally published in *Overland Monthly* 2 (May 1869): 457–62. Page citations are to the reprint edition.

Richthofen, Baron Ferdinand von. *The Comstock Lode: Its Character, and the Probable Mode of Its Continuance in Depth*. San Francisco: Sutro Tunnel Company, 1866.

Slosson, H. L. *Deep Mining on the Comstock*. San Francisco: H. L. Slosson, 1910.

Stewart, William M. *Reminiscences of Senator William M. Stewart of Nevada*. Edited by George Rothwell Brown. New York: Neale Publishing, 1908.

Sutro, Adolph. *The Bank of California Against the Sutro Tunnel*. Washington, D.C.: M'Gill and Witherow, 1874.

———. *Closing Argument of Adolph Sutro, the Sutro Tunnel*. Washington, D.C.: M'Gill and Witherow, 1872.

Sutro, Theodore. *The Sutro Tunnel Company and the Sutro Tunnel*. New York: J. J. Little and Company, 1887.

Symmes, Whitman. "The Comstock Lode." *Engineering and Mining Journal* 95, no. 2 (January 11, 1913): 129.

———. *Report: The Situation on the Comstock*. San Francisco: Blair-Murdock, 1913.

Twain, Mark [Samuel Clemens]. *Roughing It*. 1871. Reprint, New York: Harper and Brothers Publishers, 1913. Page citations are to the reprint edition.

Uhlhorn, John F. *Virginia and Truckee Railroad Directory, 1873–74*. Sacramento: H. S. Crocker and Company, 1873.

United States Bureau of Mines. *Minerals Yearbook*. Washington, D.C.: Department of the Interior, various years cited.

United States Congress. House. *Report of the Commissions and Evidence Taken by the Committee of Mines and Mining of the House of Representatives of the United States in Regard to the Sutro Tunnel*. Washington, D.C.: M'Gill and Witherow, 1872.

Waldorf, John Taylor. *A Kid on the Comstock: Reminiscences of a Virginia City Childhood*. Palo Alto: American West, 1968. Reprint, Reno: University of Nevada Press, 1991. Page citations are to the reprint edition.

Wright, William [Dan De Quille, pseud.]. *The Big Bonanza*. 1876. Reprint, New York: Alfred A. Knopf, 1953. Page citations are to the reprint edition.

———. *A History of the Comstock Lode and Mines*. Virginia City, Nev.: F. Boegle, 1889.

Newspapers

Alta California (San Francisco, California).
Comstock Chronicle (Virginia City, Nevada).
Evening Gazette (Reno, Nevada).
Gold Hill Daily News (Gold Hill, Nevada).
The Lariat (Virginia City, Nevada).
Marysville Daily Appeal (Marysville, California).

Mountain Democrat (Placerville, California).
Nevada Appeal (Carson City, Nevada).
Nevada State Journal (Reno, Nevada).
New York Herald (New York, New York).
Reese River Reveille (Austin, Nevada).
Reno Gazette-Journal (Reno, Nevada).
Sacramento Union (Sacramento, California).
San Francisco Chronicle (San Francisco, California).
San Francisco Examiner (San Francisco, California).
San Francisco Herald (San Francisco, California).
Sierra Sage (Carson City and Carson Valley, Nevada).
Silver Bow Standard (Silver Bow, Nevada).
Territorial Enterprise (Genoa, and later Carson City, Utah Territory; subsequently Virginia City, Nevada Territory and State).
Virginia City News (Virginia City, Nevada).
Virginia Daily Union (Virginia City, Nevada).
Virginia Evening Bulletin (Virginia City, Nevada).

Journals and Reports

Engineering and Mining Journal.
"First Biennial Report of the Superintendent of Public Instruction of the State of Nevada for the School Year Ending August 31, 1866." Carson City: State Printer, 1867.
Frank Leslie's Illustrated Newspaper.
Harper's Weekly.
Hutchings' California Magazine.
Mining and Scientific Press.
Nevada Territorial and State Censuses (1861, 1863, 1875).
Rolling Stone.
Sanborn-Perris Fire Insurance Maps (1890, 1907, 1923; Sanborn alone in 1907 and 1923).
U.S. Bureau of Mines. *Minerals Yearbook.* Washington, D.C.: Department of the Interior.
U.S. Manuscript Censuses and Census Reports.

Manuscript Collections

Numerous manuscripts are available for research at the Nevada Historical Society (Reno) and at the Storey County Courthouse (Virginia City) in the offices of the Recorder, the Clerk, and the Assessor. The U.S. Manuscript Censuses are available on microfilm, as are the territorial and state censuses of 1861, 1863 and 1875. In addition, the Nevada State Historic Preservation Office (Carson City) and the Comstock Historic District Commission (Virginia City) have a limited number of copies of manuscripts available for research. Several bird's-eye views of Virginia City, as well as historical photographs, are also useful in reconstructing the area's history. Some of these appear in publications, but most exist at one or more of these repositories.

Oral Histories

Bryne, Alice Mildred. Interview by Ann Harvey. University of Nevada Oral History Program, Reno, 1984.
Dolve, Carroll. Interview by Ann Harvey. University of Nevada Oral History Program, Reno, 1984.
Emrich, Duncan. "In the Delta Saloon: Conversations with Residents of Virginia City, Nevada." Recorded in 1949 and 1950. University of Nevada Oral History Program, Reno, 1991.
Flanagan, Jack. Interview by Ann Harvey. University of Nevada Oral History Program, Reno, 1984.
Gallagher, Hugh James. Interview by Lucy Scheid. University of Nevada Oral History Program, Reno, 1984.
Giuffra, John. Interview by Lucy Scheid. University of Nevada Oral History Program, Reno, 1984.
Gladding, Edward Daniel. Interview by Ann Harvey. University of Nevada Oral History Program, Reno, 1984.
Gladding, Marion. Interview by Ann Harvey. University of Nevada Oral History Program, Reno, 1984.
King, R. T. "Verne Foster and the Nevada Mining Association," University of Nevada Oral History Program, Reno, 1988.
Marks, Margaret. Interview by Ann Harvey. University of Nevada Oral History Program, Reno, 1984.
Marks, William Leslie. Interview by Ann Harvey. University of Nevada Oral History Program, Reno, 1984.
Nichols, Dorothy. Interview by Ann Harvey. University of Nevada Oral History Program, Reno, 1984.
Solaga, Inez. Interview by Lucy Scheid. University of Nevada Oral History Program, Reno, 1984.
Toll, David. Interview by Lucy Scheid. University of Nevada Oral History Program, Reno, 1984.
Totton, Kathryn M. "Comstock Memories: 1920s–1960s." University of Nevada Oral History Program, Reno, 1988.

Secondary Sources

Addenbrooke, Alice B. *The Mistress of the Mansion.* Palo Alto: Pacific Books, 1959.
Arrington, Leonard J., and Davis Bitton. *The Mormon Experience: A History of the Latter-day Saints.* New York: Alfred A. Knopf, 1979.

Bailey, Lynn R. *Supplying the Mining World: The Mining Equipment Manufacturers of San Francisco, 1850-1900.* Tucson: Westernlore Press, 1996.

Baldrica, Alice M. "Lander and the Settlement of the Pyramid Lake War." In *Frederick West Lander: A Biographical Sketch (1822–1862),* edited by Joy Leland, 151–90. Reno: Desert Research Institute, 1993.

Barnett, David. "*Virginia City* the Movie." *Nevada Magazine* 50, no. 5 (September/October 1990): 43–47.

Barth, Fredrik, ed. *Ethnic Groups and Boundaries: The Social Organization of Cultural Difference.* Boston: Little, Brown, 1969.

Basso, Dave. *The Washoe Club: The Story of a Great Social Institution.* Sparks, Nev.: Falcon Hill, 1988.

———. *The Works of C. B. McClellan.* Sparks, Nev.: Falcon Hill, 1987.

Beebe, Lucius, and Charles Clegg. *Virginia and Truckee: A Story of Virginia City and Comstock Times.* Oakland, Calif.: Grahame H. Hardy, 1949.

Bentley, G. Carter. "Ethnicity and Practice." *Comparative Studies in Society and History* 29, no. 1 (January 1987): 24–55.

Berlin, Ellin Mackay. *The Silver Platter.* Garden City, N.Y.: Doubleday, 1957.

Blackburn, George M., and Sherman L. Ricards. "The Prostitutes and Gamblers of Virginia City, Nevada: 1870." *Pacific Historical Review* 48, no. 2 (May 1979): 235–59.

Bloss, Roy S. *The Pony Express: The Great Gamble.* Berkeley: Howell-North, 1959.

Breault, William, S. J. *The Miner Was a Bishop: The Pioneer Years of Patrick Manogue, California—Nevada, 1854–1895.* Rancho Cordova, Calif.: Landmark, 1988.

Brown, Richard Maxwell. *Strain of Violence: Historical Studies of American Violence and Vigilantism.* New York: Oxford University Press, 1975.

Brown, Ronald C. *Hard-Rock Miners: The Intermountain West, 1860–1920.* College Station: Texas A&M University Press, 1979.

Burchell, Robert Arthur. *San Francisco Irish, 1848–1880.* Berkeley: University of California Press, 1980.

Butler, Anne M. *Daughters of Joy, Sisters of Misery: Prostitution in the American West, 1865–1909.* Urbana: University of Illinois Press, 1985.

———. "Mission in the Mountains: The Daughters of Charity in Virginia City." In *Comstock Women,* edited by James and Raymond, 142–64.

Chambers, S. Allen, Jr. *The Architecture of Carson City, Nevada.* Washington, D.C.: Historic American Buildings Survey, National Park Service, n.d.

Chan, Loren B. "The Chinese in Nevada: An Historical Survey, 1856–1970." *Nevada Historical Society Quarterly* 25, no. 4 (Winter 1982): 266–314.

Chrystal, William G. "The 'Wabuska Mangler' as Martyr's Seed: The Strange Story of Edward P. Lovejoy." *Nevada Historical Society Quarterly* 37, no. 1 (Spring 1994): 18–34.

Chung, Sue Fawn. "Their Changing World: Chinese Women on the Comstock, 1960–1910." In *Comstock Women,* edited by James and Raymond, 203–28.

———. "The Chinese Experience in Nevada: Success Despite Discrimination." *Nevada Public Affairs Review* 2 (1987): 43–51.

Coray, Michael S. "African-Americans in Nevada." *Nevada Historical Society Quarterly* 35, no. 4 (Winter 1992): 239–57.

———. "Influences on Black Family Household Organization in the West, 1850–1860." *Nevada Historical Society Quarterly* 31, no. 1 (Spring 1988): 1–31.

Couch, Bertrand F., and Jay A. Carpenter. *Nevada's Metal and Mineral Production (1859–1940, Inclusive)*. Geology and Mining Series, no. 37. Reno: Nevada State Bureau of Mines and the Mackay School of Mines, 1943.

Dahlgren, Carl. *Adolph Sutro: A Brief Story of a Brilliant Life*. San Francisco: Press of San Francisco, 1895.

Daley-Taylor, Andria. "Boardwalk Bons Vivants." *Nevada Magazine* 52, no. 6 (November/December 1992): 20–24, 35–37.

———. "A Feasibility Study for Piper's Opera House: Virginia City, Nevada." Report submitted to the Storey County Commissioners, 1995.

———. "Girls of the Golden West." In *Comstock Women*, edited by James and Raymond, 265–82.

———. "Piper's Opera House." Nomination to the National Register of Historic Places, 1996. On file in the Nevada State Historic Preservation Office.

D'Arcy, William. *The Fenian Movement in the United States: 1858–1886*. Washington, D.C.: Catholic University of America Press, 1947.

Datin, Richard C. *Elegance on C Street, Virginia City's International Hotel*. Reno: Privately published, 1977.

Davis, Sam P. *The History of Nevada*. Reno: Elms Publishing, 1913.

DeCurtis, Anthony, and James Henke, eds. *The Rolling Stone Illustrated History of Rock and Roll*. 2d ed. New York: Random House, 1980.

Del Papa, Frankie Sue. *Political History of Nevada*. 9th ed. Carson City: Nevada State Printer, 1990.

Dickie, G. W. "The Men and Machinery of the Comstock." *Engineering and Mining Journal* 98, series of articles (29 August–26 December 1914).

Diner, Hasia R. *Erin's Daughters in America: Irish Immigrant Women in the Nineteenth Century*. Baltimore: Johns Hopkins University Press, 1983.

Dixon, Kelly J. "The Frontier Lumbering Industry of Henness Pass Road: 1860s–1880s." Paper presented at the twenty-fifth Great Basin Anthropological Conference, Kings Beach, Calif., October 1996. On file at the Nevada Historic Preservation Office.

Dixon, Kelly J., Erika Johnson, and Juanita L. A. Spencer. "Sagehen Basin Analysis Area Historic Site Evaluation, Volume 2." Tahoe National Forest Report #05-17-816, 1997.

Duis, Perry R. *The Saloon: Public Drinking in Chicago and Boston, 1880–1920*. Urbana: University of Illinois Press, 1983.

Dwyer, John T. *Condemned to the Mines: The Life of Eugene O'Connell*. New York: Vantage Press, 1976.

Earl, Phillip I. "Brutal Times for Early-Day Virginia City Woman," *Reno Gazette-Journal*, 5 October 1986, 2E.

———. *This Was Nevada*. Reno: Nevada Historical Society, 1986.

Edwards, Sue. "Chinese Prostitution on the Comstock Lode, 1860–1880." Manuscript on file, University of Nevada, Las Vegas, Library, Special Collections, 1983.

———. "Statistical Analysis of the Chinese on the Comstock Lode, 1870–1880." Manuscript on file, University of Nevada, Las Vegas, Library, Special Collections, 1985.

Egan, Ferol. *Sand in a Whirlwind: The Paiute Indian War of 1860*. Reno: University of Nevada Press, 1985.

Elliott, Russell R. *History of Nevada*. 2d ed. Lincoln: University of Nebraska Press, 1987.

———. *Nevada's Twentieth-Century Mining Boom: Tonopah, Goldfield, Ely*. Reno: University of Nevada Press, 1966.

———. *Servant of Power: A Political Biography of Senator William M. Stewart*. Reno: University of Nevada Press, 1983.

Emmons, David M. *The Butte Irish: Class and Ethnicity in an American Mining Town, 1875–1925*. Urbana: University of Illinois Press, 1989.

Findlay, John M. *People of Chance: Gambling in American Society from Jamestown to Las Vegas*. New York: Oxford University Press, 1986.

Foster, R. F. *Modern Ireland: 1600–1972*. London: Allen Lane, Penguin Press, 1988.

Frady, Steven R. *Red Shirts and Leather Helmets: Volunteer Fire Fighting on the Comstock Lode*. Reno: University of Nevada Press, 1984.

Francke, Bernadette S. "Divination on Mount Davidson: An Overview of Women Spiritualists and Fortunetellers on the Comstock." In *Comstock Women*, edited by James and Raymond, 165–78.

———. "The Neighborhood and Nineteenth-Century Photographs: A Call to Locate Undocumented Historic Photographs of the Comstock Region." *Nevada Historical Society Quarterly* 35, no. 4 (Winter 1992): 258–69.

Galloway, John Debo. *Early Engineering Works Contributory to the Comstock*. Reno: Nevada State Bureau of Mines, 1947.

Garrison, Charles Jeffrey. "How the Devil Tempts Us to Go Aside from Christ: The History of First Presbyterian Church of Virginia City, 1862–1867." *Nevada Historical Society Quarterly* 36, no. 1 (Spring 1993): 13–34.

———. "Over 125 Years of Struggle at the Comstock's Oldest Church." *Touring the Lode: Comstock Chronicle Summer Tour Guide*, 4 April 1989.

Goldman, Marion S. *Gold Diggers and Silver Miners: Prostitution and Social Life on the Comstock Lode*. Ann Arbor: University of Michigan Press, 1981.

Goode, James. "The Making of 'The Misfits.'" *Nevada Magazine* 46, no. 6 (November/December 1986): 12–17.

Gorman, Thomas K. *Seventy-Five Years of Catholic Life in Nevada*. Reno: Journal Press, 1935.

Gray, Leslie Burns. *The Source and the Vision: Nevada's Role in the Civil War*. Sparks, Nev.: Gray Trust, 1989.

Greever, William S. *The Bonanza West: The Story of the Western Mining Rushes, 1848–1900*. Norman: University of Oklahoma Press, 1963.

Gregory, Cedric E. *Concise History of Mining*. New York: Pergamon Press, 1980.
Hafen, LeRoy R. *The Overland Mail: 1849–1869*. Cleveland: Arthur H. Clark, 1926.
Hagen, Michael. "The Mason Valley–Comstock Connection: Exploring the Moving of Old Buildings." On file at the Nevada State Historic Preservation Office, April 1996.
Hardesty, Donald L. *The Archaeology of Mining and Miners: A View from the Silver State*. Special Publication Series, no. 6. Ann Arbor: Society of Historical Archeology, 1988.
———. "Public Archaeology on the Comstock." Report prepared for the State Historic Preservation Office, 1996.
———. "Public Archaeology in the Virginia City Landmark District: The 1993 and 1994 Field Seasons." Report prepared for the Nevada State Historic Preservation Office, 1994.
Harmon, Maurice, ed. *Fenians and Fenianism*. Seattle: University of Washington Press, 1970.
Hartigan, Rachel J. "Looking for a Friend Among Strangers: Virginia City's Religious Institutions as Purveyors of Community." Undergraduate thesis, Yale University, 1993.
Hattori, Eugene M. "'And Some of Them Swear Like Pirates': American Indian Women in Nineteenth-Century Virginia City." In *Comstock Women*, edited by James and Raymond, 229–45.
———. *Northern Paiutes on the Comstock: Archaeology and Ethnohistory of an American Indian Population in Virginia City, Nevada*. Carson City: Nevada State Museum, 1975.
Hattori, Eugene M., Mary K. Rusco, and Donald R. Touhy. "Archaeological and Historical Studies at Ninth and Amherst, Lovelock, Nevada." Carson City: Nevada State Museum, 1979.
Henley, William J. "Old-time Movie Houses and the 'Virginia City' Premiere Fiasco." *Comstock Chronicle*, February 12, 1993, 7.
Hickson, Howard. *Mint Mark: "CC." The Story of the United States Mint at Carson City, Nevada*. Carson City: Nevada State Museum, 1972.
Highton, Jake. *Nevada Newspaper Days: A History of Journalism in the Silver State*. Stockton, Calif.: Heritage West Books, 1990.
———. "Salt: The Unsung 'Hero' of Nevada Mining." *Nevada Historical Society Quarterly* 26, no. 3 (Fall 1983): 172–86.
Hillyer, Katherine. *Young Reporter: Mark Twain in Virginia City*. Sparks, Nev.: Western Printing, 1964.
Historical Cultural Resources Survey (HCRS), U.S. Department of the Interior. "Greiner's Bend: A Case Study." Carson City: Comstock Project and Nevada State Historic Preservation Office, 1980.
Hobsbawm, E. J. *The Age of Capital, 1848–1875*. New York: Charles Scribner's Sons, 1975.

Hoffmann, Hemmann. *Californien, Nevada, und Mexico: Wanderungen eines Polytechnikers.* Basel: Schweighauserische, 1871.
Holliday, J. S. *The World Rushed In: The California Gold Rush Experience.* New York: Simon and Schuster, 1981.
Hulse, James W. *Silver State: Nevada's Heritage Reinterpreted.* Reno: University of Nevada Press, 1991.
Jackson, W. Turrentine. *Treasure Hill: Portrait of a Silver Mining Camp.* Tucson: University of Arizona Press, 1963.
James, Ronald M. "Defining the Group: Nineteenth-Century Cornish on the Mining Frontier." *Cornish Studies:* 2, edited by Philip Payton. Exeter: University of Exeter Press, 1994.
———. "Erin's Daughters on the Comstock: Building Community." In *Comstock Women*, edited by James and Raymond, 246–62.
———. "Knockers, Knackers, and Ghosts: Immigrant Folklore in the Western Mines." *Western Folklore* 51, no. 2 (April 1992): 153–77.
———. "On the Edge of Bonanza: Declining Fortunes and the Comstock Lode." *Mining History Association Annual* (1996): 101–108.
———. *Temples of Justice: County Courthouses of Nevada.* Reno: University of Nevada Press, 1994.
———. "Timothy Francis McCarthy: An Irish Immigrant Life on the Comstock." *Nevada Historical Society Quarterly* 39, no. 4 (Winter 1996): 300–308.
———. "Women of the Mining West: Virginia City Revisited." *Nevada Historical Society Quarterly* 36, no. 3 (Fall 1993): 153–77.
James, Ronald M., Richard D. Adkins, and Rachel J. Hartigan. "Competition and Coexistence in the Laundry: A View of the Comstock." *Western Historical Quarterly* 25, no. 2 (Summer 1994): 164–84.
———. "A Plan for the Archeological Investigation of the Virginia City Landmark District." Carson City: Nevada State Historic Preservation Office, 1993.
James, Ronald M., and John McCarthy. "McCarthy House." National Register Nomination. Carson City: Nevada State Historic Preservation Office, 1995.
James, Ronald M., and C. Elizabeth Raymond, eds. *Comstock Women: The Making of a Mining Community.* Reno: University of Nevada Press, 1997.
James, Susan A. "Queen of Tarts." *Nevada Magazine* 44, no. 5 (September/October 1984): 51–53.
Jameson, Elizabeth. "Women as Workers, Women as Civilizers: True Womanhood in the American West." In *The Women's West*, edited by Susan Armitage and Elizabeth Jameson, 145–64. Norman: University of Oklahoma Press, 1987.
Johnson, Clifton. *American Highways and Byways: The Pacific Coast.* New York: Macmillan, 1908.
Johnson, David Alan. "A Case of Mistaken Identity: William M. Stewart and the Rejection of Nevada's First Constitution." *Nevada Historical Society Quarterly* 22, no. 3 (Fall 1979): 186–98.

———. "The Courts and the Comstock Lode: The Travail of John Wesley North." *Pacific Historian* 27, no. 2 (Summer 1983): 31–46.

———. *Founding the Far West: California, Oregon, and Nevada, 1840–1890*. Berkeley: University of California Press, 1992.

———. "Industry and the Individual on the Far Western Frontier: A Case Study of Politics and Social Change in Early Nevada." *Pacific Historical Review* 51, no. 3 (August 1982): 243–64.

Jones, William R., ed. *The Murder of Julia Bulette: Virginia City, Nevada, 1867*. Golden, Col.: Outbooks, 1980.

Josephy, Alvin M., Jr. *The Civil War in the American West*. New York: Alfred A. Knopf, 1991.

Kendall, Robert E. "Henry Comstock's Offer Refused," *Comstock Chronicle*, 16 December 1994, 6.

———. "Pitfalls and Perils of Deep Mining on the Comstock." *Nevada Historical Society Quarterly* 39, no. 3 (Fall 1996): 216–31.

Lapomarda, Vincent A., S. J. "Saint Mary's in the Mountains: The Cradle of Catholicism in Western Nevada." *Nevada Historical Society Quarterly* 35, no. 1 (Spring 1992): 58–62.

Laughlin, Chandler. "Rockin' at the Red Dog." *Nevada Magazine* 55, no. 4 (July/August 1995): 80–83.

Lavender, David. *Nothing Seemed Impossible: William C. Ralston and Early San Francisco*. Palo Alto: American West Publishing, 1975.

Lewis, Oscar. *Silver Kings: The Lives and Times of Mackay, Fair, Flood, and O'Brien, Lords of the Nevada Comstock*. New York: Alfred A. Knopf, 1947. Reprint, Reno: University of Nevada Press, 1986.

Limbaugh, Ronald H. "John Muir and the Mining Industry." *Mining History Association Annual* (1996): 61–66.

———. "Making the Most of Experience: The Career of William J. Loring, Nevada Mining Engineer." *Mining History Association Annual* (1994): 9–13.

Lingenfelter, Richard E., and Karen Rix Gash. *The Newspapers of Nevada: A History and Bibliography, 1854–1979*. Reno: University of Nevada Press, 1984.

Lloyd, Christopher. *Explanation in Social History*. Oxford: Basil Blackwell, 1986.

Loverin, Janet I., and Robert A. Nylen. "Creating a Fashionable Society: Comstock Needleworkers from 1860 to 1880." In *Comstock Women*, edited by James and Raymond, 115–41.

Lowe, Sharon. "The 'Secret Friend': Opium in Comstock Society, 1860–1887." In *Comstock Women*, edited by James and Raymond, 95–112.

Lyman, George D. *The Saga of the Comstock Lode: Boom Days in Virginia City*. New York: Charles Scribner's Sons, 1951.

Lyman, Stanford M., and William A. Douglas. "Ethnicity: Strategies of Collective and Individual Impression Management." *Social Research* 40, no. 2 (Summer 1973): 344–65.

Mack, Effie Mona. *Mark Twain in Nevada*. New York: Charles Scribner's Sons, 1947.

———. *Nevada: A History of the State from the Earliest Times Through the Civil War.* Glendale, Calif.: Arthur H. Clark, 1936.
Maffly-Kipp, Laurie F. *Religion and Society in Frontier California.* New Haven: Yale University Press, 1994.
Mann, Ralph. *After the Gold Rush: Society in Grass Valley and Nevada City, California, 1849–1870.* Stanford: Stanford University Press, 1982.
Manter, Ethel. *Rocket of the Comstock: The Story of John William Mackay.* Caldwell, Idaho: Caxton Printers, 1950.
Marshall, John P. "Jews in Nevada, 1850–1900." *Journal of the West* 23, no. 1 (January 1984): 62–64.
McCormick, Jim. "Silver City: Reminiscences, Facts, and a Little Gossip." *Nevada Historical Society Quarterly* 30, no. 1 (Spring 1987): 37–55.
McCormick, Thomas J. "World Systems." *Journal of American History* 77, no. 1 (June 1990): 125–32.
McDonald, Douglas. *Camels in Nevada.* Las Vegas: Nevada Publications, 1983.
———. *The Legend of Julia Bulette.* Las Vegas: Nevada Publications, 1980.
———. *Virginia City and the Silver Region of the Comstock Lode.* Las Vegas: Nevada Publications, 1982.
McFeely, William S. *Grant: Biography.* New York: W. W. Norton, 1981.
McGloin, John B., S.J. "Patrick Manogue: Gold Miner and Bishop." *Nevada Historical Society Quarterly* 14, no. 2 (Summer 1971): 25–32.
McGrath, Roger D. *Gunfights, Highwaymen, and Vigilantes: Violence on the Frontier.* Berkeley: University of California Press, 1984.
Moehring, Eugene. "The Comstock Urban Network." *Pacific Historical Review* 66, no. 3 (August 1997).
Moody, Eric N., and Robert A. Nylen. *Brewed in Nevada: A History of the Silver State's Beers and Breweries.* Carson City: Nevada State Museum, 1986.
———. "The Comstock Brewing Industry." *OAH Newsletter* (February 1988): 4–5.
Myrick, David. *Railroads of Nevada and Eastern California.* Berkeley: Howell-North, 1963.
Nash, Gerald D. *World War II and the West: Reshaping the Economy.* Lincoln: University of Nebraska Press, 1990.
Nicoletta, Julie. "Redefining Domesticity: Women and Lodging Houses on the Comstock." In *Comstock Women,* edited by James and Raymond, 43–67.
Noel, Thomas J. *The City and the Saloon: Denver, 1858–1916.* 1982. Reprint, Niwot, Col.: University of Colorado Press, 1996.
Paine, Swift. *Eilley Orrum: Queen of the Comstock.* Indianapolis: Bobbs-Merrill, 1929.
Paul, Rodman. *Mining Frontiers of the Far West, 1848–1880.* New York: Holt, Rinehart, and Winston, 1963.
Peck, Gunther. "Manly Gambles: The Politics of Risk on the Comstock Lode, 1860–1880." *Journal of Social History* 26, no. 4 (Summer 1993): 701–23.
Pengelly, Joe. "Juan Ricardo José: Ricardo José, the Great Spanish Singer from Lanner." *Cornish World* 6 (September/October 1995): 27.

Peterson, Richard H. *The Bonanza Kings: The Social Origins and Business Behavior of Western Mining Entrepreneurs, 1870–1900.* 2d ed. Norman: University of Oklahoma Press, 1991.

———. *Bonanza Rich: Lifestyles of the Western Mining Entrepreneurs.* Moscow: University of Idaho Press, 1991.

Petrik, Paula. *No Step Backward: Women and Family on the Rocky Mountain Mining Frontier, Helena, Montana, 1865–1900.* Helena: Montana Historical Society Press, 1987.

Pickering, Lee Lukes. *The Story of St. Mary's Art Center—Now—and St. Mary Louise Hospital—Then.* Carson City: N.p., 1986.

Pomeroy, Earl S. *In Search of the Golden West: The Tourist in Western America.* New York: Knopf, 1957. Reprint, Lincoln: University of Nebraska Press, 1990. Page citations are to the reprint edition.

Pugh, William. "The History of the Baptist Church in Virginia City." *Comstock Chronicle*, 20 June, 14 July, and 28 July 1995.

Rafferty, Kevin. "Catholics in Nevada." Draft chapter for the *Nevada Comprehensive Preservation Plan* (1992). On file in the Nevada State Historic Preservation Office.

Raymond, Rossiter W. *A Glossary of Mining and Metallurgical Terms.* Easton, Pa.: American Institute of Mining Engineers, 1881.

Reps, John W. "Bonanza Towns: Urban Planning on the Western Mining Frontier." In *Pattern and Process: Research in Historical Geography*, edited by Ralph E. Ehrenberg, 271–86. Washington, D.C.: Howard University Press, 1975.

Richnak, Barbara. *Silver Hillside: The Life and Times of Virginia City.* Incline Village, Nev.: Comstock Nevada Publishing Company, 1984.

Robinson, Judith. *The Hearsts: An American Dynasty.* New York: Avon Books, 1992.

Rocha, Guy Louis. "The Many Images of the Comstock Miners' Unions." *Nevada Historical Society Quarterly* 39, no. 3 (Fall 1996): 163–81.

Rocha, Guy Louis, and Dennis Myers. "Myth #8: The Trestle of the State Seal." *Sierra Sage*, August 16, 1996, 16.

Rochlin, Harriet, and Fred Rochlin. *Pioneer Jews: A New Life in the Far West.* Boston: Houghton Mifflin, 1984.

Rosenzweig, Roy. *Eight Hours for What We Will: Workers and Leisure in an Industrial City, 1870–1920.* Cambridge: Cambridge University Press, 1983.

Rothbart, Ron. "The Ethnic Saloon as a Form of Immigrant Enterprise." *International Migration Review* 27, no. 2 (Summer 1993): 332–58.

Rule, John. *The Experience of Labour in Eighteenth-Century English Industry.* New York: St. Martin's Press, 1981.

———. *The Labouring Classes in Early Industrial England, 1750–1850.* New York: Longman, 1986.

———. "Some Social Aspects of the Cornish Industrial Revolution." In *Industry and Society in the South-West*, edited by Roger Burt, 71–106. Exeter: University of Exeter Press, 1970.

Rusco, Elmer R. *"Good Times Coming?": Black Nevadans in the Nineteenth Century.* Westport, Conn.: Greenwood Press, 1975.

Shade, Rose Marian. "Virginia City's Ill-Fated Methodist Church." *Journal of the West* 8, no. 4 (October 1969): 447–53.

Shamberger, Hugh. *The Story of the Water Supply for the Comstock.* Geological Survey Professional Paper 779. Washington, D.C.: U.S. Geological Survey, Government Printing Office, 1965.

Shepperson, Wilbur S. *Restless Strangers: Nevada's Immigrants and Their Interpreters.* Reno: University of Nevada Press, 1970.

Shepperson, Wilbur S., and Ann Harvey. *Mirage-Land: Images of Nevada.* Reno: University of Nevada Press, 1992.

Shinn, Charles Howard. *The Story of the Mine as Illustrated by the Great Comstock Lode of Nevada.* 1896, 1910. Reprint of 1910 ed., Reno: University of Nevada Press, 1980.

Simon, Paul. *Freedom's Champion: Elijah Lovejoy.* Carbondale: Southern Illinois University Press, 1994.

Sion, Michael. "Camel Race in Virginia City." *Nevada Magazine* 52, no. 5 (September/October 1992): E-17.

Smith, Duane A. "Comstock Miseries: Medicine and Mining in the 1860s." *Nevada Historical Society Quarterly* 36, no. 1 (Spring 1993): 1–12.

Smith, Grant H. *The History of the Comstock Lode: 1850–1920.* 1943. 6th rev. printing, Reno: Nevada Bureau of Mines and the University of Nevada, 1966. Page citations are to the 1966 printing.

Sohn, Anton P., M.D. "The Acceptance of Women Physicians in Nineteenth-Century Nevada." *Greasewood Tablettes* 6, no. 2 (Summer 1995): 1–2. Reno: University of Nevada School of Medicine.

———. "Chinese Doctors in Nevada and the Great Basin." *Greasewood Tablettes* 8, no. 1 (Spring 1997): 1–2. Reno: University of Nevada School of Medicine.

Stern, Norton B. "Notes on a Virginia Police Chief." *Western States Jewish Historical Quarterly* 21, no. 1 (October 1979): 89–91.

Stewart, Robert E., Jr., and Mary Frances Stewart. *Adolph Sutro: A Biography.* Berkeley: Howell-North, 1962.

Stonehouse, Merlin. *John Wesley North and the Reform Frontier.* Minneapolis: University of Minnesota Press, 1965.

Strong, Douglas H. *Tahoe: An Environmental History.* Lincoln: University of Nebraska Press, 1984.

Stroup, Dorothy. "Renaissance in Virginia City." *San Francisco Examiner,* June 18, 1989.

Terrace, Vincent. *The Complete Encyclopedia of Television Programs: 1947–1979.* New York: A. S. Barnes and Company, 1979.

Thompson, Judy Anne. "Historical Archaeology in Virginia City, Nevada: A Case Study of the 90-H Block." Master's thesis, University of Nevada, Reno, 1992.

Tilton, Cecil Gage. *William Chapman Ralston: Courageous Builder.* Boston: Christopher Publishing House, 1935.

Todd, Arthur Cecil. *The Cornish Miner in America.* 2d ed. Spokane, Wash.: Arthur H. Clark, 1995.

Totton, Kathryn D. "'They Are Doing So to a Liberal Extent Here Now': Women and Divorce on the Comstock, 1859–1880." In *Comstock Women,* edited by James and Raymond, 68–94.

Wallerstein, Immanuel. *The Capitalist World-Economy.* New York: Cambridge University Press, 1979.

———. *The Modern World-System: Capitalist Agriculture and the Origins of the European World-Economy in the Sixteenth Century.* New York: Academic Press, 1974.

———. *Politics of the World Economy: The States, the Movements, and the Civilizations.* New York: Cambridge University Press, 1984.

———. *World-Systems Analysis: Theory and Methodology.* Beverly Hills: Sage Publications, 1982.

Walsh, James P. "The Irish in the New America: 'Way Out West." In *America and Ireland, 1776–1976: The American Identity and the Irish Connection,* edited by David Noel Doyle and Owen Dudley Edwards, 165–76. Westport, Conn.: Greenwood Press, 1980.

Waters, Mary C. *Ethnic Options: Choosing Identities in America.* Berkeley: University of California, 1990.

Watson, Anita Ernst, Jean E. Ford, and Linda White. "'The Advantages of Ladies' Society': The Public Sphere of Women on the Comstock." In *Comstock Women,* edited by James and Raymond, 179–99.

Watson, Margaret G. *Silver Theatre: Amusements of Nevada's Mining Frontier, 1850–1864.* Glendale, Calif.: Arthur H. Clark, 1964.

Weisenburger, Francis P. "God and Man in a Secular City." *Nevada Historical Society Quarterly* 14, no. 2 (Summer 1971): 3–24.

West, Elliott. *The Saloon on the Rocky Mountain Mining Frontier.* Lincoln: University of Nebraska Press, 1979.

White, G. Pawley. *A Handbook of Cornish Surnames.* 2d ed. Exeter: A. Wheaton, 1981.

White, William G., and Ronald M. James. "Little Rathole on the Big Bonanza: Historical and Archaeological Assessment of an Underground Resource." Survey Report. Carson City: State Historic Preservation Office, June 1991.

Wickenden, Joseph. "History of the Nevada Militia, 1862–1912." Compiled under the direction of Brigadier General Jay H. White, Adjutant General. Microfilm copy of the 1941 manuscript; University of Nevada, Reno, Library.

Williams, George, III. *Mark Twain: His Life in Virginia City, Nevada.* Riverside, Calif.: Trees by the River, 1986.

Woods, James. *Recollections of Pioneer Work.* San Francisco: Joseph Winterburn, 1878.

Wurm, Mark, and Harry Demoro. *The Silver Short Line: A History of the Virginia and Truckee Railroad.* Virginia City, Nev.: Virginia and Truckee Railroad, 1983.

Young, George J. "History of Mining in Nevada." In *The History of Nevada*, edited by Sam P. Davis, 315–66. Reno: Elms Publishing, 1913.
Young, Otis. *Black Powder and Hand Steel: Miners and Machines on the Old Western Frontier.* Norman: University of Oklahoma Press, 1978.
———. "Philipp Deidesheimer, 1832–1916, Engineer of the Comstock." *Historical Society of Southern California* 57 (1975): 361–69.
———. *Western Mining.* Norman: University of Oklahoma Press, 1970.
Zanjani, Sally Springmeyer. *"Ghost Dance Winter" and Other Tales of the Frontier.* Reno: Nevada Historical Society, 1994.
———. *Goldfield: The Last Gold Rush on the Western Frontier.* Athens, Ohio: Swallow Press, 1992.

INDEX

Page numbers in italics refer to illustrations

Adams, John Quincy, 213
African Americans, 35, 72–73, 97–99, 115, 143, 144, 152–55, 157; marital status, 97–99; occupations, 97–98, 153–54, 165, 210, 311n45; and prostitution, 153; religion, 154, 155, 201
African Methodist Episcopal Church, 155, 201
Alemany, Archbishop Sadoc, 200
Alice in Wonderland, 136
Alsop and Company, San Francisco, 12, 50
Alta Mine, 212–13
Amazonian 601, 174
American Flats, Nev., 81
American Folklife Center, 258, 260
American Indian, 17–18, 38–41, 128, 143, 144, 156–57, 186. *See also* Paiutes, Northern; Shoshone; and Washoe (American Indian)
American Protestant Association, 162
Ancient Order of the Hibernians. *See* Irish immigrants, organizations
Anderson, "Uncle Jimmy," 221
Andreasen, Mary, 260
Anti-Chinese League, 149, 162
Antietam, Md., 70
Appomattox, Va., 71, 73, 74
Archaeology, 76, 156–57, 186–87, 204, 206
Arizona Comstock Company, 254
Arrastras. *See* Milling, arrastras
Artists, 22, 152, 210, 258, 263–64
Ash Book and Toy Store, 114
Atlantic Coast oysters, 136
Atwood, Melville, 11
Aurora, Nev., 306n45
Austin, Nev., 71, 75
Austin Brass Band, 71
Australia, 227, 237; immigrants from, 209
Austrian immigrants, 208, 217

Babcock, W. F., 55
Bagley, John, 221
Bailey, Mary, 226
Bajazette and Golden Era Mine, 61
Baldwin, Alice, 229
Bank of California, 77–80, 89–90, 104, 140, 230
Bank Crowd, the, 77–80, 88, 104–6, 148
Bannock Shoshone. *See* Shoshone and American Indians

Baptist church, 155, 201
Barbary Coast, 94, 152, 154, 177–80, *179*, 185, 186, 228, 266
Baseball, 211
Beck, H. S., 230
Beebe, Lucius, 259–60, 262–63, 272
Beecher, Henry Ward, 250
Beer Garden. *See* Van Bokkelen, Jacob L.
Belcher Mine, 58, 106
Bell, Thomas, 80
Bennett, Rev. Jesse L., 33–34, 199, 201
Best, Katherine, 260
Best and Belcher Mine, 106–7, 115
Bicknell, George, Richard, and James, 85
Big Bonanza, 91, 107–9, 236
Big Brother and the Holding Company, 263
Billiards, 180, 186, 188
Bishop, John, 6–7
Blackburn, D., 210
B'nai B'rith. *See* Jews, religious and social organizations
Bodie, Calif., 237, 306n45
Bogart, Humphrey, 259
Bonanza (television show), xix–xx, 261–63, 265, 266
Bonanza Casino, 268
Bonanza Firm, *50*, 100–110
Bonanza Inn, 259
Booth, Edwin, 250
Booth, John Wilkes, 71–72
Bootlegging, 252
Borrasca. *See* Depression
Boston, Mass., 181
Bowers, Alexander "Sandy," 7, 68–69, 229, 231–32
Bowers, Eilley Orrum, 68–69, 114–15, 229, 230, 231–32
Bow Windows, 177
Bozeman, Mont., 15
Breweries, 130, 159, 164, *183*, 188–89, 203, 307n61
Brick, the, 177, 212
British Benevolent Association, 160
Brooklyn Bridge, 124
Brown, "Fighting Sam," 32–33
Brown, Grafton T., 152, 210
Brown, Susie, 178
Brown, William A. G., 154, 301–2n25
Browne, J. Ross, xix, 22–23, 25, *31*, 59–60, 210
Brunswick Canyon, 80
Bryan, Bonnie, 268
Bryant, Edmund, 40, 103
Bryant, Marie-Louise. *See* Mackay, Marie-Louise
Buchanan, James, 42
Bucket of Blood Saloon, 261
Buckland, Eliza, 113
Bulette, Julia, 167–68, 177, 212, 260
Bullion Mine, 101, 103, 106
Bullion Ravine, 22
Burning Moscow Mine, 62–63, 65, 67
Burns, Robert, 157–58
Burton, Sir Richard, *43*, 44
Butte, Mont., 227, 237
Butte County, Calif., 110
Butter, Charles, 242–43

Caledonia Club. *See* Scottish immigrants, organizations
Caledonia Tunnel and Mining Company, 101
California, 10, 60, 71, 77, 115, 231–32; and ethnicity 34, 35; stock market, 23–24
California Foundry, 136
California Gold Rush, 1–5, 7, 19, 77, 101, 105, 201
California Mine, 57, 113, 238. *See also* Consolidated Virginia Mine
California Pan Mill, *50*
Camels, *29*, 263, *264*, 268
Camp, Herman, 18
Canada, 35, 93, 95, 143, 205, 209
Capitol Saloon, 185

Carey, E. W., 208
Carpenters. *See* Construction workers
Carroll, Lewis, 136
Carson, Tom, 187
Carson Brewery, 114
Carson City, Nev., 42, 52, 80, 83, 133, 158, 211, 222, 244, 259
Carson River, 2, 5, 21, 51, 59, 80, 234
Carson River Valley, 32–33, 88
Casteel, F. D., 7
Catholic church, 34, 200–3; cemetery, 44, 177
Census, U.S. *See* United States, Census
Centerville Mine, 61
Central Mine, 57
Central Pacific Railroad, 80, 81
Chalk mining, 252
Challenger (shuttle), 73
Chancellorsville, Va., 70
Charlatans, the, 263–64
Chattanooga, Tenn., 70
Chicago, Ill., 181
Child rearing, 191–95, 204, 206, 216–17, 221–22
Chilean immigrants, 143, 156, 187
Chin, Pooty, 227–28
Chinatown (on the Carson River). *See* Dayton, Nev.
Chinatown, Virginia City, 95–96, 99, 148–52, *151*, 205–6; and health care, 96; laundries, 95, 96, 149, 218–19; occupations, 96, 188, 201; opium use, 150, 189–90; and prostitution, 96, 150, 176, 221; social organizations, 160
Chinese immigrants, 34–35, 95–96, 109, 143, 144, 148–52, 157, 205–6, 213, 251–52, 254; occupations of, 4, 96, 149–50, 164–65, 187, 188, 217–18, 227–28, 230; and railroad, 80–81, 141, 149, 159; as servants, 112, 149, 167, 218
Ching, Charlie, 251–52
Chollar Mine, 64, 66, 67, 72, 101, 254
Chollar-Potosi Mine, 79
Chung, Sue Fawn, 96
Churches. *See under specific denominations*
Circus, 161, 209–34. *See also* Entertainment
City Restaurant, 252
Civil War, U.S., 61, 63, 70–74, 152
Clark, Walter Van Tilburg, 260
Class, 76, 215–34
Clegg, Charles, 259–60, 262–63, 272
Clemens, Orion, 25
Clemens, Samuel, xx, 25–26, 44, 68–69, 76, 117, 161, 168–69, 171–72, 208
Cobb, Ty, 253–54
Cobb, Walter, 198
Cody, "Buffalo" Bill, 250
Colorado, 34, 227, 237, 266
Columbia Quartz District, 5
Comstock, Henry T. "Pancake," 7–9, *9*, 10, 13–20, *15*, 46, 57, 67, 261, 380n41
Comstock Historic District Commission, 265, 268, 275
Comstock Pumping Association, 242–43
Consolidated Virginia Mine, 99, 106, 112, 113, 114, 116, *126*, *220*, 237, 238, 240, 241
Construction workers, 25–26, 54, 55–56, 91, 131, 179, 236
Corbett, James J., 250
Corcoran, Daniel, 178
Cornish immigrants, 30–31, 95, 121–22, 123, 143, 144, 146–48, 157, 160, 162, 165, 194, 200, 205, 227, 237, 283n23; singing, 211, 222
Cornish pumps, 58, 121–22
Cornwall, 54, 142, 162
County Cork, 144
County Hospital. *See* Storey County, hospital

Cousin Jack. *See* Cornish immigrants
Coxey, Posey, 72
Crane, Alice McCully, 112, 115
Crime, 171–75, 206
Crown Point Mine, 85–88, 102, 105–6
Crown Point Trestle, Gold Hill, *82*, 82–83, 90
Curran, Mrs. Bridget, 134

Dakins, Don, 192–92
Daley, Andria, 259–60
Darmstadt, Hesse, 55
Daughters of Charity of Saint Vincent de Paul, 93–94, 145, 177, 198–99, 222, 241–42; hospital, 198–99, 241; school and orphanage, 75, 198–99, 221–22
Davis, Jefferson, 70
Davis, R. D. W., 61
Dayton, Nev., 21, 51–52, 169, 213
Dean, Walter S., 110
DeGroot, Henry, 12, 14, 16
Deidesheimer, Matilda, *56*
Deidesheimer, Philipp, 36, 55–57, *56*, 62, 160
Delaware, 259
DeLongchamps, Frederick J., 249
Delta Saloon, 185, 260, 268
Denver, Colo., 306n45
Depression, 4–5, 24, 65, 74–79, 89–90, 105, 116, 118, 138, 140, 219, 222, 235–57. *See also* Great Depression
De Quille, Dan. *See* Wright, William
Deutsche Union, 159
Devil's Gate, 6, 22, 40, 51
De Voto, Bernard, 260
Dildine, Abram S., and "Old Mother" (Nancy Nelson), 221
Dini, Speaker of the Assembly Joe, 268
Divide, the, 101, 161, 254, 257
Divorce, 194–95, 207, 218, 260
Donner Party, 52
Donohue, Ralston, and Company, 77
Doolittle, Jimmy, 248
Dosch, Col. Henry E., 169
Doten, Alfred, 71, 72, 76, 158
Dressmakers. *See* Needlework
Drought, 5
Drury, Wells, xix, 205
Duis, Perry, 181
"Dutch Nick's" saloon, 4
Dwellings, 4, 12, 23, 30, 98–99, 122, 244, 252–53, 254, 265

Edgington, Abram M., 226–27
Edgington, Lillie, 226–27
Egyptian, 42
El Dorado Saloon, 185
Elliott, Russell, 141
Emmet Guard, 72, 144, 146–47, 205
Empire Mine, Grass Valley, Calif., 10
Emrich, Duncan, 258–59, 260–61
England, 46, 93, 162, 205, 222
English immigrants, 34, 95, 143, 165, 194, 205
Entertainment, 4, 92–93, 160, 208–10
Episcopal Church, 113, 201
Ethnicity, xx–xxi, 34–37, 67, 95–99, 143–66, 193–94
Eureka Benevolent Society, 72
Euro-Americans, 4, 34–35, 37, 38, 41, 143, 144, 149, 162–65, 189–90, 193, 206; and fraternal organizations, 162; and occupations, 165, 184, 187, 208–9, 227, 230

Fair, James G., 100–10, *104*, 207–8, 230, 231, 244
Fair, Theresa, 207–8
Fenians. *See* Irish immigrants, social organizations
Fijians, 161
Finney, James "Old Virginny," 6–7, 12, 13–14, 16–20, 57, 67, 280n34
Firefighters, 86–87, 111–17, 198
Fire of 1875, 111–17, 149, 177–78, 191, 207, 232

Fitzsimmons, Bob, 250
Flood, James C., 105, 231
Flynn, Errol, 259
Folklore, xix, xx, 3–4, 8, 28–29, 57, 58, 67–69, 89–90, 101, 222, 232, 234, 244, 253, 260–61, 263, 276; Bank of California, 77–80; Big Bonanza, 108; Bonanza kings, 207–8; early discoveries, 8, 13–20; ghosts and tommyknockers, 121–22; Great Fire of 1875, 112–13, 117, 198, 294n39; Great Seal of Nevada, 83; and health care, 198; and Father Patrick Manogue, 201; and prostitution 32, 168, 176–77, 185, 260–61; and saloons, 182, 188; Spanish, 12; and violence, 32–33, 168, 171; and women, 192
Folklore Section, Library of Congress. *See* American Folklife Center
Fort Churchill, 113, 140
Fort Homestead, 72, 81, 110, 147, *148*
Fortunetellers, 114–15, 229
Foundries, 133–34, 136
Fourth Ward School, 178, 191–92, *192*, 254, 267
Francis, Jim, 197–98
Frederica, Sister, 198
Frederick House, 182, 251
Fredricksburg, Va., 70
Freemasons (Free and Accepted Order of Masons), 72, 162, 205, 213–14, 227
Freiburg School of Mines, 35
French immigrants, 35, 93, 95, 143, 161, 168, 221
Friends of Poland, 160
Fulton Foundry, 134, 136

Gable, Clark, 263
Gallagher, Fr. Hugh, 34, 200
Gambling, 167, 186–88, 204
Gans-Nelson fight, 251
Geiger, Dr. D. M., 52, 230
Geiger Grade, 52–53, 169, 248
"General Grant" (cannon), 81, 110
Genoa, Nev. (previously Utah Territory), 1
Gentry Mine, 61
Georgia, 35
German immigrants, 34–35, 88, 95, 110, 143, 159–60, 161; occupations of, 93, 159–60, 164, 184, 189, 208, 227, 250, 303n39; social organizations, 72, 159; Turnverein societies, 159
Germany, 162
Gettysburg, Penn., 70
Gobey and Keely's Saloon, 182
Gold Canyon road, 52–53
Gold Hill, Nev., *38*, 81, *82*, 169, 269; and depression, 240, 254; and ethnicity, 147; and founding, 7, 67; and saloons, 240
Gold Hill Brass Band, 81
Gold Hill Foundry, 134, 136
Gold Hill Miner's Union, 140
Gold Hill Theater, 208
Gold Hill Water Company, 60
Golden Jubilee (Virginia City), 235, 249–50
Goldfield, Nev., 243–44, 251
Goldman, Marion, 220
Goodman, Joe, 76, 160
Gould, Alva, 7
Gould and Curry Mine and Mill, *47*, 50–51, 72, 77, 129
Government, Comstock, 9, 11, 41–42, 171, 172, 196–97
Grant, Ulysses S., 70, 205, 231, 238
Grass Valley, Calif., 10, 11, 129
Grass Valley Mine (Comstock), 61
Great Depression, 76, 252–56. *See also* Depression
Great Fire of 1875. *See* Fire of 1875
Greek immigrants, 143
Greeley, Horace, 28–29

Greiner, John, 134
Greiner's Bend, 51, 209
Gridley, Reuel Colt, 71, 75
Grosh, Hosea Ballou and Ethan Allan, 3–4, 6, 13
Grunwalds (actors), 208

Hahnlen, Jacob F., 159–60
Hale and Norcross Mine, 103–5, 254
Hallidie, Andrew Smith, 124
Hall's Pioneer Laundry, 168
Hamlin, William, 134
Haong, Dr. Son, 227–28
Hardesty, Donald L., 186–87, 204
Harris, Elias B., 48
Harris, Waterman, and Haynes (gang), 169
Harrison, B. A., 11
Hart, Miss Sallie, 204
Hayes, Rutherford B., 238
Hayward, Alvinza, 79, 105–6
Health care, 197–99
Hearst, George, xxi, 69
Heffernan, Arthur Perkins, 170–71
Helvetia Society, 160
Henderson, Alexander, 6–7
Herrick, H. S., 71
Hibernia Brewery, 186–87, 204
Hillyer, Katherine, 260
Hispanics, 35–37, 50, 95, 96–97, 143, 144, 155–56, 157; demography of, 96; mining technology, 12, 36–37, 46, 279n24; occupations of, 4, 36–37, 97, 156, 164, 165, 187, 208, 210
Hobart, W. S., 110
Hobart, W. W., 115
Hobsbawm, Eric, xv, 277n3
Hoffmann, Hemmann, 203
Honey Lake Rangers, 39
Houseworth, V. A., 9
Houston Oil and Mineral, 269, *270*
Hughes, Lynn, 264–65
Hungarian immigrants, 34, 143
Hurdy Gurdy Girls, 185
Huston, John, 263
Hutchings, James, 134

Idaho, 14, 75, 103, 237
Illinois, 34, 213
Independent Order of Good Templars. *See* Temperance
International Hotel, 75, 113, 116–17, 182, 185, *210*, 232, 251, 252
International Order of Odd Fellows, 72, 116, 162, 205, 213–14
Ireland, 144
Irish Fenian Brotherhood. *See* Irish immigrants, social organizations
Irish immigrants, 34–35, 95, 100–1, 103–4, 122, 143, 144–48, 157, 178–79, 182–83, 201, 215, 221, 230, 237; and children, 165, 194; demography of, 163–65; military companies, 72, 144, 146–47, 158; occupations of, 93, 145–46, 149–50, 164, 184, 201, 208, 219; social organizations, 44, 72, 144–45, 147, 160, 162, 205
Irish Land League. *See* Irish immigrants, social organizations
Irish Republican Army. *See* Irish immigrants, social organizations
Italian Benevolent Society. *See* Italian immigrants, social organizations
Italian immigrants, 34–35, 93, 95, 143, 161, 210, 217; social organizations, 159, 160

Jackson, Robert D., 242
Jacobsen, Aileen, 276
James, Isaac E., 80
Japanese immigrants, 143
Jews, 34, 88, 143, 160; occupations of, 160, 226–27; religious and social organizations, 72, 160, 201
Johntown, Nev., 21
Jolson, Al, 250

Jones, John P., 105–6, 231
Joplin, Janis, 263
Jose, Richard, 222, 251
Jumbo Grade, 51, 53

Kee, Chung, 254
Kelly, Chief of Police, 175
Kenitzer and Raun (architectural firm), 116
Kennedy, John F., 73
Kentuck Mine, 85–88, 102–3
Kentucky, 230
Kesey, Ken, 263
Keystone Mine, 61
King, Clarence, 55
Kirk, George B., 171
Knight, William, 7
Knights of Pythias, 205

Lake Tahoe, 137
Lake's Crossing. *See* Reno, Nev.
La Plata Hall, 208
Larkin, Peter, 178
Larson, Karl, 272
Las Vegas, Nev., 244
Latter-day Saints, 1–2, 17–18
Lawyers, 24, 92
Lee, Robert E., 70, 71
Leitch, R. P., *38*
Leslie, Miriam, 174, 185
Lester, Jessie, 177
Liberal Mexican Club, 156
Library of Congress. *See* American Folklife Center
Limitation Order L-208, 256
Lincoln, Abraham, 42, 63–64, 66, 70, 71–73, 88, 152
Lloyd, Christopher, xv
Lobb buckwheat, 53
Lobo, Frank, 210
London, Eng., 231
Lord, Eliot, 4, 18–19
Loring, William J., 254
Loring Cut, 254
Lovejoy, Edward, 213–14
Lovejoy, Elijah, 213
Lynch, Philip, 76, 230

Mackay, John W., xxi, 100–10, *102*, 112–13, 117, 148, 207–8, 222, 231, 236, 241, 275
Mackay, Marie Louise, 103, 199, 207–8, 231
Magnolia Saloon, 182
Maguire, Tom, 208
Maldonado brothers, 36–37, 50, 54–55, 230
Manogue, Fr. Patrick, 86–87, *87*, 103, 113, 201
Massachusetts, 34
Mathews, Mary McNair, 94, 149, 155, 174–75, 188, 206, 222; and occupations, 195, 198, 228; and temperance, 204
Maximilian, Emperor, 156
McCarthy, Timothy Francis, 201, 297n27, 210n31
McClellan, Cyrenius B., 210
McCone, John, 134
McCone Steel, 134
McEntire, Mary, 215, 221
McLaughlin, Patrick, 8–9, *9*, 10–11, 46, 67, 261
Melodeon Theater, 208
Melton, Greg, 268
Menlo Park, 231
Merchants, 24–25, 34, 91–92, 188–89, 222; and ethnicity, 36
Mercury poisoning, 134–36
Meredith, Henry, 40
Methodist Church, 34, 113, 199, *200*
Metropolitan Brass Band, 72
Mexican-American War, 37
Mexican Benevolent Society, 156
Mexican Mine, 36–37, 50, 54–55, 57, 74, 156, 230
Mexico, 4, 37, 46, 75, 131; immigrants from, 34–37, 72, 91, 93, 143, 156,

160, 164, 187, 230. *See also* Hispanics
Millian, John, 168, 212
Milliners. *See* Needlework
Milling, 6, 12, *49*, 50, *51*, 74, 78–79, 84, 88, 132–36, *135*, 241, *243*; arrastras, 11, 46, *47*, 279n24; cyanide, 242–43; work force, 25–26, 91, 133; innovations, 45–51, 242
Mills, Darius Ogden, 77, 80, 101, 241
Milton Mine, 101
Miner's League, 140
Miner's Protective Association, 140
Miner's Union. *See* Mining, unions
Miner's Union Hall, 155
Mining, 119–42; engineers, 91; and ethnicity, 36–37, 122; and fire, 84–88; hardrock, 6, 10, 12–13, 67, 119–42, *126*, *127*; injuries related to, 84, 119–21, 125, 129, 130; innovations, 55–56, 120–21, 123–24, 128–30; legal issues, 60–67; Law of 1872, 197, 231; open pit, 53; placer, xix, 1–20, 67, 118; pumping, 57–60, 212–13, 241, 242–43, 248; rathole, 36–37, 67, 102, 131–32, *132;* supports, 53–57, 128; unions, 81, 88, 140–42, 196, 204; work force, 25–26, 91, 131, 137–39. *See also under specific mines*
Misfits (film), 263
Missouri, 42, 77
Molinelli Hotel, 182, 232–33, 251
Monk, Henry James "Hank," 28–29
Monroe, Marilyn, 263
Montana, 75
Montgomery Guard, 144, 146–48
Morgan Mill, 226
Mormon Church. *See* Latter-day Saints
Moroccan immigrants, 143, 161, 209
Mott, Judge Gordon N., 62, 64, 65
Moundhouse, Nev., 133
Mount Butler, 22
Mount Davidson, xx, 2, 4, 6–7, 22, 76, 211
Mount Pleasant, 11
Muckle, Hugh, 211
Mundall, John H., 76
Murphy, John, 85
Myth. *See* Folklore

Nagel, F., 210
National Broadcasting Corporation, 261
National Guard, 71–72, 114, 144
National Guard Hall, 161, 170
National Park Service, 268
Native Americans. *See* Paiutes, Northern; American Indian
Nelson, Nancy. *See* Dildine, Abram S., and "Old Mother"
Nevada, 75, 109; constitution, 63, 64–66, 74; historic preservation office, 267; legislature, 29, 66, 267; state census, 109; state prison, 171; statehood of, 66, 70; superintendent of public instruction, 195; territorial census, 37; territory, 25–26, 39, 42
Nevada Brewery, *183*, 189
Nevada City, Calif., 40, 46, 129
Nevada Humanities Committee, 267
Nevada Opera, 258
Nevada Pioneer, 159
Nevin, Joe, 237
New Almaden Mine, 136
New Mexico, 34, 35
Newspapers, 76–77
Newsweek, 259
New York, 34, 42, 108, 159, 207, 230, 231, 267
New York Mine, 211
Niagara Concert Hall, 208
Nile River, 44
Nixon, Richard M. and Patricia, 261
Nobel, Alfred B., 129

North, Judge John Wesley, 62–66
Northern Paiutes. *See* Paiutes, Northern
Noyes, M. J., 71
Numaga ("Young Winnemucca"), 39–40
Nye, James W., 42, 140
Nye's Store, 225

O'Brien, Jack, 101, 222
O'Brien, William S., 104–5
O'Brien and Costello's Saloon and Shooting Gallery, 186–87
Occupations, 24–32, 91–95, 222–23, 226–30. *See also under specific categories*
Odd Fellows. *See* International Order of Odd Fellows
Ohio, 77, 230
Ophir, Town of, 11–12
Ophir Mill, 50, 51
Ophir Mine: biblical, 1, 11–12; Comstock, 11–12, 36, *43*, 44, 54–55, 57, 62–63, 65, 67, 74, 77, 101, 103, 106, 129, 237; and fire, 112, 114, 116; and lynching, 170
Ophir Pit, 255
Opium, 150, 189–90
Order of the Caucasians, 162
O'Riley, Peter (early miner), 8–9, 10–11, 46, 167
O'Riley, Peter (hotel keeper), 40
Ormsby County, 80, 226
Orrum, Eilley. *See* Bowers, Eilley Orrum
Osbiston, Frank, 148, 227
Osborn, John A. "Kentuck," 9, 11, 46
Overton, Captain J. B., 241

Packers, 36, 97, 164. *See also* Teamsters
Page, Joe and Carol, 265, 275–76
Paint Your Wagon, (film), 67
Paiutes, Northern, 156–57, *158;* and depression, 247; food gathering, 156–57; gambling, 157, *158*, 253–54; occupations of, 157, 217–18; Pyramid Lake War, 38–41, 103
Palmer, Louise, 188, 195, 225
Panama, 35
Paris, France, 207, 231
Paul, Almarin B., 46–48, 67
Payne, Amanda, 154, 302n26
Pearl Harbor, Hawaii, 73
Peerless Mine, 61
Pennsylvania, 230
Penrod, Immanuel "Manny," 9, 10–11, 46, 67
Philadelphia Brewery, 189
Pioneer Quartz Company, 6
Piper, John, 250–51
Piper's Old Corner, 185
Piper's Opera House, 113, 207, 208, 250–51, 276. *See also* Entertainment
Pittsburgh, Penn., 58
Placerville, Calif., 11, 27
Plato, Joseph, 7
Polish immigrants, 34, 143, 160
Pony Express, 41–42
Porter, Cole, 260
Portuguese immigrants, 34, 95, 143, 210
Potosi Mine, 64–66, 67, 72, 254
Powder River, Calif., 14
Presbyterian church, 201, *202*
Prostitution, 31–32, 93–94, 165, 176–78, 185, 211–12, 220–21, 223, 257, 260–61, 272, 273; and Barbary Coast, 177–80, *179;* and Chinese, 93, 165, 176, 251; and ethnicity, 93; and red-light district, 93–94, 176; and violence, 167–68, 177
Provost Guard. *See* National Guard
Puebla, Battle of, 156
Pullman, George M., 182
Pyramid Lake, Nev., 3, 39–41
Pyramid Lake War. *See* Paiutes, Northern

Quong Li Hoy and Company, 228

Railroad. *See* Virginia and Truckee Railroad
Ralston, William C., 77–80, 90
Raymond, R. W., 162
Red Dog Saloon, 264
Red-light district. *See* Prostitution, and red-light district
Reno, Nev., 52, 80, 81, 84, 133, 244, 259, 260
Reno Number 11 (V&TRR), *256*
Risdon Iron Works, 110
Rising, Judge, 171
Rising Star Mine (Idaho), 103
Robinson, Henry, 217
Roebling, John August, 124
Rogers, James F., 7, 9
Rolling Stone, 264
Roman Empire, 42
Rooney, S. B., 199–200
Roop, Isaac, 39
Roosevelt, Franklin D., 73
Rosener's Store, 225
Russell, Majors, and Waddell, 41
Russian immigrants, 34–35, 143, 160
Ruthbart, Ron, 181

Sacramento, Calif., 42, 77
Sacramento Saloon, 182
Saloons, 4, 22, 92–93, 105, 167, 169, 180–87, 203–4, 223, 240; and Barbary Coast, 177–79; and ethnicity, 36, 154, 161, 182–83
Salt, 29, 137
Salt Lake City, 42
Sanborn-Perris Fire Insurance Map, 177, *179*, 180, 228
Sands, Clarence, 154, *155*
Sands, David and Laura, 154–55
San Francisco, Calif., 12, 27, 67, 88, 110, 231, 235, 263–64; cable cars, 124; Great Fire of 1875, reaction to, 114; and milling 46, 47, 50; theater, 208, 227
San Francisco Catholic Diocese, 200
Sanitary Commission, 71
Sarsfield Guard, 144, 147, 205
Savage, Jim, 133
Savage Mine, 72, 106, 129, 235, 254
Sawdust Corner, 251, 268
Sayers, Nellie, 178
Sazerac Lying Club, 205
Sazerac Saloon, 185
Scandinavian immigrants, 34–35, 95, 143; organizations, 160
Schussler, Herman, 110, 160
Scott, Randolph, 259
Scottish immigrants, 34, 143, 157–59, 160; organizations, 157–59
Seattle oysters, 136
Servants, 92, 94
Shannahan and O'Connor Saloon. *See* Hibernia Brewery
Sharon, William, 77–81, *78*, 83, 90, 103–6, 110, 140, 229–30
Shea, "Crazy Kate," 111
Shenandoah Valley, Va., 181
Sherman, William T., 70
Shooting galleries, 186
Shoshone, 38–41
Sierra Nevada Mine, 124
Sierra Nevada Mountains, 1–4, 55–56, 75, 110
Silver City, Nev., 21, 239, 249, 254, 264, 270–71; militia, 40–41
Silver Palace, the, 185
Singer Sewing Machine sales office, 204
Sisters of Charity. *See* Daughters of Charity of Saint Vincent de Paul
Six Mile Canyon, 6, 8, 50, 51, 60, 234, 242, *243*, 276
"601" committee. *See* Vigilantes
Skae, John, 110
Sky Deck Saloon, 259

Smith, Duane, 198
Smith, Grant, 74, 238
Society of Pacific Coast Pioneers, 241
Sons and Daughters of Temperance. *See* Temperance
Sousa, John Philip, 250
South American immigrants, 35, 143
South Carolina, 35
South Sea Islanders, 34, 143
Southern Baptist church. *See* Baptist church
Spanish immigrants, 35
Sparks, Nev., 244
Spite houses, 253
Square-set timbering, 55–57
Stagecoaches, 28–29, 53
State of Deseret, 1
Steamboat Hot Springs, Nev., 52–53, 169
Steel, O. C., 182
Stephenson, Dr. W. H. C. and Jane, 153–54
Stewart, William M., *63*, 63–67, 77–78, 197, 231, 244
St. Louis Brewery, 189
St. Mary in the Mountains Catholic Church, 71, 112–13, 116, 117, 199, *200*, 201, 207, *253*
St. Mary Louise Hospital. *See* Daughters of Charity of Saint Vincent de Paul, hospital
Stocks, 23–24, 74–75, 88, 104–7, 132–33, 232, 235–37, 241; manipulation of, 175–76, 243
Stone Ground, 264
Storey County, Nev., mentioned throughout, 80; courthouse, 115, 116, 267–68; hospital, 215, 221; jail, 170–71; register of death, 120
St. Patrick's Catholic Church (Gold Hill), *82*
St. Patrick's Day, 144, 146, 268
St. Paul's Episcopal Church, 201, 227
Strouse, Mark, 226–27
Strout, J. A., 134
Sugarloaf Mountain, 276
Sullivan, Sandy, 237
Sun Mountain. *See* Mount Davidson
Sunderland Mill, 134
Sutliff's Music Hall, 208
Sutro, Adolph, xx, 1, 58–59, 88–89, 90, 160
Sutro Baths, 235
Sutro City, 59, 238, 268
Sutro Tunnel, 58–59, 88–89, *89*, 90, 129, 160, 235, 238, 241, 248
Sweeney Guard, 144, 205
Swiss immigrants, 33, 35, 72, 143, 203

Tahoe Basin, 56
Tamkin, William and I. K., 187–88
Teamsters, 12, 26–30, *28*, 36, 41, 53, 83–84, 92, 97, 132, 189
Telegraph, 41–42
Telluride, Colo., 266
Temperance, 72, 188, 203–4
Temples of Comedy, 208
Tennesee, 66
Territorial Enterprise, 25–26, 69, 76, 241, 260, 262
Theaters. *See* Entertainment
Thompson, Cad, 177, 212
Tiffany and Company, 108
Tilton, J. H., 52
Tilton, Virginia, 193
Toiyabe Range, 234
Toll roads, 27–28, 51–53
Tommyknockers. *See* Folklore
Tonopah, Nev., 243–44
Topliffe's Theater, 208
Tourism, 258, 260, 261–62, 265–66, 269
Townsend, James W. E. "Lying Jim," 76
Transatlantic cable, xxi
Treasure Hill, Nev., 80

Trinity County, Calif., 213
Truckee Meadows, Nev., 11, 52, 169, 244
Tucker, J. H., 134
Turkish immigrants, 143
Tyler, Jenny, 177
Twain, Mark. *See* Clemens, Samuel

Uncle Sam Mine, 61
Union Brewery, 189
Union Iron Works, 136
Union Mill and Mining Company, 79, 266
Union shaft, 58
Union Tunnel, 57
United States: Bureau of Land Management, 271; census of 1860, 25–27; census of 1870, 91–99, 201; census of 1880, 238–39, 244; census of 1900, 144–46; census of 1910, 244–47; Mint, 83, 137, 241; Postal Service, 261; War Department, 257; War Production Board, 256; Works Progress Administration, 254
University of Nevada, Reno, 108
Utah Territory, 35, 42

Van Bokkelen, Gen. Jacob L., 72, *73*, 130, 134, 159, 183, 199, 272, 303n38
Van Duzen, John "Yank," 213
Van Sickle, Henry, 33
Varga, Billy, 272
Victoria (Queen), 68
Vigilantes, 170–71
Violence, 32–33
Virginia, 17, 230
Virginia and Gold Hill Water Company, 110
Virginia and Truckee Railroad, 80–84, 88, 90, 99, 113, 132–34, 137, 158, 239–40, *256*, 259, 261
Virginia Board of Brokers, 72
Virginia, Carson, and Truckee River line. *See* Virginia and Truckee Railroad
Virginia City, (film), 259
Virginia City, Nev., throughout; government, 42; naming, 17; in 1860, 31, 43; in 1865, *100*
Virginia City, Mont., 266
Virginia City Historic District Commission. *See* Comstock Historic District Commission
Virginia City Miners' Union, 140
Virginia City Water Company, 59–60
Virginia Daily Union, 76
Virginia Evening Bulletin, 76
Virginia Foundry, 136
Virginia Range, 3

Wages, 5, 7, 30, 108, 133, 140–41, 216, 225, 240, 283n23
Waldorf, John, 195, 207, 209, 216–17, 232, 242, 247–48
Wales, 162. *See also* Welsh immigrants
Walker, J. M., 101–3
Walker River, 3
Wallerstein, Immanuel, 277n3
Walsh, Judge James, 11, 12, 27
Warner Brothers, 259
Washington (state), 34
Washington, D.C., 154
Washington Guard, 155
Washington Monument, 42
Washington Saloon, 182
Washoe Brewery, 189
Washoe City, Nev., 84
Washoe Club, 185, 233, 242
Washoe County, 84
Washoe Gold and Silver Mining Company No. 1, 46, 48
Washoe Indians, 34
Washoe Lake, Nev., 50, 51
Washoe Pan Process, 47–48, *49*
Washoe Seeress. *See* Bowers, Eilley Orrum

Washoe Typographical Association, 72
Washoe Valley, Nev., 50, 51, 68, 231–32
Washoe zephyrs, 34, 100, 257, 267
Water (drinking), 5, 59–60, 110–11. For mine water, *see* Mining, pumping
Watt, Secretary of the Interior James, 269
Weldon, Richard, 188
Wells, Fargo and Company, 169
Welsh immigrants, 105, 143, 160
Werrin Building, 122
Whalen, Martin, 134
White Pine County, 90
Wild West. *See* Folklore
Wiley, Anna, 229
Williams Station, 39
Wilson, W. F., 120
Winchester, Jessi, 271–72
Winnemucca, "young." *See* Numaga
Wing, Dr. Song and Choney, 227–28
Winters, John D., Jr., 11, 46
Women, 185–86, 189, 192–94, 244–45; in 1860, 31–32, 35–36; in 1870, 92–95; in 1880, 238–39; occupations of, 131, 145, 184
Woodworth, Joseph, 11
World War II, 73
Wright, Judge, 115
Wright, William, 4, 16–19, 76, 117–18

Yee family, 228
Yellow Jacket Mine, 61, 79, 83, 84–88, 90, 102, 106, 112, 119, 129, 198
Yerington, Nev., 244
Yosemite Saloon, 182
Young, Otis E., 36–37
Yount, John, 6–7

Zanjani, Sally, 32–33